Geothermal energy

Frontispiece: White Island Volcano, Bay of Plenty, New Zealand. (By courtesy of The New Zealand High Commissioner, London.)

GEOTHERMAL ENERGY

Its past, present and future contributions to the energy needs of man

H. Christopher H. Armstead

B.Sc., C. Eng., F.I.C.E., F.I.Mech.E., F.I.E.E., A.C.G.I.

E. & F.N. SPON LTD
LONDON
A Halsted Press book
John Wiley & Sons New York

First published 1978
by E. & F.N. Spon Ltd
11 New Fetter Lane
London EC4P 4EE
© 1978 H.C.H. Armstead
Photosetting by Thomson Press
(India) Limited, New Delhi
Printed in Great Britain by
J. W. Arrowsmith Ltd, Bristol

ISBN 0 419 11240 5

Distributed in the USA by Halsted Press, a division of
John Wiley & Sons, Inc., New York

Library of Congress Cataloging in Publication Data
Armstead, H. Christopher H.
  Geothermal energy.

  "A Halsted Press book."
  Bibliography :   p.
  Includes index.
  1.   Geothermal engineering.   2.   Geothermal resources.
I. Title.
TJ280.7.A74   1978   553.7   78–2545
ISBN 0–470–26337–7

Dedication

To all those who feel profound unease at the profligate squandering of the earth's dwindling resources and at the wanton fouling of our once beautiful planet this book is dedicated, in the hope that it may at least contribute to the solution of one of the grave crises confronting Mankind – that of finding abundant energy and of simultaneously reducing the terrifying degree of pollution associated with the combustion of huge quantities of fossil fuels on which we are now so dependent for our social needs.

Contents

viii   Contents

xii   Contents

Unit conversion table

Physicists, professors and engineers are usually accustomed to working in different units, and despite the tendency to encourage the use of the SI system, it is a fact that old customs die hard. Engineers, especially of the middle and older generations, are unaccustomed to thinking in terms of newtons, pascals or even joules: they prefer to talk of atmospheres (ata and atü), psia and psig, btu, foot-pounds and watts. To most engineers a statement that the world's annual energy consumption is so many millions of terajoules is more or less meaningless until he has converted it into kilowatt-hours. Even amongst engineers of different nationalities customs vary. New Zealanders usually refer to geothermal bore yields in terms of kilopounds/hour, Americans in tons/hour, Continentals in kilograms/second. The author has generally tended to use the metric system and the Celsius temperature scale except where custom, or quotation from other authors' works makes the use of other unit systems more convenient. On many of his graphs he has used multiple scales to suit different readers. However, to avoid ambiguities and to facilitate comparisons, the following conversions should serve a useful purpose.

Class	To convert	into	Multiply by
Length	centimetres	inches	0.394
	metres	feet	3.281
Volume	cu centimetres	cu inches	0.061 02
	cu metres	cu feet	35.31
	cu metres	litres	1000
	cu metres	US gallons	264.2
	cu metres	Imp gallons	220.5
	US gallons	Imp gallons	0.833[1]
	US gallons	cu feet	0.1337
	Imp gallons	cu feet	0.16

[1] 1 imperial gallon of water at 4° C weighs 10 lb.
  1 US gallon of water at 4° C weighs 8.33 lb.

Class	To convert	into	Multiply by
Mass	kilograms	pounds	2.205
	kilograms	long tons	0.000 9842[2]
	kilograms	short tons	0.001 102[3]
	kilograms	metric tons, } or tonnes (t) }	0.001
Density	grams/cm^3	lb/in^3	0.03613
	kg/m^3	lb/ft^3	0.062 43
Pressure	atmospheres	cm of Hg	76.0
	atmospheres	kg/cm^2	1.033
	atmospheres	lb/in^2, or psi	14.70[4]
	bars	atmospheres	0.987
	bars	newtons/m^2	100 000
	bars	psi	14.5
	bars	kg/cm^2	1.02
	atmospheres } absolute (ata) }	atmospheres } gauge (atü) }	subtract 1.0
Energy and Calorific value	British thermal } units, or btu }	ergs	1.055×10^{10}
	btu	ft lb	778.3
	btu	joules	1054.8
	btu	kg cal	0.252
	btu	kWh	0.000 292 8
	btu/lb	cal/g, or } kcal/kg }	0.555 56
	tonnes coal equivalent (tce)	kWh	8130
Concentration	parts/10^6, } or ppm }	grains per } US gallon }	0.0584
	ppm	grains per } Imp gallon }	0.070 16
	ppm	grams/litre	0.001
	ppm	% by weight	0.000 01
Enthalpy	joules/gram, } or J/g }	btu/lb	0.43
	kcal/kg	btu/lb	1.8
Heat flow	kilowatt (thermal)	btu/h	3413
	kilowatt (thermal)	kcal/h	860

[2] 1 long ton = 2240 lb
[3] 1 short ton = 2000 lb
[4] Absolute pressure in psia = gauge pressure in psig + 14.7
   Absolute pressure in ata = gauge pressure in atü + 1

Temperature conversion $°F = (1.8 \times °C) + 32$
$°C = (°F - 32) \div 1.8$
Absolute Celisus temperature $= °C + 273 = K$ (Kelvin)
Absolute Fahrenheit temperature $= °F + 460$

Prefixes

Tera (T)	10^{12}	
Giga (G)	10^9	
Mega (M)	10^6	
Kilo (k)	10^3	
Hecto (h)	100	
Deca (da)	10	
Deci (d)	10^{-1}	SI system
Centi (c)	10^{-2}	
Milli (m)	10^{-3}	
Micro (μ)	10^{-6}	
Nano (n)	10^{-9}	
Pico (p)	10^{-12}	
Femto (f)	10^{-15}	
Atto (a)	10^{-18}	

Reciprocals It is perhaps scarcely necessary to add that inverse relationships in the above table can be derived from the reciprocals of the last column of figures.

An interesting conversion *chart* devised by Professor P.M.C. Lacey of the University of Exeter appeared in an article published in the 'Chartered Mechanical Engineer', p. 62, June, 1977. It is understood that Professor Lacey is preparing more extended charts to cover a wider range of conversions.

Preface

The harnessing of earth heat to the service of man calls for the collaboration of a team of specialists, each versed in a different discipline. Geologists, geohydrologists, geophysicists, geochemists, petrologists, engineers, metallurgists, economists, industrialists, environmentalists, financiers, lawyers, – even medical men, politicians and others – will or may become involved. To embrace a full description of the activities of all these specialists and of the theories and hypotheses on which they work would virtually require an encyclopaedia, and would certainly be too great a task for any single author. This book makes no claim to being encyclopaedic, nor is it a 'popular' work for the man-in-the street. It is addressed to the well educated reader who has enjoyed the benefits of a scientific training, who may perhaps be one of the specialists mentioned above but whose activities have not hitherto been slanted towards the geothermal aspects of his own particular discipline, and who wants to have a fairly broad idea as to 'what geothermal energy is all about'. There is many a well trained specialist who knows virtually nothing of geothermics but whose knowledge could well be channelled profitably into geothermal work did he but know how to apply it in that direction. In short, he lacks the background information that would illuminate the scene sufficiently to show him how to direct his talents most usefully towards the science of geothermics, which is of rapidly growing importance and which will soon urgently need all the qualified man-power that can possibly be recruited in its service. This book seeks to provide that necessary background.

As the author is an engineer, it is only to be expected that the engineering aspects of geothermal development will here receive perhaps rather more than their fair share of attention. For this he tenders his apologies, but at the same time he would point out that had someone of a different provenance written the book it seems probable that rather greater stress would have been placed upon some other facet of what is essentially a many-sided subject. He, nevertheless, hopes that he has done justice to those aspects of geothermics in which he cannot claim to be a specialist.

Since the early 1950s there has been a spate of technical, economic, legalistic and other writings on geothermal matters. They have appeared in the proceedings of international conferences and symposia and of learned societies, and also in the technical and lay press. Since, as is to be expected, such writings have been partly good, partly bad and partly indifferent, it follows that too much rather than too little 'literature' is available. However, this very plethora of writings is apt to be confusing to the would-be student of geothermics, who cannot see the wood for the trees and who (to change the metaphor) needs guidance in separating the chaff from the wheat. One of the purposes of this book is to provide that guidance by drawing the reader's attention to some of the more apposite writings by means of an extensive bibliography which will enable him to pursue any aspect of geothermics in which he may be particularly interested, as far as he wishes; for most of the writings to which the reader will be referred will in turn contain their own bibliographies for still more detailed study ... and so *ad infinitum*. The author is very conscious of the fact that in selecting works of reference for inclusion in his reference list he will undoubtedly have omitted many excellent writings that would equally, and in some cases perhaps even better, have illustrated or amplified the various points made in the text of the book. He offers his apologies to those authors whose works have been omitted – not by intent, but by his inability to be familiar with *all* the relevant writings and by necessity of limiting the length of his reference list.

With a science that is developing as rapidly as geothermics, it is inevitable that a book of this type is overtaken by events between the time of submission of the manuscript and the date of publication; (e.g. see the note accompanying Fig. 1). The information given here, except where specifically stated otherwise was broadly valid in 1976.

Finally, the author would like to thank those many co-operative people from several countries who have so helpfully responded to his persistent, and probably troublesome, requests for statistical and other information. He hesitates to mention them by name for fear of inadvertently omitting any.

Rock House,   H.C.H.A.
Ridge Hill,
Dartmouth,
South Devon

April 1977

Introduction

γη Earth　　　θερμη Heat

Within a few years earth heat – or to use its more scientific but ponderous title, geothermal energy – may well become a factor of supreme importance for the survival of industrial man. Its resources are gigantic, could we but gain easier access to them, and it is vitally necessary that the gospel of its study and exploitation be propagated as widely and as rapidly as possible. Hence this book.

It will be shown in Chapter 20 that the world is in serious danger of approaching an energy famine within a few decades, if not earlier. In fact it is maintained by some that the first harbingers of such a famine are already with us. However, public complacency is such that very few people can read the writing on the wall. Although alarmist preachings are generally to be deplored, a few well-placed jeremiads can be salutary. By recognizing in good time the potential dangers that lie ahead we can perhaps take timely action to avoid them; but by ignoring them we shall be inviting disaster.

Two of the most promising sources of long-term salvation, of greater ultimate importance even than that of fossil fuels and nuclear fission, are controlled thermo-nuclear fusion (which has so far eluded us) and geothermal energy which has progressed rapidly and for which the promise of impressive new developments can now be discerned. Active research into thermo-nuclear fusion will doubtless be pursued and if successful that would be a 'consummation devoutly to be wished'. However we must not, we *dare* not, rely solely upon that problematic success. It is essential that we have an alternative and it is suggested that that should be the exploitation of earth heat – not in the relatively puny quantities hitherto won, but on an altogether vaster scale.

There are, of course, other sources available to us, such as hydro-, tidal, wind and wave power, and while it is important that none of these should be neglected, it must be recognized that they could perhaps have little more than a palliative effect. For example, if all the estimated exploitable hydro-resources of the world were developed (and at present only about one-seventh of those resources have been touched), they could satisfy only 13 or 14% of the world's *present* annual energy requirements. It is also true that we continuously receive immense quantities of solar energy, but its diffuseness at present renders it difficult to harness on the required grand scale, though it can be of considerable local small-scale importance.

The rewards of exploiting earth heat appear to be far more easily and quickly attainable, so we must lose no time in seeking to grasp them before it is too late.

Even today, geothermal energy is regarded by most people, if they think about it at all, merely as an interesting freak of nature that serves as a tourist attraction in certain parts of the world, either for its spectacular or for its alleged medicinal properties. It would scarcely be an exaggeration to say that tomorrow it may well become one of the best hopes for the survival of civilization.

There are promising signs of a belated, but rapidly growing awareness of the great potentialities of earth heat. Since the United Nations first convened in Rome an international conference on new sources of energy in 1961 [1], at which geothermal energy had to share the platform with wind and solar energy and attracted less than one-third of the total number of papers submitted, there has been an acceleration in the frequency and scale of geothermal symposia and seminars. By the late 1960s the geothermal sciences had advanced sufficiently for the United Nations to consider it justifiable to hold an international symposium at Pisa in September, 1970 [2], devoted solely to these sciences. This was so successful that a further United Nations Geothermal Symposium was held less than five years later in San Francisco in May, 1975 [3], at which nearly 400 papers were accepted for publication and something between 1300 and 1400 participants attended. Since there must be many others actively engaged in geothermics who neither attended nor contributed papers at these various conventions, it is apparent that a large number of workers is now occupied with geothermal exploration and exploitation: but more will be needed.

The United Nations should be given due credit for the great contributions they have made towards geothermal development, not merely by organizing international conventions but by providing technical assistance and funds for exploration projects in several countries that lacked the resources to proceed without aid. UNESCO too have helped by sponsoring specialized training courses in the geothermal sciences and by publishing a *vade mecum* in the form of a Review of geothermal research and development [4]. National and international committees and study groups have recently been established to examine the prospects of winning earth heat in countries that have hitherto been regarded as geothermally worthless.

From all this it is clear that throughout the world there is a marked growth of interest in earth heat. The oil crisis of 1973 undoubtedly provided a great stimulus to this interest, which will no less certainly gather impetus in the years to come. Each successful geothermal project engenders confidence that encourages the development of further projects, and the science of geothermics, born at the turn of the twentieth century, will undoubtedly have grown to an impressive maturity before the century closes. Exciting new developments are now taking place that may well increase the rate of geo-thermal exploitation to a staggering degree, so that earth heat may assume an importance comparable with that accorded during the last hundred years to fossil fuels.

In spite of a handful or so of very impressive installations, it must be admitted that so far we have scarcely scratched at the surface of geothermal potentialities. At present (1977) earth heat is contributing only a small fraction (less than 0.2%) of the world's electrical needs and a far smaller fraction of its non-electrical energy requirements. Clearly we are a very long way from being able to claim that geothermal energy is yet a major factor in the world economy. However, with the techniques now available we can exploit earth heat only in what are known as 'geothermal fields' (see Chapter 5), which are confined by nature to certain favoured parts of the earth. At last, however, we seem to be within reach of *creating* such fields where none exist in nature (see Chapter 19) and of ultimately tapping immense reserves of heat that have hitherto been regarded as inaccessible.

Wherever earth heat has been exploited, it has proved to be cheap by comparison with alternative sources of energy (see Chapter 15). Why then, it may reasonably be asked, has its development not been much more rapid since the 1930s when the Italians had so successfully exploited the Larderello field in Tuscany for power generation and the Icelanders had made such great progress in the heating of Reykjavik's homes by means of subterranean heat? The explanation lies partly in the comparative rarity of geothermal fields, some of which are topographically remote from potential energy markets. Others occur in countries well endowed with cheap alternative energy souces that can be developed without risk by well established techni-ques involving no more than conventional engineering methods. New Zealand and Iceland are cases in point; for both are rich in hydro-power resources, and New Zealand also possesses coal and natural gas. It is true that New Zealand *has* developed geothermal power to some extent and that Iceland *has* used low grade earth heat for space heating and domestic hot water supplies, but had either of these countries been poor in other energy resources it is highly probable that by now they would have exploited their earth heat to a far greater extent.

But these are not the only reasons for the relatively slow past development of geothermal energy. There has also been a philosophical conflict of ideas between 'risk capital' on the one hand and 'safe development' on the other hand. Although there are many hot spots in the world where the existence of geothermal fields is suspected, there can be no advance certainty that the right geological, hydrogeological and chemical conditions occur, without which exploitation would be impossible. Geothermal development must be preceded by exploration to establish whether a truly exploitable field exists. Exploration requires a fairly large outlay of risk capital which, if successful, can be amply repaid by results; but which, if unsuccessful, will represent just so much money wasted (except, perhaps, for any purely scientific knowledge that may have been gained). Governments – particularly those of developing countries – tend to be reluctant to invest risk capital if an alternative energy source is clearly obtainable without risk, even at a higher price than that at which geothermal energy could *perhaps*, but by no means *certainly*, be won if the outcome of the risk were successful. In short, governments are seldom willing to take a gamble. Private enterprise, on the other hand is often very willing to invest risk capital (e.g. the oil companies) but, not unnaturally, it expects a fairly high return on its money if the results of the risk should be successful. Now the commonest application for geothermal energy has hitherto been electricity generation, with district heating second. But electricity and domestic heat supply are industries which, being essential public services, are apt to be intolerant of high profits; so there is less inducement to stake risk capital. Moreover, whereas a successful oil strike produces a commodity that is readily marketable and easily shipped all over the world, an enterprise that wins geothermal energy must find a local market, which may be more limited.

Reluctance to take financial risks has, moreover, been reinforced by a certain conservatism and unwillingness on the part of electricity supply undertakings to take geothermal energy seriously: there has been a tendency to regard it as 'new fangled', and to cling to more conventional sources of energy that have for longer stood the test of time. Prejudicial though this attitude may have been, it has undoubtedly been an adverse psychological factor that has retarded the rate of geothermal development, and one which even now has not been fully eliminated.

Examples can of course be cited both of governments and of private enterprises that have been venturesome and willing to invest risk capital in geothermal exploration that has paid off handsomely. In many countries, however, the philosophical gap between risk and caution remains, despite the valuable work done by the United Nations in helping to bridge this gap by partly financing geothermal exploration programmes through the medium of their United Nations Development Programme Special Fund, thus reducing the element of risk that has to be taken by client nations.

Certain trends, however, are already doing much, and will undoubtedly do more in future, to remove these barriers of reluctance. First, many countries have by now 'skimmed the cream' off their readily available indigenous energy resources and must now perforce seek less conventional supplies. Secondly, the spectacular increases in fuel prices since 1973 are helping to create a greater willingness to face the required risks inherent in rather costly exploration work, in view of the great prizes that may be won in the form of relatively cheap energy and, equally or even more important, self-sufficiency and independence from the political whims and economic policies of other nations. Thirdly, there have been great advances in recent years in exploration techniques that have gone far in *reducing* the risk element. In the early days of geothermal development, before we knew much about what is happening below-ground, exploration was a rather chancy hit-or-miss business largely based on 'wild-cat' drilling in the neighbourhood of visible surface thermal manifestations. Drilling is a very costly operation, and much money was sometimes spent before any positive results could be shown. Now, by using more refined techniques and by applying the accumulated knowledge gained during the last two or three decades, we are able to collect much more reliable information from relatively inexpensive measurements made at the surface, from which we can build up a reasonably probable 'model' of the structure of the field so that we may postpone expensive drilling operations until we can choose bore sites with a much better chance (though still without *certainty*) of success. Lastly, whereas at one time there was a tendency to regard earth heat as being suitable almost exclusively for electricity generation or district heating, there is now a growing awareness of its industrial and farming potentialities which can thus widen the choice of markets.

A useful rule-of-thumb which graphically illustrates the potential value of earth heat is that *every kilowatt of geothermal base load power is capable of saving about two tons of oil fuel per annum*, or its equivalent, in an integrated power system. Thus if the price of oil were, for example, US $80 per ton, one kilowatt of geothermal base load could save about US $160 p. a. which, if capitalized at 16% p.a., would be worth US$1000 per kilowatt! It is of interest to note that in California the capital cost of a 106 MW geothermal power plant (excluding the steam supply) in 1976 was about US $200 per kW: with the cost of the steam supply included, the gross costs would probably be less than double this figure. Now, as will be shown in Chapter 15, capital costs per kilowatt can vary greatly from country to country, from time to time and from plant capacity to plant capacity. Nevertheless, this example will suffice to show that in terms of fuel saved, investment in geothermal power can be very attractive. It can be still more attractive for low grade heat.

The example also emphasizes the importance of *time* in an oil-importing country (or for that matter also in an oil-exporting country, for oil not burned at home is oil available for export). For example, with oil fuel at US $80 per ton, a year's delay in constructing a 100 MW geothermal power

Table 1   Growth of installed geothermal power capacity.

Year	Geothermal MW installed	Average annual compound growth rate
1904	0	
1942	130	
1958	293	5.22% (1942–1958)
1961	420	
1963	536	
1970	675	7.20% (1958–1970)
1976	1362	12.42% (1970–1976)
1981	2828 (estimated)*	15.74% (1976–1981) (estimated)*

*See note on Fig. 1.

plant would effectively burden the national balance of payments by about US $16 millions, even if the possible effects of inflation upon the price of the plant during the year's delay were disregarded. This underlines the importance with pressing ahead as rapidly as possible with geothermal development in countries possessing potentially exploitable earth heat, especially in those lacking cheap indigenous fuels.

It is of interest to study the past and present growth trends of geothermal development, and also the expected growth during the next few years. Only power generation will here be considered – not because other applications of earth heat are in any way to be despised, but simply because heat statistics are less readily available than kilowatt figures. It took more than half a century, from 1904 to 1958, for 293 MW of geothermal electric power to be developed, but the growth of installed capacity of such plants since then has shown a marked upturn, as can be seen in Table 1, which is illustrated graphically in Fig. 1.

From 1950 to 1970 the world demand for electricity grew at about $8\frac{1}{2}$% p.a., so that until 1970 geothermal power development failed even to keep pace with that growth, but thereafter it has been taking upon itself an ever increasing share of the world electricity market. From 1970 to 1973 the world electrical growth had declined to an average of 7.35% p.a. It is still too early to assess the full effects of the 1973 oil crisis upon the world electricity market, as statistics take a few years to prepare and publish, but there can be little doubt that geothermal power is already beginning to play a far larger role in the world power market – partly, no doubt, due to the very occurrence of that crisis. Moreover the growth figures shown above are not entirely fair to geothermal energy, for whereas the average world load factor on generating plants lies between 45 and 50%, geothermal power plants generally operate at annual load factors of about 90% – certainly 85%. Hence the *energy* (in kilowatt-hours) contributed by geothermal power plants is likely to be relatively more than might be deduced from the installed megawatt capacities. This point, however, should not be overstressed, as

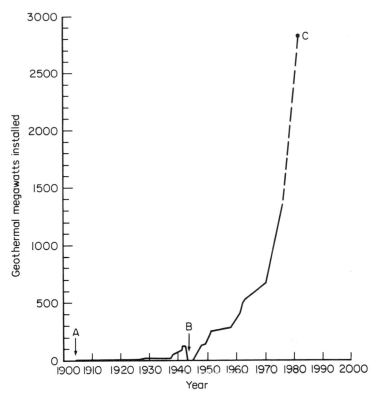

A. Prince Conti starts to develop power at Larderello
B. Destruction of Italian plant by war damage
C. 1976 estimate of situation in 1981

Figure 1 World capacity of geothermal power plant installed.

Note: Since this book went to press a survey has been published [197] which puts the estimated installed geothermal power capacity at nearly 12 000 MW by 1985. This would imply a mean annual growth rate of about 27% from 1976 to 1985 – a far more optimistic forecast than deducible from Table 1 and from this figure, the data for which was obtained from a questionnaire sent out early in 1976.

much would depend upon the types of alternative plants that would have been installed if the geothermal plants had not been adopted. However, it would generally be true to say that a geothermal kilowatt is worth more in terms of energy than any other kind of kilowatt – except perhaps for run-of-river hydro-plants.

To judge from the growing interest in non-power applications of earth heat, it seems probable that the growth in such applications is also accelerating rapidly. Encouraging though these obvious signs of growing geothermal development may be, it is necessary to preserve a proper perspective. Even by 1981 the geothermal contribution to the world's electrical needs is still likely to be only a fraction of 1%. Moreover, with existing exploitation techniques, it seems improbable that the present accelerating growth trend in geothermal development can be sustained for more than perhaps two or three decades, simply because of the limited natural occurrence of hyperthermal fields in which substantial power projects could be developed. Like fossil fuels, the resources of hyperthermal fields are not unlimited – though it would at present be too hazardous to make a guess at what those reserves might be. Non-power applications of earth heat could probably expand for longer, with the use of the techniques available, as vast semi-thermal fields are known to occur without the same geographical limitations.

From all this the impression might be gained that we are enjoying a short-lived geothermal 'boom' that will gradually peter out – perhaps in the earlier part of the next century. However, the operative words that may have given this misleading impression were *with existing exploitation techniques*. Why should we assume that greatly improved techniques should not be effected for winning earth heat? Why should such improvements, if achieved, lie in the remote future? Necessity is the mother of invention. Great and exciting developments are even now the subject of experiment in the USA which hold out reasonable hope that within a decade or so we may possess the means of vastly extending the areas of the world where earth heat may be won – perhaps even of making it universally available at any convenient point on the earth's surface.

There is a common tendency amongst writers and commentators to write off geothermal energy as little more than a resource which, though sometimes of great local importance, could certainly never make a major contribution to the solution of the world's energy problems. The author believes this to be an unwarranted defeatist attitude, and that we are, on the contrary, on the eve of a geothermal 'renaissance' that will assume a significance comparable with, and perhaps exceeding, that of fossil fuels at the present day. After all, less than half a century ago there was probably only a handful of dreamers who seriously believed it would ever be possible for man to reach the moon and planets, because, *with the techniques then available*, such an achievement could clearly never have been realized.

It is useful to summarize the 'virtues' of earth heat as an energy source:

 (i) Its reserves are immense – virtually infinite on the scale of history (see Chapter 4).
(ii) It is highly versatile (see Chapter 2).
(iii) It is far less guilty of pollution than fuel combustion or nuclear fission (see Chapter 17).
(iv) It has the merit of *security*.

This question of security requires amplification. Unlike hydro-power it is not at the mercy of seasonal rainfall. If indigenous geothermal resources can be developed they can liberate a country from political and economic dependence upon other countries and can help the national balance of payments. The continued exploitation of earth heat, once won, is not labour-intensive and is therefore less vulnerable to industrial disputes than, say, coal mining. A geothermal installation, unlike a marine oil rig, is relatively in little danger from sabotage which, in these lawless days, is a threat that cannot be ignored. For all these reasons earth heat may be regarded as having the merit of security.

In 1967 in Washington DC, the author publicly expressed the opinion that humanity might have reaped a far greater reward had the vast expenditure hitherto lavished upon space exploration been devoted instead to activities in a downward direction. He still stands by that opinion.

Historical note

Visual evidence of earth heat is provided by such natural phenomena as volcanoes (Frontispiece and Plate 1), geysers (Plate 2), fumaroles, hot springs and pools of boiling mud (Plate 3). These manifestations are to be found in various parts of the world and are often associated with the occurrence of earthquakes. It is not surprising that the ancients should have regarded the depths of the earth with horror, as the seat of Hell and of malignant gods – as do some primitive peoples to this day. Fortunately we have now come to view them as the seat of hope. The transition from horror to hope will now be described in this historical note, which may conveniently be treated for the most part under headings corresponding with the various applications of earth heat. Some of these applications have developed collaterally with others.

1.1 Applications of earth heat

1.1.1 *Balneology.* Although the ancients who inhabited the thermal regions of the world tended to hold in fear the more violent manifestations of earth heat such as volcanoes, and therefore to shun them—and after the destruction of Pompeii and Herculaneum who can really blame them? – they were not slow to exploit for their comfort the gentler thermal phenomena such as hot springs, which have been used for centuries for bathing purposes by the Etruscans, the Romans, Greeks, Turks, Mexicans, Japanese, Maoris and no doubt others. Sometimes the waters from these springs have a rather foul taste and smell, suggestive of rotten eggs or worse; and as there is a curiously perverse human tendency to think that that which is unpleasant must be good for you, a belief grew up that to drink one's bath water was commendable and beneficial to the health! Certainly it is true that some thermal waters act as powerful aperients. Furthermore, as hot water is undoubtedly soothing to tired limbs, the cult of 'taking the waters' was built up over the centuries. Thermal waters were alleged to possess healing

Plate 1 Mount Ngauruhoe Volcano, the North Island, New Zealand. (By courtesy of The New Zealand High Commissioner, London.)

and prophylactic properties when applied externally, as in bathing, or internally when taken orally or used for douches. Thus was born the balneology industry – the oldest application of earth heat. It flourished in the days of Imperial Rome where the *thermae* became an institution, not only as centres for health and hygiene but also as foci for social intercourse – rather like the eighteenth century coffee houses in London. Later, in the eighteenth and nineteenth centuries the custom of taking the waters underwent a tremendous revival, enthusiastically supported by the jaded world of fashion whose members sought, for a few weeks annually, to recoup from the effects of excessive port-drinking during the rest of the year. 'Spas', 'hydros' and 'watering places' sprang up all over Europe and elsewhere, and were frequented by invalids, hypochondriacs and the merely affluent. Not all, but most of these spas were of thermal origin. The balneology industry thrives to this day, as will be shown in Chapter 12, (though without its erstwhile aristocratic overtones) and millions of bottled 'mineral waters' from thermal springs are consumed every year all over the world. The *hamams* of Turkey,

Plate 2 'Old Faithful' geyser, Yellowstone National Park, USA. (By courtesy of the United States Information Service, American Embassy, London.)

Plate 3 Boiling mud pool, Rotorua, New Zealand. (By courtesy of the New Zealand High Commissioner, London.)

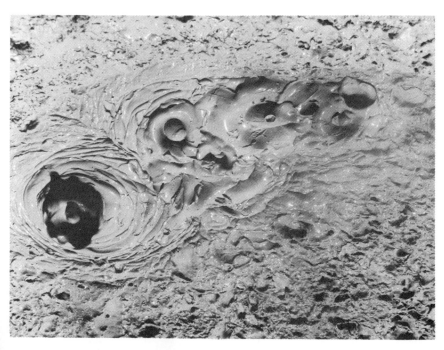

from which originated the westernized 'turkish bath', were patronized by the Ottomans, and even today they are frequented by women who squat over jets of hot water in the hopes of inducing fertility or of curing unmentionable diseases.

1.1.2 *Domestic services.* The Maoris have adapted geothermal phenomena to their domestic needs ever since they settled in New Zealand in the fourteenth century, and their traditional rural way of life in the thermal areas may be witnessed today. At a village near Rotorua in the North Island one may see a fisherman catch his trout in a cool river and drop it into a nearby pool of boiling water to cook it. A few yards away his wife may be seen administering a geothermal bath to the baby, while his daughter is doing the household laundry in a hot spring and the vegetables are cooking over a fumarole. Each of these simple domestic chores is performed with natural fluids at appropriate temperatures and it is important not to confuse the trout with the baby. Although the hidden hand of organized tourism may perhaps be detected behind this scene of rural bliss, the activities nevertheless have a factual historical basis.

1.1.3 *Mineral extraction.* Another very early application of natural thermal activity was for the procurement of minerals. The Etruscans extracted boric acid from the boiling springs – later known as *lagoni* – to the south of the ancient town of Velatri (modern Volterra), and used it for making the splendid enamels with which they decorated their vases. From the mid-thirteenth to the end of the sixteenth century these *lagoni* were exploited for the extraction of sulphur, vitriol and alum. In the early nineteenth century Francesco Larderel, Count of Montecerboli – a Frenchman who had emigrated to Italy – built up a prosperous boric acid industry in Tuscany which survived until well after the middle of the present century, by which time it was no longer profitable. The count gave his name to the now famous township of Larderello.

Another ingenious exploiter of geothermal minerals was none other than Hernan Cortez, who used them for military purposes. The Fates seem to have favoured this Conquistador by spreading superstitious dismay among the Aztecs. Not only did these doomed Mexicans believe Cortez to be the reincarnation of the mythical Quetzalcoatl who, centuries earlier, had come out of the East and had later sailed away into the Atlantic, promising to return one day; not only did they identify the hitherto unknown horse with its rider so as to produce in their imagination a sort of synthetic centaur; but Mount Popocatépetl, after centuries of dormancy, decided to stage a miraculous eruption at the precise time when the small Spanish force was advancing upon the Emperor Moctezuma's capital city. With such supernaturally loaded dice, the conquest of a mighty empire by a small band of determined adventurers was perhaps not so surprising. Cortez was not only a soldier: he was a great opportunist. He decided to earn a double dividend from the eruption of Popocatépetl. Having burned his fleet to ensure that his comrades-in-arms could not defect, he was still able to maintain his supplies of ammunition by making his own gunpowder, using as an ingredient the sulphur deposited near the crater of the volcano after its obliging activity.

The extraction of chemicals from geothermal fluids continues. Elemental sulphur is recovered from fumaroles in Japan and Taiwan, sulphuric acid is manufactured from thermal fluids in Japan, common salt and calcium chloride are produced from hot brines beneath the Salton Sea area in Southern California; and the possibilities of winning valuable trace elements from geothermal fluids are being closely studied.

1.1.4 *Electric power generation.* The most spectacular advance to be made in the exploitation of earth heat was heralded when Prince Piero Ginori Conti first promoted electric power generation at Larderello in 1904. Attempts were first made to use reciprocating steam engines fed with natural steam, but these were short-lived owing to virulent chemical attack. Then, at first by using heat-exchangers with which to raise 'clean' steam from 'dirty' natural steam, and later by improving the quality of materials used in the manufacture of prime movers so that natural steam could be used directly without incurring the losses inherent in the use of heat-exchangers, the chemical problems were gradually overcome. A 250 kW power station was put into service in 1913; and thereafter a steady expansion in the sizes and numbers of generating units took place until, by the early 1940s, some 130 MW of geothermal power plant were feeding the electrified Italian railway system. This plant, then the only installation of its kind in the world, was totally destroyed by enemy action during the latter part of World War II, but shortly after the return of peace to Italy reconstruction was begun – not only in Larderello, but also at several other places in the vicinity. Now, a complex of several geothermal power plants having a total installed capacity exceeding 400 MW is supplying the integrated power network of the Italian state organization *Ente Nazionale per l'Energia Elettrica*, commonly known as 'ENEL'. Although larger geothermal power plants are now in service in America, it is to the eternal credit of the Italians that the first really impressive break-through in geothermal power exploitation was achieved.

About half a century was to pass before any other country followed Prince Conti's early pioneering work in geothermal power generation. New Zealand started serious exploration in the Wairakei area in North Island in about 1950, and in 1955 they decided in conjunction with the United Kingdom Government to embark upon a dual purpose chemical/ power project which was to have produced heavy water and 47 MW of electric power. While the plant was under construction the market price of heavy water fell to a level at which its production would no longer have been economic, so the installation was modified to become a 'power only' project. Meanwhile, successful drilling operations had proved a lot more steam, and the project was stepped up in scale to a total installed capacity of 192 MW. The first power unit was commissioned in 1958 and by 1963 the entire project was completed. The discovery of natural gas in North Island and the construction of the 600 MVA Cook Strait cable which enabled large quantities of cheap hydro-power to be transmitted from the South Island to the more populous North Island caused a postponement of the tentative plans that had been drawn up for additional geothermal power development in New Zealand. After the oil crisis of 1973, however, the New Zealand Government decided to proceed with further exploitation of natural steam. As the Wairakei field had meanwhile shown signs that its ultimate potential might fall rather short of earlier expectations, it was decided to develop an entirely new field at Broadlands – also in the North Island – where a 165 MW installation is now planned for commissioning in about 1981.

California was the next country to produce geothermal power. After a very cautious start – 12 MW in 1960 and a further 14 MW in 1963 – the pace of development accelerated rapidly as more and more steam was proved. By 1976 the total geothermal generating plant capacity installed in the Geysers field, some 60 miles or so North of San Francisco, had reached a figure of 522 MW, and the rate of further expansion is now limited more by the need to comply with the stringent environmental standards than by the ability to win steam or to build the plant. Estimates ranging from 2000 to 25 000 MW have been talked about as the possible ultimate potential of the Geysers field alone.

In a historical note such as this it would be tedious to make a full catalogue of every geothermal power development. Suffice it to say that after the great economic success of geothermal power had been clearly demonstrated by the three 'geothermal giants' – Italy, New Zealand and California – a spate of other power plants using earth heat began to appear from about the mid-1960s in Japan, the USSR, Mexico, El Salvador, Iceland and Turkey. By 1976 a total installed geothermal power capacity of 1362 MW had been achieved (Fig. 1) and a far greater quantity of further plant had been definitely planned. Other installations are now tentatively contemplated, while active exploration is being undertaken in a dozen or so countries other than those mentioned above, with power generation in view. The production of electric power from earth heat is now a well established economic activity, but so far (with one exception, namely Paratunka in Kamchatka, USSR) it has been confined to hyperthermal fields only (see Chapter 5).

It is of interest to note that Iceland, one of the most thermally active countries in the world (and whence the world 'geyser' is derived from an

Icelandic word *geysir*, which can most nearly be translated as 'gusher') was a late starter in the development of geothermal power. It is true that Iceland pioneered large scale geothermal district heating, and that a 17 MW plant power was planned in the early 1960s to exploit the thermal field of Hveragerdi [5], but this power plant was cancelled when a large hydro-power installation which could not at that time have competed economically with Hveragerdi for normal demand growth, suddenly became justified by the unforeseen emergence of a large new power-intensive industry which changed the economic perspective. It was only in 1969 that a small non-condensing 3 MW geothermal power unit was first installed at Namafjall. This was rather an anti-climax, but one which was quite natural in a country of so small a population. Iceland will only become a fully qualified member of the community of geothermal power generating nations when the planned 70 MW of plant is commissioned at Krafla in 1977.

1.1.5 *District heating, domestic hot water supply and air-conditioning.* More or less simultaneously with the major development of the Larderello field in Italy for power generation, the Icelanders were initiating large scale exploitation of their earth heat for distict heating and domestic hot water supplies, not on the small scale that had been practised for centuries in various parts of the world, but as a large scale public supply system similar to the electricity and gas distribution systems of big cities. After some modest beginnings at the turn of the twentieth century a pilot district heating scheme was established in 1930 in the capital city of Reykjavik to supply about 70 houses, two public swimming pools, a school and a hospital. The scheme was so successful that the Municipality of Reykjavik started to drill for hot water about 15 km outside the city in 1933 with the intention of ultimately heating the entire capital. By 1943 no less than 2300 houses were thus sup-plied. New areas were then drilled, some within the city boundaries, and gradually the system grew until by 1975 all but 1% of the buildings in Reykja-vik were supplied with domestic heat – an amenity enjoyed by about 90 000 people. While the Reykjavik system had been extended to outlying suburbs, other smaller public heating supply systems were springing up in various parts of Iceland. More than half of the entire population of Iceland were enjoying the benefits of geothermal heating by 1975 and the proportion was rising rapidly.

In the USA too there have been smaller public heating systems utilizing earth heat. At Boise, Idaho, the Warm Springs residential area was geo-thermallly heated as long ago as 1890, while at Klamath Falls, Oregon, a similar scheme has been in operation since the turn of the century. In Japan and New Zealand geothermal domestic heating has developed more on individualistic lines for single homesteads, though recently a group of public buildings in Rotorua, North Island, New Zealand, was collectively supplied with natural hot waters. In the late 1960s space heating began to be developed in Hungary, where a large low grade thermal reservoir occurs below-ground, and to a lesser degree similar activities have been pursued in Kamchatka, in the Eastern USSR. In the early 1970s public heat supplies were made available on a modest scale at Melun, in the Paris Basin, by circulating water at depth through bore-holes originally sunk for oil exploration purposes. The success of the Melun scheme led to the initiation in 1976 of several other geothermal district heating projects in suburban areas around Paris: far more ambitious plans for district heating by means of earth heat in other parts of France are now being contemplated (see Section 11.1).

As a complementary development to space-heating, house-*cooling* (or air-conditioning) was first developed in Rotorua for a hotel building, by making use of a single geothermal bore [6]. Einarsson suggested in 1975 that the destruction of Managua, Nicaragua, in the devastating earthquake of 1972, offered an excellent opportunity for providing a geothermally activated 'public cooling system', [7] but this proposal was not pursued.

1.1.6 *Farming.* Another thriving application of earth heat was first practised in Iceland in the 1920s, when natural hot waters were used to heat greenhouses in which vegetables, fruit, mushrooms and flowers were raised in a country where such produce could not be farmed under natural conditions owing to the prevailing inhospitable climate (Plates 4 and 5). Before then, virtually all food other than fish and potatoes had to be imported into Iceland, but by using geothermal heat under glass or plastic roofs, the country's economy was greatly improved and the population was enabled to enjoy foods that were formerly unobtainable except at exorbitant prices. By 1975 140 000 m² were under cultivation in Iceland in geothermally heated greenhouses. More recently the Hungarians have exploited their natural hot water fields to an even greater extent. By the end of 1975 no less than 1.7 million m² of greenhouse area were being geothermally heated for the cultivation of tomatoes, paprikas, cucumbers, other vegetables and flowers. The USSR, the USA, Italy, Japan and New Zealand have all developed this kind of cultivation to some extent in more recent years, and the farming application of earth heat has been extended to animal husbandry, soil heating, fish hatcheries, dairy farming, egg incubation, poultry raising and the biodegradation of farm wastes. At Agamemnon, near Ismir in Turkey, waste geothermal waters from *hamams* are used to irrigate the fields, where crop yields are alleged to be substantially increased as a result of the warmth. In the State of Oregon, USA, the application of geothermal soil heating is said to have raised the output of corn by 45%, tomatoes by 50% and soya beans by 66%, as well as showing improvement in the *quality* of certain crops [8]. Experiments in geothermal greenhouse heating are being conducted in Ladakh in the Himalayan districts of India where the

Plate 4 Grapes grown in geothermally heated greenhouse, Iceland. (Photo by Icelandic Photo and Press Service, Reykjavik.)

Plate 5 Tropical fruit grown in geothermally heated greenhouse, Iceland. (Photo by Icelandic Photo and Press Service, Reykjavik.)

exceptionally severe climate virtually precludes the possibility of natural farming.

Farming applications require only low grade heat, ranging from about 20° C for fisheries, 40° C for soil warming to 60° C, or perhaps as much as 80° C, for greenhouse heating.

1.1.7 *Industry.* The word 'industry' of course embraces a very wide range of activities, including some of those already referred to in this chapter. With a growing realization that many industries are very heat-intensive – some requiring high grade, but many more requiring only low-grade heat – there has been a steady expansion in the industrial applications of earth heat since the end of World War II, before which such application was more or less confined to the borax industry in Italy, referred to in Section 1.3. The first large scale industrial application of geothermal heat was initiated in the 1950s at Kawerau, North Island, New Zealand, where the Tasman Pulp and Paper Mills have for many years now been producing paper – particularly newsprint. At the Kawerau plant about 200 tons of natural steam are consumed hourly for processing purposes, and incidentally for small scale power generation also. Other examples of geothermally activated industries that have sprung up in recent years are a salt factory, a brewery and a sulphuric acid plant in Japan; timber seasoning and veneer fabrication in New Zealand; the extraction of ammonium bicarbonate, ammonium sulphate and sulphuric acid from geothermal steam in Italy; a curing plant for light aggregate cement building slabs in Iceland; the extraction of elemental sulphur from volcanic fumes in Japan and Taiwan; and the recovery and upgrading of diatomite at Námafjall in Iceland (with some incidental small scale power generation). The application of earth heat to many other industries has been, or is being, considered.

1.2 Scientific precursors

No historical note of this description would be complete without mention of certain great scientists who, while not having directly contributed to the harnessing of earth heat in the service of man, have nevertheless influenced thought that has promoted geothermal development.

First there was *Goethe* (1749–1832) who, though principally renowned as a poet, was a man of great versatility. He might well be regarded as the Father of Geology, for it was he who developed the study of rocks into a serious science worthy of systematic treatment – a science that figures so largely in geothermal exploration. Under Goethe's initial impetus, geology developed rapidly during the nineteenth century, particularly after Charles Darwin endowed the whole of pre-history with an evolutionary perspective, and it advanced further in the twentieth century when isotopic studies became possible.

Lord Kelvin (1824–1907) was fascinated by theories to account for the origin and nature of the interior heat of the Earth, and hoped that in the pursuit of these theories he would be able to determine the age of our planet. Although he outlived by about ten years the discovery of radioactivity by Becquerel in 1896, his theorizing was mainly based upon the classical nineteenth century conceptions of heat. He was unable to reconcile the observed rates of outward heat flow from below-ground with any calculable store of 'original' heat with which the earth could have been endowed at the time of its formation, without conflicting with well established geological evidence as to the age of the earth. But he was largely responsible for directing much thought towards the search for an explanation of the mysteries of subterranean heat.

Sir Charles Parsons (1854–1931), the eminent British engineer whose name is more generally associated with the development of the steam turbine and with astronomical telescopes, also took an interest in subterranean heat. In 1904 he proposed that £5 million – a very considerable sum in those days – should be spent in sinking a shaft twelve miles deep into the earth. He estimated that the work would take about 85 years to complete, and on the basis of 'normal' temperature gradients he expected to encounter temperatures of the order of 600° C at the base of his shaft. The practical way in which this exposed heat was to be reached and used seems to have been rather left to the imagination, and the fact that he referred to his proposal as 'The Hellfire Exploration Project' suggests that Sir Charles's tongue was in his cheek. Nevertheless, he again referred to this proposal in 1919 in his presidential address to the British Association.

Another outstanding scientist who influenced geothermal thought was the German, *Albert Wegener*, who was a pioneer in developing the theory of 'floating continents' (see Chapter 3). It was in 1915 that Wegener first suggested that the earth's crust was behaving in a peculiar manner. His theories aroused fierce scepticism, but modern theory has vindicated his ideas, which have had a profound influence upon geothermal thinking.

There can be little doubt that if Goethe, Kelvin, Parsons and Wegener were alive today they would all be geothermal enthusiasts.

1.3 Development of the environmental conscience

It is fortunate that the rapid acceleration of geothermal development in recent years has coincided with the growing public awareness of the dangers threatening our human environment. The fantastic and self-evident growth of pollution from fuel combustion has at last frightened the more responsible elements of society into a realization of the terrible things we are doing to our planet, and belated recognition of the awesome problems to be faced is now becoming evident. Geothermal energy was at first rather unjustifiably hailed as 'clean', and altogether innocent of pollutive side effects:

it was sometimes argued that its development should be encouraged even for that reason alone. But on closer examination it was found that the exploitation of earth heat has its pollutive problems too, but that these problems are far more manageable than with fuel combustion. A good cause can be harmed by exaggerated claims, but fortunately (as will be seen in Chapter 17) the pollutive aspects of geothermal development are being attacked with vigour and success. So although earlier irresponsible claims of entirely non-pollutive earth heat were not well founded, the balance is now being redressed. So long as we remain alive to the pollutive risks, no serious difficulty need arise in combatting them.

1.4 Conclusion

This brief historical note will serve to show that the exploitation of geothermal energy, apart from one or two earlier minor applications, is really a child of the twentieth century, and that it entered the period of adolescence after the Second World War. It is still quite juvenile, but it may be expected to attain adulthood by the early years of the twenty-first century.

By definition, 'history' stops at the present moment. Nevertheless, the discerning eye can detect, embedded in the present, certain 'arrows' that point to future possibilities – even probabilities. These future prospects will be dealt with more fully in Chapter 19, but it may here be conjectured with reasonable confidence that many more *natural* geothermal fields will be discovered and exploited within the next few years, that it will before long become possible to *create* artificial geothermal fields (where none now exist) in places where impermeable hot rock is to be found at moderate depths, and that ultimately we shall be able to win earth heat from very great depths *wherever we wish*.

The versatility of earth heat

In Chapter 1 the principal applications of earth heat have been broadly classified in the process of their historical development. Moreover, in Chapters 10, 11 and 12 these applications will be dealt with in greater technical detail; but the purpose of this chapter is to emphasize the great *versatility* of geothermal energy.

In the earlier part of the twentieth century there was a strong tendency to think of electricity generation as by far the most important application of earth heat. This was understandable. In the first place electricity seemed to be one of the world's most marketable commodities, since the demand for it was growing at an accelerating pace and since it was easily transported over considerable distances. Furthermore, the geothermal fields that were then attracting attention all provided fluids at fairly high temperatures at which the steam turbine could operate at acceptable (though not high) efficiencies. Apart from these factors, electricity itself is so versatile that it can be adapted to almost any application – lighting, heating, motive power, etc. – so that its generation was analogous to the minting of coin, a commodity having the great merit of exchangeability into whatever energy form we may need. Hence the interest in discovering geothermal fields was chiefly concentrated on finding fairly high temperatures and pressures suitable for power generation. The occurrence of low-grade earth heat was then thought to be chiefly of interest to the balneologist or, if sufficiently hot, to the suppliers of district heating and of domestic hot water. But district heating and domestic hot water supply usually become commercially attractive only for sizeable communities which are often not sufficiently obliging as to be located close to a low grade geothermal field (and the transmission of low-grade heat over long distances is not economically attractive); it is also suited to cold, or fairly cold, climates only, which somewhat restricts the choice of available fields. This is not intended to belittle the importance of district heating and domestic hot water supply, it is only intended to show that it lacks some of the ease of marketability possessed by electricity. In point of fact it has been pointed out by Kunze *et al.* [9] that nearly 20% of the total energy consumed in the USA is accounted for by space heating.

This former bias towards electricity generation as the 'natural' application for earth heat tended to overlook the fact that power generation is essentially an inefficient operation owing to inescapable thermodynamic constraints; and the lower the temperature of the heat source the lower will be the efficiency of energy conversion. On the other hand, the direct use of geo-thermal heat for space heating, industrial processes or husbandry can be highly efficient, since the losses incurred are not imposed by the laws of thermodynamics but only by such imperfections as must inevitably arise from insulation losses, drains and terminal temperature differences in heat exchangers. These imperfections can, moreover, be controlled within econo-mic constraints: they are not dictated by natural laws.

Another important fact is that the sources of high enthalpy natural heat which are both accessible to man and suitable for power generation are believed to be far less abundant than those of lower enthalpy fluids that can be used for other purposes. Kunze *et al.* [10] underline the importance of moderate temperature geothermal fluids (e.g. 150° C) on the interesting proposition of a probable Poisson distribution curve, from which they draw the conclusion that what lower temperature fluids lack in enthalpy they make up for in quantity. As utilization techniques improve, the authors argue, lower temperature fluids will become of increasing economic impor-tance. Moreover, low enthalpy fluids are likely to contain less dissolved solids and gases than high enthalpy fluids and are therefore less likely to cause corrosion troubles. On the other hand the authors admit that a low temperature geothermal field would require more land for a given thermal scale of development than a high temperature field, and that subsidence (see Chapter 17) could perhaps present problems.

The advantages of low grade heat for farming and animal husbandry were realized fairly early in this century though, as with district heating, most of these applications are of greater interest to cold than to hot countries. It was not until well into the middle of the twentieth century that the enormous *industrial* potential of earth heat began to be appreciated – for industries cover a wide spectrum of uses that require heat of all grades from about 20° C upwards, and some industries are very heat-intensive.

Lindal [8] published an interesting chart (Fig. 2) which sets out a large number of uses of heat over a temperature range of 20–200° C. He would probably not claim to have shown every conceivable industry or application on his chart, and he emphasizes that many of the applications shown are practised over a *range* of temperatures rather than at a single value. But the diagram clearly shows that heat of almost any grade obtainable from geo-thermal fluids can be put to good use. Clearly 'fire heat' – i.e. the very high

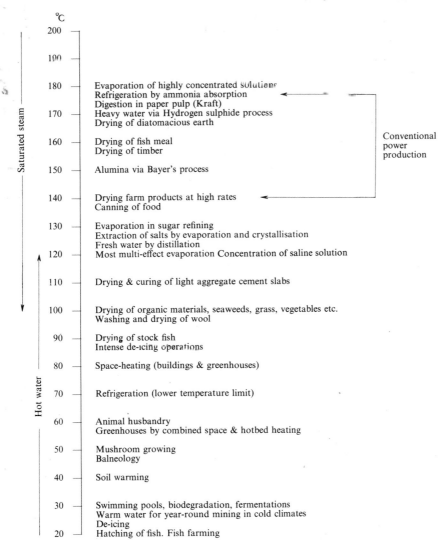

Figure 2 Approximate temperature requirements of geothermal fluids for various applications. (From Lindal [8], p. 146.)

temperatures needed for the pottery industry, smelting, etc. – could not fall within the geothermal 'net' (at any rate not until or unless we become capable of utilizing hot lavas or magma in the conceivable remote future) but the scope of using natural steam and hot water is clearly very great. It should also not be forgotten that something like 90% of the world's consumption of energy is for non-electric purposes. In short, geothermal energy is far too versatile an asset to be used for power generation only.

In a most interesting paper [11] Reistad emphasizes some of the points made here and presents a most revealing breakdown of the different uses of energy in the USA (which may be taken as fairly typical of most highly industrialized countries) under various headings – electricity generation, residential, commercial, industrial and transportation. The electricity is then re-allocated to the other four sectors according to the requirements of each. Each sector's energy needs are further broken down into the various applications such as space heating, cooking, refrigeration, process steam, etc. and a final allocation is then made to one or other of two groups – 'potential geothermal use' and 'non-geothermal use'. The first group excludes all uses requiring temperatures exceeding 250° C: it also excludes transportation and such applications as cooking, which have intermittent demands. Reistad's final conclusion is that about 41% of the USA's energy requirements could be provided by geothermal energy *were it available*. If the upper temperature limit were dropped from 250° C to 200° C the proportion of energy that could theoretically be supplied geothermally would scarcely be affected – a drop merely from 41% to 40% – because very few processes use temperatures between 200° C and 250° C. If the temperature limitation were further dropped to 150° C and 100° C, the percentage of national energy requirements falling within the 'potential geothermal use' group would become about 30% and 20% respectively. This interesting analysis clearly shows that the problem of geothermal application is one of *availability* rather than *applicability*, as the uses to which geothermal energy could be put, if it were available, would be enormous. Reistad's exclusion of transportation (which could partly be provided by electrified railways and – perhaps in future – by battery cars) and of cooking on the grounds of intermittency of demand shows that his conclusions tend to err towards the conservative side. There is growing evidence of a much keener awareness of the wideness of applicability of earth heat – of its immense versatility. The present imbalance between applicability and availability will, it is hoped, be ultimately rectified.

One of the most promising economic potentialities of earth heat is the establishment of dual or multi-purpose plants. Different processes often require different grades of heat. Thus a power plant, for example, taking fairly high pressure steam could exhaust (either wholly from a back-pressure turbine, or partially from a pass-out turbine) low pressure steam that could be used to supply a district heating system, a farm, a distillation plant, a heat-intensive industry, or at least a swimming pool. Conversely, an industry requiring high pressure steam could exhaust low pressure steam to supply condensing turbines. Or again, with a 'wet' geothermal field the steam could be used for one purpose and the hot water for another. The possible combinations of processes are very great. Sometimes more than two end products could be obtainable – e.g. if minerals could be extracted from the thermally spent fluids after they had fulfilled two or more preliminary functions. In this way the most efficient practicable use could be made of earth heat (and perhaps its chemical by-products) in a manner similar to that used in 'total energy' plants that are rapidly gaining favour in more conventional contexts.

Dual or multi-purpose plants are not without their own peculiar problems, the chief of which is the matching of variable demand patterns of the two or more end products; but such problems are not necessarily insuperable, particularly if storage facilities are available, either for hot fluids or for the end products.

The structure of the earth[*]

Before attempting to study or to understand geothermal phenomena, the reader would be well advised first to learn something of the main facts and theories concerning the physical structure of the earth. Although we are rapidly becoming familiar with the conditions of outer space we have, until fairly recently, been surprisingly ignorant of what is going on only a mile or two beneath our feet. That the nether regions of the earth are hot is a fact that has long been known to every schoolboy. This fact, no doubt, was originally deduced from the visual evidence of volcanoes and other surface thermal manifestations that are to be found in several parts of the world. It was given non-scientific, but highly emotional support by the widespread belief, entertained by many of the world's religions, in the existence of a Hell.

3.1 Concentric layers of the earth

The only parts of the earth with which we are directly familiar are the crust and the atmosphere. But these together account for less than half of 1% of the total mass of the planet. The remaining $99\frac{1}{2}$% or more lies concealed beneath the crust, and our knowledge of its nature is indirect, largely deduced from the study of earthquake waves and of lavas and from measurements of the outward flow of heat from the interior towards the surface. Nevertheless, this indirect knowledge has enabled us to build up a fairly clear and consistent picture of the structure of the earth, which is now believed to consist of five distinct concentric spheres, assembled in 'onion fashion'. These, passing from the outside towards the centre of the earth, are the *atmosphere* (itself a complexity of sublayers), the *crust* (including the land masses, the seas and the polar icecaps), the *mantle*, the *liquid core* and finally the *innermost core* which is probably solid. Both temperatures and densities

---

[*] For fuller information on the structure of the earth, and on the bearing of this subject upon geothermics, see [14] and [15].

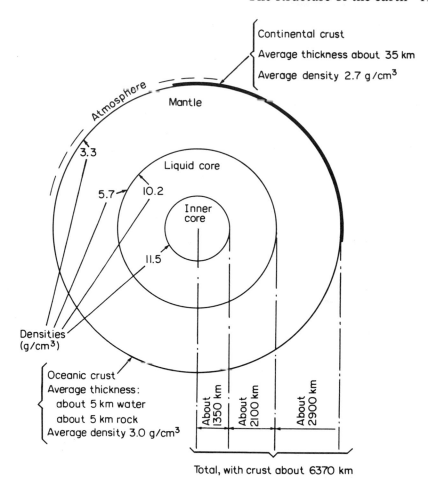

Continental crust
Average thickness about 35 km
Average density 2.7 g/cm³

Atmosphere

Mantle

Liquid core

Inner core

3.3

5.7

10.2

11.5

Densities (g/cm³)

Oceanic crust
Average thickness:
  about 5 km water
  about 5 km rock
Average density 3.0 g/cm³

About 1350 km

About 2100 km

About 2900 km

Total, with crust about 6370 km

Figure 3   Concentric layers of the earth. (After Bullard, p. 20 [15]).

rise rapidly as the centre of the earth is approached (Fig. 3). Thus we have a picture of a very hot planet, efficiently lagged by a thin crust of low thermal conductivity. Thanks to this insulating crust, the conditions on the land surfaces and in the seas are sufficiently temperate to support life. Although the vast quantities of heat stored within the entire planet are of course 'geothermal' it is only with the crust and the upper layers of the mantle that we need concern ourselves for practical purposes.

3.2 The crust
Let us take a closer look at the crust (Fig. 4), which bears to the rest of the planet a similar relationship to that of an eggshell to an egg, except that its

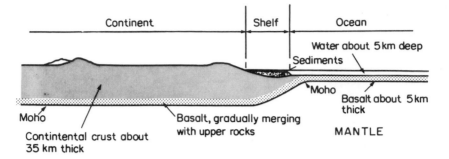

Figure 4 Schematic section through the earth's crust at an inactive coastline (e.g. that of eastern North America). (After Bullard, Fig. 1 [15].)

Note: Under the continent the proportion of basalt probably increases with depth, but it does not necessarily form a separate layer. A great thickness of sediments usually underlies the continental shelf.

thickness is far from uniform. This illustration shows a section (not to scale) through the earth's crust at an 'inactive' coastline such as the eastern seaboard of North America. (The meaning of the word 'inactive' will later be explained). The crust forming the bed of the ocean is very thin (about 5 km). The continental crust is about seven times as thick – even more in mountainous districts – and is fringed with a sedimental 'shelf' of varying width. It is in this shelf that off-shore oil and gas may sometimes be found. Relatively shallow waters, up to about 200 m, cover the shelf, but the depth of the water in mid-ocean averages about 5 km and reaches about 8 km in certain 'trenches'. The boundary that separates the mantle from the crust is known as the Mohorovičić discontinuity (after its Yugoslavian discoverer), or 'Moho' for short. At this boundary there is a sudden change in the velocity of earthquake waves, which indicates a corresponding change in material composition and physical state. The temperature at the Moho is believed to be generally of the order of 600° C, while at the centre of the earth it may perhaps reach about 6000° C.

3.3 The seismic zone

This simplified model here described is not, however, everywhere in a state of equilibrium. High stresses can be built up in the crust and upper mantle, and when these stresses are relieved by material movements the released forces give rise to earthquakes. Seismologists are now able to detect many thousands of earthquakes every year, varying in intensity from the lightest of tremors, imperceptible to the human senses and detectable only by the most delicate devices, to violent shocks capable of devastating cities – such as those that destroyed Lisbon in 1755, San Francisco in 1906, Tokyo in 1923 and Tangshan in 1976. These particularly severe earthquakes

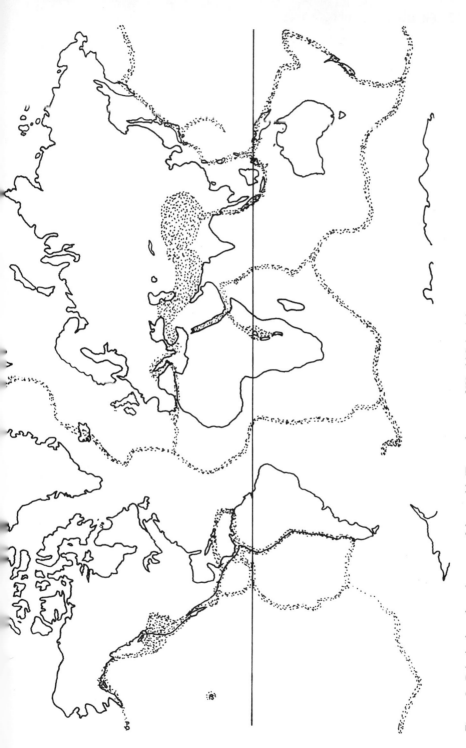

Figure 5 Principal earthquake zones of the world. (From the map of World Seismicity, 1961–9, published by the National Earthquake Centre, Washington, D.C.)

release enormous quantities of energy, incomparably greater than that emitted by the largest of man-made nuclear bombs. It is now possible to map with precision the intensities and locations of earthquakes and the depths at which they originate: it is found that with very few exceptions they are confined to ribbon-like zones that spread over the face of the globe more or less in the manner shown in Fig. 5. These zones are collectively called 'The seismic belt', much of which lies beneath the oceans.

3.4 Plate tectonics

The explanation of this zone pattern baffled investigators for a long time, but since the late 1950s a body of very impressive evidence has been assembled by Sir Edward Bullard of Cambridge University, and others. It now seems almost certain that the crust of the earth consists of several discrete 'plates' – six major and a few smaller ones – that are in a state of continuous relative motion at rates of a few centimeters *per annum*. These relative motions imply that in some places, where the plates are moving apart, the crust must be splitting; while at others, where plates are colliding, the crust must be crumpling. Fig. 6 shows a rough sketch of the two processes. That part of the seismic belt that runs down the middle of the Atlantic Ocean is typical of crust-splitting: the gap formed between the two separating plates is continuously being replenished by the upward flow of hot 'magma' from the mantle to form a steadily spreading ocean floor. The Andes, on the other hand, are typical of colliding plates, where the crumpling has thrown up great mountain ranges. This crumpling effect is modified by the fact that one

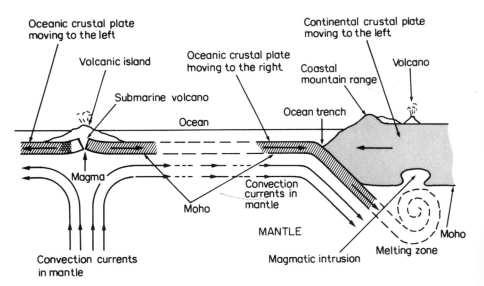

Figure 6   Crustal plate movements (not to scale).

of the colliding plates yields to the other and slides beneath it at a downward angle, finally merging into the mantle whence it originated many millions of years ago. A collision zone, such as that shown to the right of Fig. 6, is 'active', as distinct from the 'inactive' coastline of Fig. 4.

According to the theory of Continental Drift (or Plate Tectonics as it is sometimes known) the seismic belt follows the lines along which the plates are either separating or colliding. For the motion of the plates is believed to occur in a series of jerks, large or small, giving rise to major and minor earthquakes respectively. These jerks occur whenever the built-up stresses exceed the frictional and tensile resistances to motion. The motive power behind the plate motion is not understood with certainty but could well arise from convection currents in the mantle which, though acting as a solid when transmitting seismic waves, behaves like an extremely viscous liquid under the influence of forces maintained over very long periods. In fact the material of the mantle 'creeps' like other materials under the combined influence of high temperatures and mechanical stresses. In some places, contiguous plates are neither colliding nor separating, but sliding past one another in a shearing motion. Such is the San Andreas fault in California which, being kinked or 'stepped' instead of having a plane boundary, can give rise to such dangerous situations as that which caused the destruction of San Francisco in 1906. The kink tends to resist the sliding motion and thus causes the build-up of very high stresses until it finally gives way to the accompaniment of a very severe seismic shock.

A feature common to crust-splitting and plate-colliding zones is the presence of volcanoes – active (Frontispiece and Plate 1), dormant or extinct. The mid-Atlantic seismic belt coincides with a long range of submarine volcanoes, while the presence of visible volcanoes round the periphery of the Pacific Ocean has caused that zone to be known as 'The Belt of Fire'. In a few places the mid-Atlantic Rift rises above sea level, appearing as volcanic islands such as Iceland, The Azores, St. Helena, Tristan da Cunha and others, as shown to the left of Fig. 6. Crust-splitting zones sometimes appear also in the continental land masses—e.g. the Great African Rift Valley, along which there is a line of volcanic cones, including Mt Kenya and Mt Kilimanjaro. Common features of plate-collision zones are the presence of a deep ocean trench on one side of a mountain range and of volcanoes at some distance behind the range on the other side. For example, the Andean volcanoes of Chimborazo and Cotapaxi lie to the East of the main sierra, while the Peru/Chile ocean trench lies at a comparatively short distance to the West of the South American continent. As a corollary to the presence of volcanoes there must be, or have been, magmatic penetrations of hot material from the mantle into the crust as a result of the disturbances at the edges of the plates. Such penetrations may be complete, when lava issues from an active volcano, or they may be partial, when they take the form of magmatic intrusions, or pockets of mantle material bulging into the crustal rocks to within perhaps a few kilometres of the surface. Partial intrusions are probably volcanoes in embryo.

The explanation of the observed pattern of crustal structure at collision zones may perhaps be somewhat as illustrated to the right of Fig. 6. The descending plate, forcing its way into the reluctant mantle, generates intense frictional heat and causes a melting zone to form at some distance inland. The hot, relatively fluid mantle material would be fairly buoyant and could penetrate the base of the continental mass where the high temperature may have rendered the crustal rocks more or less plastic. In the course of time these intrusions could work their way to the surface and erupt as volcanoes.

The theory of Continental Drift dates from 1915 when Alfred Wegener (mentioned in Chapter 1) drew attention to the fact that the east coast of South America would fit snugly against the west coast of Africa (Fig. 7) like pieces of a jigsaw puzzle. He also pointed out that geological and botanical similarities were to be found along the two coasts, and accordingly suggested that the two continents had once been joined together and had later split and drifted apart. He even carried the idea to an extreme by proposing that at some time in the remote past there had existed only a single, very large continent, which he named Pangeia, embodying all the land masses of the world. The whole surface of the Earth, according to his theory, was either ocean or part of Pangeia: there were no islands. Then

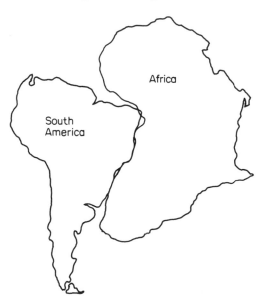

Figure 7   Wegener's observation of continental 'fit'.

at some date, of the order of 100 million years ago, Wegener alleged that unexplained internal convulsions within the core(s) of the earth must have caused this super-continent to break up, with its disintegrated fragments drifting apart until they formed the present configuration of land masses that we know today.

Wegener's theories were at first received very sceptically, but nevertheless our modern conceptions of Continental Drift and Plate Tectonics stem from his original rather vague hypotheses. It has now been discovered from very accurate geodetic measurements that the Americas are in fact moving away from Europe and Africa at a slow, but detectable average rate of the order of 10 cm per annum. Similar relative motions have been detected between other crustal masses. Moreover, if the continental *shelf* be taken instead of the coastline, the apparent 'fit', not only between South America and Africa but also between North America and Europe, is too good to be explained away by mere chance (Fig. 8). Similar good fits are to be found elsewhere; for example, between the west coast of the Arabian peninsula and the African Red Sea coast.

Wegener's postulation of Continental Drift posed some awkward questions at first; for the rates of movement, although they seem slow in absolute terms, are relatively fast on a geological time scale. The separation of the Americas from Europe and Africa to their relative positions at an average rate of 10 cm a year could have been achieved in about 50 million years, while the movement of the Tonga Isles from the Eastern Pacific Rift would have taken about twice as long. Geologists tell us that the age of the earth is of the order of 4500 million years. Why, after surviving for more than 4000 million years, should Pangeia suddenly have started to disintegrate? What cataclysmal event could have accounted for it? Will the dispersing land masses ultimately coalesce again into some new Pangeia within a few hundreds of million years from now, and will the whole process be repeated? Had disintegration and re-coalescence in fact occurred several times before the last Pangeia started to break up?

It is now generally believed that the crustal plates have been in a state of relative motion ever since an early stage of the earth's history, and that these plates are continuously being created at splitting zones and simultaneously destroyed at collision zones. The plates, as it were, 'float' upon the mantle, and in the course of the ages they may re-form and change their directions, like ice floes in an eddying pool, under the influence of magmatic forces as yet not properly understood. There was probably never a Pangeia.

Modern theories of plate tectonics, as broadly described in this chapter, are not without critics; but the evidence in support of these theories is very impressive. Apart from the improbably good fit of the continents as shown in Fig. 8, it is highly significant that whereas earthquakes at splitting zones

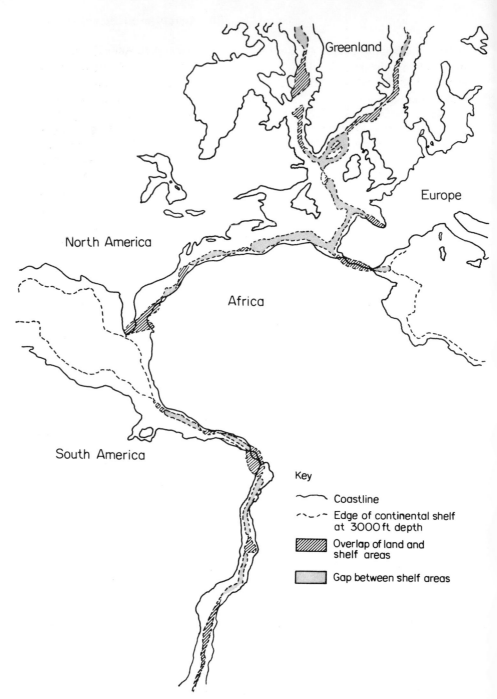

Figure 8 Probable arrangement of continents before the formation of the Atlantic ocean. (After Bullard, p.68, [12].)

generally originate at shallow depths, those that occur on one side of collision zones are often very deep-seated (at depths of up to about 650 km), as would be expected where a downward moving plate is forcing its way into a resisting mantle (Fig. 6). The existence of deep ocean trenches along one side of what are believed to be collision zones, as shown in Fig. 6, is also significant; and what better explanation can be offered of the presence of coal seams in temperate and even Arctic regions (e.g. Spitzbergen) that have clearly been formed by the decay of *tropical* vegetation, than that such parts of the crust have wandered in the course of millions of years far from the torrid latitudes where the primaeval forests once grew? Finally there is further evidence provided by what Bullard aptly describes as 'a kind of geological tape-recorder' to be found in the directional magnetization of the magnetic elements in the ocean floor. It is known that the orientation of the earth's magnetic field is not only constantly shifting but that it has reversed its polarity at various times during the history of the planet. As the newly formed ocean floor cools at the sides of the mid-ocean rift, the magnetic elements of the solidifying rock become magnetized in the direction of the earth's magnetic field. Records of this directional magnetization, obtained quite simply by towing a magnetometer behind a ship, show symmetrical bands of similarly magnetized ocean bedrock on either side of, and more or less parallel with, the rift; thus suggesting a fairly steady outward movement of the sea floor away from the rift [13].

The most critical of detectives, judges and juries could hardly fail to be impressed by this imposing mass of evidence or to reject 'coincidence' as a satisfying explanation. While much remains to be explained, it seems reasonable to accept that the crust and outer layers of the mantle have behaved, and continue to behave, more or less in the manner here described. The earth appears to be a gigantic heat engine, whose precise method of working is not yet fully understood but whose internal energy motivates the crustal creep and the changing patterns of magmatic convection within the mantle.

The importance of the theory of plate tectonics to the geothermicist lies in the fact that, as far as is known, hyper-thermal fields (see Chapter 5) are to be found only within the seismic belt; that is, within zones of crustal weakness where volcanic activity is, or has been, intense. It is therefore within the seismic belt that exploration for high temperature fields at moderate depth should be confined. It is true that lower temperature geothermal fields have been found outside the seismic belt – e.g. in Hungary, in the Russian Steppes, in the Arzacq Basin of France, in Switzerland, Australia and doubtless elsewhere. It is also true that fairly hot zones which do not constitute geothermal fields (see also Chapter 5) occur in places lying outside the seismic belt. The existence of these zones of relatively low grade heat could perhaps be attributed to any or all of the following conditions:

(i) There could be parts of the crust that bear abnormally high proportions of radioactive rocks (see Section 3.5 below).
(ii) The crust could well be relatively thin in places, so that the Moho is at shallower depth than in more normal 'non-thermal' areas. (This emphasizes that Fig. 4 is somewhat stylized: there is no reason to suppose that the Moho is uniformly horizontal beneath the continental masses).
(iii) The temperature of the mantle just beneath the Moho is not necessarily everywhere the same: there could be local magmatic disturbances that give rise to hot spots in certain places.

The fact that the known high temperature fields are all found in or very close to the seismic belt does not imply that such fields are to be found anywhere and everywhere within that belt. They occur only at certain places within the belt where the geological and hydrogeological conditions favour the formation of high temperature fields: hence their comparative rarity.

3.5 Temperature gradients and heat flow

Apart from the visual evidence of a hot interior to the earth, as provided by volcanoes and other natural thermal phenomena, we have the less spectacular, but even more convincing evidence provided by the fact that as we penetrate the earth's crust by drilling or by mining operations we invariably experience a rise in temperature. Observed thermal gradients vary widely from place to place, being as low as about $10°$ C per kilometre in some places; but a fair average in the non-thermal regions of the earth would be of the order of 25 to $30°$ C per km. In a few places, specially favoured by nature, very much higher gradients may be encountered. The mere existence of a thermal gradient implies a continuous outward flow of heat from below-ground to the surface.

Thermal gradient and *heat flow* must not be confused with one another. Both are found to vary from point to point in the earth's crust. The outward *conductive* heat flow at any point in the solid crust will, of course, be proportional to the product of the thermal gradient and the thermal conductivity of the rock at that point. Thus a high thermal gradient could be associated with low heat flow if the rock conductivity is poor. But in permeable zones containing fluids, the outward heat flow is no longer purely conductive but is aided by *convection*, so that this simple relationship is no longer valid. Differences in thermal gradient in impermeable zones are accounted for partly by differences in thermal conductivity and partly by the presence of local hot spots at depth.

Accurate measurements of heat flow have been made in many parts of the world – even in the ocean bed – and it is found to vary quantitatively from place to place, being generally lowest in the ocean bed far from submarine rifts, and highest in parts of the seismic belt. The worldwide average value is about 1.4μ cal/cm^2 s or 0.06 Watts (thermal)/m^2. The escape of heat through active volcanoes, though locally very impressive, forms a negligible fraction of the total outflow of heat from the earth – this total amounting to something like $2\frac{3}{4} \times 10^{16}$ cal/h, or the thermal equivalent of about 30 000 million kilowatts. This is about 17 times the present total capacity of installed power plant in the world, and probably represents about 35 times the rate of energy output of those plants. But it is of little use to lament this apparently wasteful loss of energy, because the natural outward heat flow from below-ground is far too diffuse and of far too low a grade to be of any value to us.

It might be thought that this vast natural outward heat flow originates entirely from the huge quantities of stored heat within the mantle and cores of the earth beneath the crust. It is true that part of it does in fact arise from this source and is conveyed outwards mainly by the thermal conductivity of the rocks, aided in places by fluid convection in permeable zones. Nevertheless, in non-seismic areas this usually accounts only for part of the total heat flow: nearly all of the balance is believed to be derived from radioactivity in the crustal rocks. It is the presence of this radio-activity that largely confused Lord Kelvin when he attempted to account for the age of the earth in terms of original stored heat and of other factors such as shrinkage energy. Most of the earth's radioactive elements are believed to occur in the crust – a fact that may cause some surprise since the radioactive elements are the heavier ones, which could reasonably be expected to gravitate towards the central core of the earth. But as, for the most part, these elements have been dissolved in relatively light-weight rocks, they have tended to concentrate in the outer layers of the planet.

There are three other internal sources of heat which, though of relatively minor importance, could collectively account for appreciable quantities of energy. These are:

(i) The heat released by exothermic chemical reactions within the crust.
(ii) The friction generated in faults, where the sliding action of huge masses of rock exerting enormous gravitational and tectonic pressures against one another is caused by the readjustment of tectonic stresses.
(iii) The latent heat released by the crystalization or solidification of molten rocks on cooling.

3.6 The Mohole project

For many years there has been a project mooted, but repeatedly postponed for one reason or another, known as the 'Mohole' enterprise. This was a proposal advanced in the 1950s to sink a very deep drill-hole right through the crust of the earth, through the Moho into the mantle beneath, choosing as a drilling site a point deep in the ocean where the crust was believed to be at its thinnest. The Mohole Project could be regarded as an updated and restricted version of Sir Charles Parsons's 'Hellfire Project' (see Chapter 1) and would undoubtedly cost a great deal more than the sum which that earlier project was expected to absorb. The underlying idea of the Mohole operation was to bring to the surface direct evidence of what had hitherto been surmise. Its execution could perhaps help to reveal the true nature of the upper mantle and advance our knowledge of the working of the planet. However, the urgency of the project has now waned and interest has declined for two reasons. In the first place certain formations are to be found (e.g. in Cyprus) where the mantle has actually extruded through the crust and outcropped at the surface. Secondly, the great success of the enterprise known as JOIDES (Joint Oceanographic Institutions for Deep Earth Sampling) – an American operation for drilling shallow holes in the ocean floor – has contributed greatly to our knowledge of the structure of the earth and to the study of heat flow. It seems probable that Mohole will never materialize.

4 The heat energy resources of the earth

4.1 Stored heat

The total heat potential of the earth is not easily assessed. In the first place we do not know how much of the internally stored heat could safely be extracted without risking cataclysmal consequences of a seismic or climatic nature. Then we really have no direct knowledge of the total quantity of radioactivity contained in the crustal rocks, nor of the physical properties of the extremely dense matter deep within the mantle and cores of the earth, since the conditions prevailing there cannot be reproduced in the laboratory at the surface.

We do know, as mentioned in Section 3.5, that the continuous natural outflow of heat from the earth is about 30 000 million kW (thermal). But the *natural* outflow through a crust of low thermal conductivity is not to be compared with what could be artificially extracted from the interior of the planet if we could bypass much of the insulating crust – partly with the natural assistance of magmatic intrusions from below, and partly by drilling deep into the crust from above.

A point worth noting is that the natural outward heat flow from below-ground is only a tiny fraction of the heat we regularly receive from the sun – perhaps only about one-fortieth of 1%. Of course much of the total energy received at the earth's surface, from below-ground and from the sun, is radiated into space, particularly at night; but some of the balance is absorbed by photosynthesis in the formation of organic matter in the creation of plant and animal life, some of which is ultimately transformed into fossil fuels. Large amounts of the net energy receipts are just shuffled about through the weather cycle, by the lifting of water vapour and the creation of winds; but the energy thus absorbed is later again degraded into heat as the water vapour is condensed and falls back into the sea, and as the wind energy is dissipated in frictional heat.

Despite all these complicated aspects of the problem it is nevertheless possible to gain a very rough idea of the order of magnitude of the earth's heat resources, in the following manner:

The following facts about the earth are known:

Mean diameter	12 740 km, or 1.274×10^7 m.
Mean radius	6 370 km, or 6.37×10^6 m.
Volume	$4/3 \times \pi \times 6.37^3 \times 10^{18} = 1.083 \times 10^{21}$ m³.
Density	5.52 (mean)

Hence the mass of the earth $= 5.52 \times 1.083 \times 10^{21} \times 1000$ kg
$$= 5.978 \times 10^{24} \text{ kg.}$$

The next factor of importance, and one of which we have no direct knowledge, is the average specific heat of the earth. It may be noted that the following specific heats of various materials apply to conditions near to the earth's surface:

Water	1.0
Ice	0.504
Rocks	from 0.19 to 0.22 (average about 0.2)
Iron	0.13
Nickel	0.109.

Now the first three items are the principal ingredients of the crust, which accounts for less than 0.5% of the earth's mass; and the rocky part of the crust greatly outweighs that of the oceans and ice-caps, forming more than 90% of the mass of the crust. Nevertheless, the very high specific heats of water and ice, despite their small proportions by weight, could probably effectively raise the average specific heat of the crust to perhaps something of the order of 0.25. The composition of the mantle, which accounts for about 68% of the earth's total mass, is very complex, but its outer layers are presumably of a 'rocky' nature since they are the source of volcanic emissions. As a rough approximation an average specific heat of 0.2 will here be assumed for the mantle. The two innermost cores of the earth, which together account for about $31\frac{1}{2}$% of the mass of the earth, are believed to consist largely of iron and nickel, so that an average specific heat of about 0.12 would not be unreasonable for these cores. Hence a *very rough* first estimate of the mean specific heat of the earth could be reckoned as follows:

$$0.005 \times 0.25 = 0.00125$$
$$0.68 \quad \times 0.2 \ = 0.136$$
$$0.315 \times 0.12 = 0.0378$$

$$\overline{} \qquad \overline{}$$

$$1.000 \qquad 0.17505$$

i.e. Average specific heat $= 0.175$

No one can say, however, what the effect would be of the immense pressures and high temperatures at depth upon the specific heats of the composite materials. Nor can we by any means be sure of the accuracy of the average specific heats assumed above for conditions of normal pressures and temperatures. Little confidence can therefore be placed upon the accuracy of the mean figure of 0.175 deduced above. To be on the cautious side, let it be assumed that *the mean specific heat of the earth is 0.15.*

It is now possible to make a rough estimate of the amount of heat that would be released if the average temperature of the earth were reduced by a thousandth part of one degree centigrade (0.001°C), namely:

$$\frac{5.978 \times 10^{24} \times 0.15}{1000} = 8.967 \times 10^{20} \text{ kgcal.}$$

Or, since 860 kgcal are equivalent to 1 kWh (thermal), 1.043×10^{18} kWh (th).

Or again, since 1 t of coal equivalent (1 tce) is defined by the United Nations Statistical Department as 8130 kWh

$$\frac{1.043 \times 10^{18}}{8130} = 1.283 \times 10^{14} \text{ tce} = 1.283 \times 10^{8} \text{ Mtce.}$$

Now it is of interest to note that the probable world energy consumption in 1976 was about 9250 Mtce (actual statistics not yet available). Hence, by cooling the earth through 0.001°C we could theoretically supply the world with its energy needs *at the 1976 level* for ... $1.283 \times 10^{8}/9250$, or 13 870 years (say about 14 000 years).

Alternatively it may be noted that the world's *electrical* production in 1976 was probably about 6.625×10^{12} kWh. Let it be assumed that all the heat removed from the earth by cooling it through 0.001°C were converted into electricity at an efficiency of 16% – a low figure deliberately chosen on account of the relatively low temperatures and pressures normally associated with good quality geothermal fields by comparison with those adopted in large modern fuel-fired power plants. It would then be possible to generate $0.16 \times 1.043 \times 10^{18}$, or 1.669×10^{17} kWh. Hence, by cooling the earth through 0.001°C we could keep the world supplied with electricity at the 1976 level for $(1.669 \times 10^{17})/(6.625 \times 10^{12})$, or 25 192 years (say about 25 000 years).

It must be admitted that prognostications over such very long periods are of little real value, since even the *survival* for so long of the species flatteringly known as *home sapiens* is by no means certain, and even if mankind should still walk the earth after all those thousands of years, his way of life, his energy requirements and the population will almost certainly have changed beyond all recognition. In any case, it may justifiably be asked, what is there that is magical about the arbitrarily assumed temperature drop of 0.001°C? A very small average temperature drop has here been deliberately chosen because it is most improbable that we shall ever succeed in extracting heat except from the upper layers of the earth, so a very small *average* temperature drop for the whole planet could be achieved by cooling only the upper layers through quite a substantial temperature drop. Nevertheless, the drop chosen was entirely arbitrary and has no special significance: it was intended merely to give a broad idea of scale. The figures, even in the form expressed, could be inaccurate on such counts as the assumed mean value of specific heat, and the fact that only the *stored* heat of the earth has been taken into consideration without regard to the heat that is being, and will continue to be, generated by radioactive decay and by the other minor sources mentioned at the end of Section 3.5. (It has even been suggested [104] that the earth may be getting *hotter* on account of radioactivity. This, however, seems improbable for, as stated in Section 3.5, it is believed that most of the earth's radioactivity is located in the crustal rocks. If the mantle and cores were getting hotter as a result of this radioactivity, *negative* temperature gradients would be implied in the lower levels of the crust; and so far no evidence of such a phenomenon has been detected. Any heating of the upper crustal rocks by more deeply seated radioactivity would simply appear as part of the observed heat flow averaging 1.4 μ cal/cm²s which, as already stated, is very small by comparison with the solar energy reception and which is mostly radiated to the sky at night.)

Quite apart from all these considerations, the figures quoted above are misleading in that they ignore the effects of demand growth – effects which, as will be shown in Chapter 20, can be startling. For example, at 5% annual compound growth the periods of 14 000 and 25 000 years deduced above would shrink to less than $1\frac{1}{2}$ centuries. As, however, it is certain that we cannot in future tolerate growth at rates anywhere approaching those to which we have become accustomed, the curtailment of the earth heat durability times would be far less drastic than this, though it could still be considerable.

There has been one purpose, and one purpose only, in bandying these rather fanciful figures about, and that has been to show that the heat reserves of the earth are gigantic. If we hope to arrive at figures with any semblance of meaning it would be more realistic to confine our attention to the *crust* only, and moreover just to that part of the crust that lies beneath the land masses, where accessibility offers fewer problems. Even then, the value of any derived figures would be strictly limited.

The thermal capacity of that part of the crust that lies beneath the land masses (excluding Antarctica) is about 1/300th of that of the entire planet. Since we are now dealing with such a small fraction of the earth, it would be reasonable to assume a larger (though still arbitrary) standard temperature drop as a datum of reference, because it would be easier to spread heat extraction activities more evenly over the land-covered crust than throughout the whole globe. One-tenth of a degree Celsius will now be assumed.

Hence, by cooling only the land-covered crust through 0.1°C we could supply the world with its energy needs *at the 1976 level* for

$$\frac{14\,000}{300} \times \frac{0.1°\,C}{0.001°\,C}, \text{ or about 4700 years,}$$

and its electricity requirements, also *at the 1976 level* for

$$\frac{25\,000}{300} \times \frac{0.1°\,C}{0.001°\,C}, \text{ or more than 8000 years.}$$

Even these figures are still impressive, especially when we take into account a fact that will be brought out in Chapter 20, namely, that growth rates *must* be substantially and rapidly reduced. Moreover, whilst it may be prudent to assume that we could never recover heat from elsewhere than the crust beneath the land areas of the world – which incidentally amounts to about 7 Mtce/sq. mile of land surface per degree C of cooling – there is no reason why we should not ultimately succeed in extracting heat from beneath parts of the continental shelf or extend our range of heat capture into the upper layers of the mantle. The quoted figures could therefore in some respects be regarded as conservative.

The modest temperature drop of 0.1°C here assumed is in recognition of the fact that it would never be practicable to extract heat uniformly from all parts of the land-covered crust. Even in the remote future heat extraction would necessarily have to be confined to a limited, though perhaps large, number of places scattered over the land surface, at each of which quite substantial temperature drops might be attained, but the crustal average would be small. The permissible *local* temperature drop would be limited partly by the hazard of creating high seismic stresses and partly by considerations of crustal shrinkage, which would amount to about 4 ft/°C of average temperature drop beneath the affected locality. These factors must be studied empirically, but a certain amount of work has been, and is being, done in this direction. As to the permissible world total of heat extraction, this might ultimately have to be limited by climatic considerations. Regardless of conversion efficiency, nearly all the extracted heat would ultimately find its way back into the atmosphere and oceans. Could we continue to consume energy, even at the present level (disregarding possible future growth) for centuries without adversely affecting the climate? More will be said of this in Chapter 17, but here it may be broadly stated that this is a problem with which we need not probably concern ourselves – at least for a very long time.

The geographical density of heat extraction is likely to vary widely from place to place for all time. For example, at the rate of 7 Mtce/sq. mile/°C quoted above, the 1976 energy consumption of the United Kingdom (325 Mtce) could have been supplied by cooling the crust beneath 464 sq. miles through 0.1° C, and the entire land area of that country could support heat extraction at that level for 183 years. For the whole world's land area (excluding Antarctica) the 1976 energy demand (9250 Mtce) would require about 13 200 sq. miles of crust to be cooled through 0.1° C and, as already shown, extraction could be sustained at that level for about 4700 years. The discrepancy between the 183 years and the 4700 years is of course due to the fact that the United Kingdom is densely populated and highly industrialized. A larger temperature drop would probably have to be tolerated in such countries.

It is of interest to turn to the World Energy Conference 'Survey of Energy Resources (1974)', in Table IX-2 of which the following figures are given of the world's *measured* energy reserves (Table 2).

Now uncertain though these figures may be, it is worth noting that the estimated grand total of all measured resources of fossil and fission fuels could theoretically be matched by cooling the whole earth through only $(3.472 \times 10^{12})/(1.283 \times 10^{17})°$ C, or 0.000 0271° C (or the crust beneath the land masses alone by about 0.01° C).

It is only fair to point out that these theoretical cooling figures are of course *average* world values. On a national basis the equivalent figure could be very different. For example, it has been suggested that the gross *coal* resources in Britain may be as much as 16×10^{10}t, including coal-fields inferred but not yet proved and ignoring the fact that only a fraction of them may ever be commercially exploitable. However, if the whole of this coal could conceivably be recovered and burned it would be equivalent to cooling that part of the crust that lies beneath Britain by something of

Table 2   The world's measured energy resources (1974).

Resource	Millions of tera-joules	Millions of tonnes of coal equivalent
Solid fuels	15 111	531 000
Crude oil	3 958	139 000
Natural gas	1 933	68 000
Shales and tar-sands	10 921	384 000
Total fossil fuels	31 923	1122 000
Uranium (non-breeders)	824	29 000
Total	32 747	1151 000
Uranium (breeders)	49 463 ⎫	2321 000
Thorium (breeders)	16 602 ⎭	
Total measured reserves	98 812	3472 000

the order of $\frac{1}{4}$°C–perhaps about 30 times as much as the world average 'land-covered crust cooling figure'. This is indicative of a very high concentration of coal reserves beneath a very small area of the crust: it also emphasizes the rather nebulous implications of such figures, which should therefore not be treated too seriously.

Imprecise though such figures may be, they serve at least to show that, unless the problem of thermonuclear fusion is ever mastered, there would seem to be no other energy source that could possibly compete with geothermal heat. It is true that in a little more than a week the amount of *solar* radiation falling upon the earth's surface would equal all the measured reserves listed above, but the collection of such a very diffuse source would pose immense problems, whereas it is known that large local concentrations of earth heat can be won merely by using existing technology.

To descend from the global to the local scale, it is of interest to note that Decius [16] has estimated that one cubic mile (rather more than 4 cubic kilometres) of magmatic intrusive material, in cooling from 1800° F to 100° F – i.e. through 944° C – would liberate enough heat to generate 28 million kilowatt-years, or about $3\frac{1}{2}\%$ of the world's electrical production in 1977.

4.2 Renewability of earth heat

Is geothermal energy a *renewable* source? With the gigantic reserves that have been disclosed above it could, in one sense, be regarded as almost infinitely renewable. Nevertheless, on a local scale it is possible for a field to become exhausted through excessive exploitation. The Italian field in Tuscany continues to maintain its output of about 400 MW, and may even expand in capacity, but only by drilling more and more wells over an ever-widening area. A day must surely come when the boundaries of the field are reached and the stored heat begins to run down, though that may not happen for quite a long time. In Wairakei, New Zealand, the rate of hot fluid recharge is known to be a high proportion of the rate of draw-off, but this could perhaps be due in part to the extent of the field being wider than was at first believed and in part to the pick-up of residual rock heat from inflowing cooler waters from further afield. If a field were run down to an uneconomically low temperature for the purpose for which it is being exploited, it might in theory recover in the course of a very long time; but this would almost certainly involve periods of no commercial interest – perhaps centuries or millennia – for thermal conduction through rock is an incredibly slow process, while other forms of heat infeed (see Chapter 5) are quantitatively so individual to each field as to be quite unpredictable. If a field were tapped only to the extent of the rate of infeed from depth, it would be truly renewable – or at any rate for as long as the infeed were maintained; but although rates of infeed are not easily deduced, they are generally believed to be fairly small in relation to the rates of withdrawal at which a field becomes commercially attractive. It has been noted in Sections 3.5 and 4.1 that the total natural outward heat flow from the earth is of the order of 3×10^{10} kW. Even if it be assumed that the heat flow from those parts of the surface that are covered by the seas is irretrievably lost, the remaining heat flow passing through the land-covered areas is still substantial – about 9×10^9 kW, or about 9700 Mtce p.a. which is *of the same order of magnitude as the present total world energy consumption.* Can nothing be done to recover this renewable heat flow? Even though it must very slowly decline owing to the gradual loss of stored heat and the exponential decay of the radioactivity, the rate of decline would be virtually imperceptible for many centuries. It might therefore at first be thought that, leaving aside the question of future demand growth, it should be possible to renew energy from this heat flow at a rate more or less sufficient to make good our *present* consumption: that is to say, that earth heat is largely, though not completely, renewable. But on reflection it will be realized that although some of this heat could be recovered at extraction points where the crust is being cooled, most of it would be lost where it passes between extraction points; and even at the latter the underground temperatures would remain fairly high so that there also much of the natural heat flow would continue to pass outwards. The answer to the posed question is therefore regrettably 'No. Earth heat is *not* a renewable energy source except perhaps in a very few places such as Iceland, where there are grounds for believing that magmatic convection beneath a thin crust is exceptionally high'.

5 The nature and occurrence of geothermal fields

5.1 The comparative rarity of geothermal fields

From what has been said in Section 3.4 it will be understood that geothermal fields are rarities. Those of higher temperature are to be found only in a few parts of the seismic belt, which itself covers only a very small fraction of the earth's surface. Fields of lower temperature, however, are not confined to the seismic belt, but again they are to be found only in certain favoured places. By far the greater part of the earth's surface is non-thermal.

5.2 Classification of 'areas'

It is convenient to classify all the areas of the earth's surface into three broad groups, so:

(i) *Non-thermal areas*, having temperature gradients ranging from about 10 to 40° C per kilometre of depth.

Thermal areas. These are of two kinds:

(ii) *Semi-thermal areas*, having temperature gradients of up to about 70° C per kilometre of depth.

(iii) *Hyper-thermal areas*, having temperature gradients many times as great as those found in non-thermal areas.

(The absence of specific temperature gradients in the case of hyper-thermal fields is deliberate, for there are places – e.g. Lanzarote in the Canary Islands – where gradients can more conveniently be expressed in terms of degrees per centimetre rather than per kilometre [17]. In fact on Montaña del Fuego in Lanzarote it is possible to burn one's fingers by scraping away a few inches of the ground surface with one's hands. It is for this reason that the rather vague expression 'many times as great' has been used).

It could be argued that a more logical basis for classifying the areas of the world would be in terms of *heat flow* rather than temperature gradient; but whereas heat flow must be deduced after making allowance for the thermal conductivity of rocks and for fluid escape, temperature gradient is an easily measured and simple function that can usefully serve as a preliminary basis for this very broad and rough classification which, it is emphasized, has no sanction of classic authority whatsoever: it is purely a suggestion of the author's. Others might well prefer some different means of classifying areas. It should also be understood that the gradients used in the above definitions are only approximate figures, and not clearly defined demarcations.

5.3 Classification of 'thermal fields'

It should be noted that up to this point no mention has been made of the word 'field', but only 'area'. Every point on the earth's surface lies within some *area*, but by no means does every point lie within a *field*. It is now necessary to make a distinction between a 'thermal area' and a 'thermal field'. A thermal *area*, if associated only with impermeable rock below the surface, is, with the techniques at present available to us, incapable of exploitation. A good example of a thermal area which can scarcely claim to be a thermal field is to be found at Pathé in México. Here there is a temperature gradient of about 550° C/km, but despite the drilling of 17 bores, some of which are of diameters up to 12-inches, the amount of steam won is so feeble that only about 150 kW of power can be generated (despite a 3500 kW installation). This is due to inadequate subterranean permeability.

A thermal *field* may be defined as a thermal area (either semi- or hyper-) where the presence of permeable rock formations below ground allows the containment of a working fluid, without which the area could not be exploited. The working fluid – water and/or steam, normally associated with certain gases – serves as a medium for the conveyance of deepseated heat to the surface. Since geothermal fields require both water and heat for their existence, they are often referred to as 'hydrothermal systems'.

Like 'areas', geothermal fields may also be conveniently classified into three types, as follows:

(i) *Semi-thermal fields*, capable of producing hot water at temperatures up to about 100° C from depths of 1 or 2 km.
Hyper-thermal fields. These are of two kinds:
(ii) *Wet fields*, producing pressurized water at temperatures exceeding 100° C, so that when the fluid is brought to the surface and its pressure reduced, a fraction is flashed into steam while the greater part remains as boiling water.
(iii) *Dry fields*, producing dry saturated, or slightly superheated steam at pressures above atmospheric.

Hyper-thermal wet and dry fields are sometimes referred to as 'water-dominated' and 'steam-dominated' fields respectively.

5.4 Distribution and potential values of areas and fields

There seems to be some kind of perverse law of nature whereby the world distribution of areas and fields is more or less in inverse ratio to their actual or potential economic worth, as shown in Table 3.

As already stated, hyper-thermal fields, both wet and dry, appear to be confined to the seismic belt. Semi-thermal fields may occur elsewhere. Whereas hyper-thermal fields of either type can supply comparatively high grade heat that can be exploited for a very wide range of applications, including electric power generation, the use of semi-thermal fields (which supply low grade heat) is more restricted, being suitable for district heating, certain industries and husbandry (though sometimes they could perhaps be used for power generation in conjunction with binary fluids, as will be explained in Chapter 10). This explains why hyper-thermal fields have been shown in Table 3 as of greater economic worth than semi-thermal fields. The reason why dry hyper-thermal fields have been shown as still more valuable than wet is that they are free from certain technical problems associated with the latter, as will later be seen.

Hyper-thermal areas, like hyper-thermal fields, also appear to be confined to the seismic belt : they are assumed to be due to the presence of hot impermeable rocks at quite moderate depth. Semi-thermal areas are not confined to the seismic belt and are assumed to be due to the presence of hot impermeable rocks at greater depths – say 2 to 3 km.

One interesting thermal occurrence worth noting is in the island of Lanzarote in the Canary Islands. As already mentioned in Section 5.2, the temperature gradients experienced there are fantastically high – in fact as much as about 100° C/metre near the surface. Whether the Montaña

Table 3 Distribution and potential values of areas and fields.

Approximate order of frequency of occurrence	Areas and fields	Economic worth	Remarks
Commonest	Non-thermal areas	At present worthless	Perhaps exploitable after many years
	Semi-thermal areas	At present worthless	Perhaps exploitable after a decade or so
	Hyper-thermal areas	At present worthless	Perhaps exploitable fairly soon
	Semi-thermal fields	Valuable	Exploitable now
	Wet hyper-thermal fields	More valuable	
Rarest	Dry hyper-thermal fields	Most valuable	

del Fuego is a hyper-thermal *area* or a hyper-thermal *field* is still the subject of discussion, for the exact nature of the subterranean phenomena is not yet properly understood; but none can dispute the fact that it is hyper-thermal. And yet it does not lie on any clearly defined part of the seismic belt, but some 700 km or so to the south of the East–West spur that branches off from the mid-Atlantic Rift and passes close to Gibraltar (Fig. 5). Nevertheless, Lanzarote has been subject to volcanic eruptions in the earlier part of the eighteenth century and again in 1824; so perhaps it lies on a dormant 'tributary' of the more clearly defined seismic belt to the north.

Although, for convenience of grouping areas and fields, semi-thermal fields have been shown in Table 3 as rarer than hyper-thermal areas, it is quite probable that the reverse is true, since the former are not confined to the seismic belt. Although comparatively few semi-thermal fields have been clearly defined, and still fewer actually exploited, it is now believed that they are likely to be very widespread. Some of them are thought to be of huge extent. For example, Tikhonov and Dvorov [18] have claimed that 50–60% of the territory of the USSR contains economically exploitable thermal waters.

Fig. 9 shows the occurrence and grouping of fields and areas. Considering the fact that the seismic belt forms such a small fraction of the total surface of the earth, and that much of it lies beneath the oceans, it is not surprising that hyper-thermal fields – wet or dry – are comparatively rare phenomena. It will be noted that in this figure all 'areas' not associated with permeable formations are marked as 'not at present exploitable'; but this does not imply that they never will be exploitable. More will be said about this in Chapter 19, but meanwhile it is well to remember that even in non-thermal areas there is a great deal of heat some few kilometres below the surface. These hot rocks at great depth, which will almost certainly be impermeable, will not necessarily always remain inaccessible. It is probable that in non-thermal areas there is low grade heat at about 100° C only 3 or 4 kilometres below ground, and high grade heat at about 250° C at 9 or 10 kilometres depth. Even now we are technically capable of reaching these depths, though it would not be economic to do so; but the problems of exploitation at such great depths involve more than mere penetration.

The presence of a hyper-thermal field is usually, but not always, accompanied by surface manifestations of thermal activity, such as geysers and fumaroles; but the converse is not necessarily true. Surface manifestations may simply be heat leakages through occasional cracks in a generally impervious formation. Likewise, excellent fields have been detected – e.g. at Monte Amiata in Italy – in places totally devoid of any visible surface thermal phenomena whatsoever.

Evidence of semi-thermal fields is seldom betrayed at the surface at all, unless perhaps by warm springs; so there is nothing inherently improbable

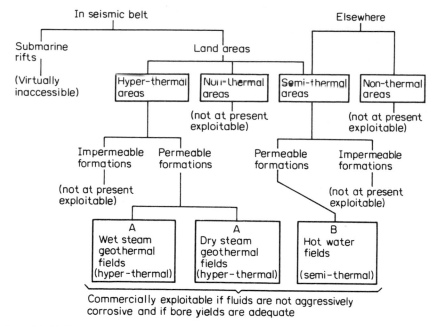

A, exploitable for power generation or for other applications under B below.
B, exploitable for district heating, industrial or other purposes, and perhaps for power generation if binary fluids are used.

Figure 9   Classification of 'areas' and 'geothermal fields'.

in the modern belief expressed earlier in this Section, that such fields may be far more widespread than was formerly thought.

5.5 Speculation concerning the nature of geothermal fields

The structure of geothermal fields, both hyper- and semi-thermal, and the mechanism of the upward transfer of deep-seated heat to higher levels are still subject to a certain amount of speculation. As underground fields are invisible, it is necessary to rely more upon our powers of deduction than upon direct observation. Nevertheless, from the evidence collected over the years in the course of exploration and exploitation it is possible to construct a 'model' consistent with the observed behaviour of the field and capable of describing the underground processes which could cause that behaviour. The probable validity of such a model must stand or fall by the accuracy with which predictions drawn from it are fulfilled. The most obvious and practical predictions of importance are the locations of promising drilling sites for gaining access to hot exploitable geothermal fluids. Hence if drill sites chosen according to the model show a high success ratio, the implication may reasonably be drawn that the hypothetical model fairly closely resembles the actual physical features of the field. If the success ratio is low, then clearly the model must be faulty. The validity of the model can be further checked in the course of time against predictions of downhole changes of pressure and temperature, and other symptoms of field behaviour. So much exploration and research has now been carried out in the last two or three decades that the fog of uncertainty is gradually lifting. Working hypotheses have been advanced that have generally fitted fairly well with the observed facts; and the present concensus of opinion is that the nature of thermal fields is likely to be somewhat as broadly described in the next two sections of this chapter, though new theories and models may well be advanced in future to cause some of the previously held theories to be modified.

5.6 Model of hyper-thermal field
A hyper-thermal field requires five basic constituents:

  (i) a source of heat.
 (ii) a layer of bedrock.
(iii) an 'aquifer', or permeable zone of fractured and fissured rock capable of entraining a large quantity of water and/or steam.
(iv) a source of water replenishment to make good the fluid losses induced by nature or by artifice from the aquifer.
 (v) a 'cap-rock' to prevent the wholesale loss of heat and vapour from the field into the atmosphere.

Such a combination of five widely different features cannot be expected to occur very often; hence the comparative rarity of hyper-thermal fields. The relative arrangement of these features is probably somewhat as illustrated in Fig. 10.

In seeking to account for the first requirement – a source of heat – it surely cannot be without significance that hyper-thermal fields occur only within the seismic belt (with the possible exception of Lanzarote which, while lying to one side of the belt, has nevertheless been subjected to recent volcanic activity). If the theory of plate tectonics outlined in Chapter 3 is broadly correct, then hyperthermal fields will occur only at those places of crustal weakness where adjacent plates are in relative motion and at which the intrusion of magma into the crust is liable to occur (Fig. 6). It is therefore generally believed that the main source of heat beneath a hyper-thermal field is a *magmatic intrusion*. Such an intrusion, bulging into the body of the crust, has the effect of reducing the thickness of the thermally insulating crustal cover. Moreover, magmatic intrusions will tend to occur where the temperature of the magma is higher than the general temperature level of the mantle just beneath the 'Moho' in undisturbed places, because of the high frictional forces beneath plate-colliding zones and because of gaping cracks penetrating deeply into the mantle at splitting zones. Hence

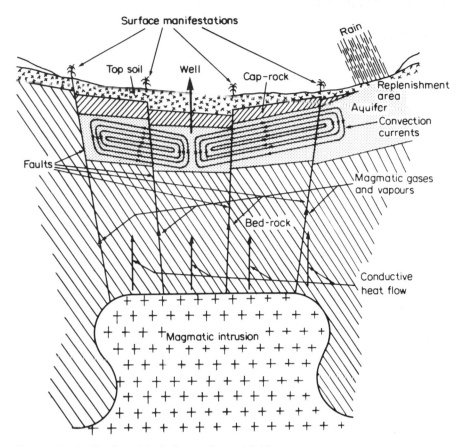

Figure 10   Stylized model of a hyper-thermal field.

the high outward heat flow in hyper-thermal fields may chiefly be due to these two causes – local hot spots in the mantle and local thinning of the crust. Other possible contributory sources of heat are:

(i)   high concentrations of local radioactivity in the crustal rocks.
(ii) exothermic chemical reactions.
(iii) frictional heat due to the differential motion of rock masses sliding over one another at geological faults.
(iv) latent heat released on crystallization or solidification of molten rocks.
(v)   the direct ingress into the aquifer of intensely hot magmatic gases forcing their way through faults in the bedrock.

All but the first and last of these possible contributory sources are probably of minor importance.

The second requirement of a hyper-thermal field – a layer of bed-rock – needs no special explanation, as it will usually form part of the mainly basaltic crustal rocks. As the bedrock generally occurs at great depth – probably from 2 to 5 km at its upper boundary – the prevailing lithostatic pressure will usually be so high as to render it more or less impermeable by squeezing out of existence any fissures, especially horizontal ones, that may have been temporarily formed at moments of high mechanical stresses. Nevertheless, despite this re-closing tendency, it is possible that vertical or near-vertical faults of very narrow width could perhaps survive at these great depths under the influence of extremely high pressure magmatic gases and vapours flowing upwards from far greater depths.

The third requirement of a hyper-thermal field is the aquifer, or permeable zone of rock. The word 'aquifer' of course means 'water-bearer', but the term is also used to describe *steam*-bearing permeable zones occurring in hyper-thermal fields. The permeability of the aquifer is probably accounted for by the incidence and variations of frequent and intense thermal and mechanical stresses that have been induced at times of volcanic activity, resulting in cracks and fissures in rocks of relatively low mechanical strength.

Before discussing the fourth requirement of a hyper-thermal field – a source of water replenishment – it is well to mention that there are three types of water with which the hydrogeologist and 'field model-maker' are concerned:

(i)   *meteoric waters*, which is merely a 'snob' term for rainwater or the waters released from melting snows and ice at the earth's surface.
(ii) *magmatic waters*, which originate from the released vapours from the water of crystallization when magmatic material fully solidifies on cooling.
(iii) *connate waters*, as implied by the adjective, are waters 'born with' some rock formation – e.g. sea-water trapped in marine rock formations, or water of crystallization released from certain rocks when their stability has been disturbed by chemical or physical changes.

The water within a hyper-thermal field, and its associated vapour, could be any of these types. First, the permeable aquifer may outcrop at some distance from the field, as shown in Fig. 10, in which case meteoric waters will continually be flowing into the aquifer to make good any *natural* losses occurring at points of surface activities (e.g. hot springs, geysers, etc.) together with any *artificial* fluid losses arising from the draw-off through boreholes when a field is exploited. Secondly it is possible that magmatic water vapour under very great pressure and at very high temperature may rise through the bedrock from the magmatic intrusion beneath and enter the fissures and voids of the aquifer. Finally, connate waters could occur within a field, but as they would lack the means of replenishment they play only a minor part in most thermal fields.

The last requirement of a hyper-thermal field – an impermeable cap-rock to form the 'lid of the kettle' that prevents the wholesale escape of heat and vapour into the atmosphere – has caused much speculation in past years. The presence of such a convenient 'lid' in all known hyper-thermal fields seemed to be too fortuitous. In a few cases, of course, layers of impervious rock may have been deposited quite naturally in the course of geological evolution. But in others it seems probable that in the remote past permeability was not confined to the aquifer alone, but extended right up to the surface, in which case there would have been a continuous escape of geothermal fluids and of heat to the surface for a very long time by way of fissures in the upper layers of the rock formation. If this were so, then the very escape of fluids could have caused a self-sealing process that could ultimately block the fissures and render the upper layers of the rock impermeable, thus forming a cap-rock that would prevent further escape of heat and fluids. Such self-sealing action could be caused by two chemical processes; first, the deposition of minerals (mainly silica) from solution, and secondly the hydrothermal alteration of the upper rock causing kaolinization. This theory was advanced in 1964 by Facca and Tonani [19]. The evidence of the suggested processes can clearly be seen in certain fields, and the theory enables the far-fetched assumption of coincidence to be dispensed with.

Thus we have a plausible general model of a hyper-thermal field in its simplest form, as illustrated in Fig. 10. Some years ago it was commonly believed that magmatic vapours were probably responsible for supplying both the source of heat and a large fraction of the water or vapour held in the aquifer; but isotopic measurements have now largely disproved this theory. Water of magmatic origin differs from water of meteoric origin in its deuterium content. The ratio of hydrogen to deuterium (H/D) in rain water is approximately 6800 : 1, whereas in magmatic steam (or 'juvenile' steam as it is sometimes called) the ratio is about 6400 : 1; hence it is a simple matter to determine the relative proportions in a mixture of meteoric and magmatic waters. It has now been clearly demonstrated that the magmatic water content of all known hyperthermal fields is very small – perhaps as much as 5% but never more than 10%. The bulk of the water appears to be of meteoric origin, having penetrated from the surface through permeable formations to depths where it has been heated, mainly by conduction through the bedrock from below, though perhaps sometimes assisted by the upward infiltration of small quantities of magmatic vapours.

A field such as that illustrated in Fig. 10 would almost certainly be 'wet', yielding steam/water mixtures up bores sunk through the cap-rock into the pressurized aquifer beneath. Convection currents would be expected to flow within the aquifer; and in the process of rising to the under side of the cap-rock, hotter water from below would tend to boil and release steam. But with a configuration such as that shown in Fig. 10 the steam could not accumulate to any great extent: it would tend to slide along the under side of the cap-rock towards cooler, outer zones where it would be condensed and rejoin the convection water stream. Although only two convection loops are shown in Fig. 10, there could well be several such loops or 'circulation cells'.

Sometimes, if the aquifer is slightly dome-shaped, hot water may collect in the lower part while steam collects in the upper part. Where this happens, as in Tuscany, the field would be 'dry' (although by drilling very deeply into the aquifer boiling water might be encountered). Dry hyper-thermal fields are less common than wet.

The cap-rock will often be faulted in a few places, and it is by way of such faults that small quantities of thermal fluids will sometimes escape upwards and appear as surface manifestations.

5.7 Model of semi-thermal field

A semi-thermal field could sometimes take much the same form as that of a hyper-thermal field as illustrated in Fig. 10, but with the heat flow from depth insufficiently intense to induce the relatively high temperatures associated with hyper-thermal fields. However, as the hot water entrained in the aquifer of a semi-thermal field seldom reaches a temperature of $100°$ C, no cap-rock is needed to retain the heat and pressure within the aquifer; nor does it normally occur. The thermal gradient and depth of the permeable aquifer of a semi-thermal field should be sufficient to maintain convective circulation, but the temperature in the upper part of the reservoir is unlikely to exceed $100°$ C, partly because there is probably no cap-rock to hold down a pressure build-up above atmospheric, and partly because there may be mixing with cool ground waters. Re-charge with meteoric ground waters can of course occur at any point in the field except where there may be fortuitous formations of impermeable layers of rock above the aquifer. The problems of accounting for the source of heat in semi-thermal fields that do not occur near the boundaries of tectonic plates is not simply answered. It could be due to one or more of the three causes mentioned near the end of Section 3.4.

5.8 Actual hyper-thermal fields

It is emphasized that Fig. 10 is stylized. Individual fields may well be of a more complicated structure. Nevertheless, a close examination of the geology of known hyper-thermal fields enables the five basic features (see Section 5.6) to be identified. Two of the simplest fields will now be described.

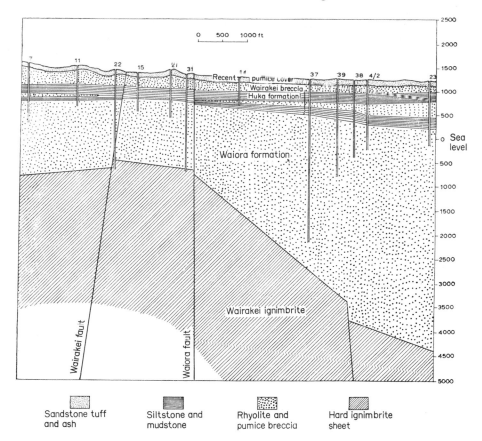

Figure 11   Geological section of the Wairakei thermal field, New Zealand.
   Note: Some of the bores, sunk to various depths, which lie approximately along the plane of the vertical section are shown (with their local identification numbers). (Reproduced from Fig. 1 of [68].)

5.8.1 *Wairakei*. Fig. 11 shows a simple geological section through this field, which occurs in the Taupo volcanic zone in the North Island, New Zealand. In this case there are believed to be two layers of bedrock. The upper layer, which has been partially but not fully penetrated by drilling, is an almost impermeable formation of ignimbrite, extensively faulted as shown in the figure. Beneath the ignimbrite, but not shown in the figure, and at a depth of about 4000 metres, a sedimentary basement of greywacke is believed to exist. The aquifer is a clearly defined layer of permeable pumice breccia known as the Waiora formation, of thickness ranging from about 350 to 1400 m. The cap-rock is no less clearly defined, and consists of the impermeable 'Huka formation', composed of siltstones, mudstones, pumiceous sandstones and diatomites of thicknesses varying from about 70 to 170 m. The cap-rock is covered with 80 to 160 m or so of pumice and breccia. Water replenishment of the field is believed to originate from two sources. Mostly it is probably from rainwater precipitated upon outcropping parts of the Waiora formation beyond the confines of the cap-rock, but to some extent also it is believed to come from below, through the faults in the ignimbrite. Some of this second source of water may be of meteoric origin, having penetrated from far away into the boundary between the greywacke basement and the ignimbrite, while some of it may be of magmatic origin having forced its passage through faults in the greywacke. The source of heat is probably for the most part conduction through the bedrock from a deep-seated magmatic intrusion, though a fraction may be conveyed through the bedrock with magmatic vapours.

5.8.2 *Cerro Prieto.* This field, situated a few kilometres to the south of Méxicali – a township near the Western end of the US/Mexican frontier – is related to the San Andréas fault system which extends from the Gulf of California to Cape Mendocino between San Francisco and the Oregon State boundary. A cross-section of the field is illustrated in Fig. 12. Here there is a granitic basement, faulted in many places, lying at a depth of about 2500 m beneath most of the field but rising to the West to outcrop in the

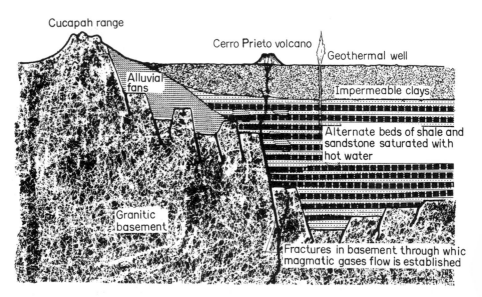

Figure 12 Geological section of the Cerro Prieto thermal field, Baja California, Mexico. (Reproduced by kind permission of the Comision Federal de Electricidad, Mexico, from their illustrated brochure describing the Cerro Prieto venture, 1971, [189].)

Cucapah mountain range. A layer of impervious plastic clays, about 700 m thick, serves as a cap-rock, while between this layer and the granitic basement lie alternate beds of shale and sandstone which are sufficiently permeable to form an aquifer. A re-charge area for replenishing the aquifer with meteoric water is provided by the alluvial fans that form the eastern slopes of the Cucapah range, while the presence of the extinct Cerro Prieto volcano shows that magma has at some time in the past penetrated to the surface, and is suggestive of the present existence of a magmatic intrusion at depth, to provide the source of heat. It is also possible that conductive heat through the granitic basement may be supplemented to a small extent by very hot magmatic vapours rising through the faults in that basement from below.

These two fields closely follow the conventional pattern of a hyper-thermal field as illustrated in Fig. 10. Space does not permit the description of other fields in this chapter, but Facca [20] examines the structures of three other fields – The Geysers (California), Otake (Japan) and Larderello (Italy). The first and third of these three fields yield dry superheated steam, while Otake (like Wairakei and Cerro Prieto) is a wet field. Although the structure of these three fields lacks the extreme simplicity of Wairakei and Cerro Prieto, an underlying resemblance to the archetype of Fig. 10 can be detected in every case.

5.9 Blowing of wet hyper-thermal bores

It is of interest to consider what would happen if a very long vertical tube, having walls of thermally insulating material and blocked off at the base, were filled with water and steadily heated at the bottom of the tube with high grade heat. The presence of convection currents in such a tube may virtually be ruled out if the heat is uniformly applied across the whole area of the base. The temperature of the lowest layer of the water would steadily rise, and although a small amount of heat would be conductively transferred to higher layers, the lowest layer would ultimately boil and give off steam. This steam, being buoyant, would rise into a slightly higher and cooler layer and would condense, thereby raising the temperature of this second layer until it, in turn, would boil. However, the hydrostatic pressure in this second layer would be slightly lower than that in the bottom layer, so it would start to boil at a very slightly lower temperature than that at which the bottom layer boiled. Steam from this second layer would then rise into a still higher layer, condense, and cause boiling at a rather higher level and at a rather lower temperature. The process would continue until the entire column of water is boiling; the temperature at each level in the tube being at the boiling point corresponding with the hydrostatic pressure at that level. The resulting temperature/depth relationship would then be as shown in Fig. 13 (which allows for the falling density of water with rising temperature and depth), and the continued addition of further heat from

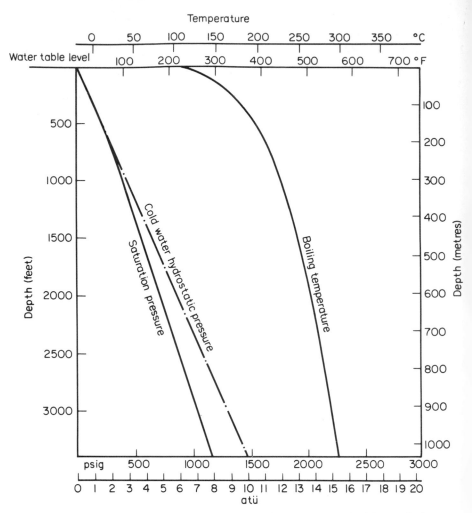

Figure 13   Variation of boiling temperature, saturation pressure and cold water hydrostatic pressure with depth below water table.

below would cause the emission of steam from the top of the tube at atmospheric boiling point.

A water-filled aquifer receiving a steady supply of heat from below would tend to imitate this curve to some extent; but the relationship between depth and temperature could be disturbed by the cooling influence of ground waters near the surface and by other complex factors. Near the outer boundaries of the convection loops (Fig. 10) the temperatures of the descending convection currents would almost certainly be lower than those of the boiling point/depth curve, as the waters would not be descending unless they had cooled appreciably. Elsewhere the relationship could approximately hold. In the middle of the field, where a 'plume' of heated water is rising more or less vertically at the point where two convection loops are contiguous, the falling hydrostatic pressure would cause the water to boil and the temperature to drop as the fluid rises; and the steam bubbles so formed would give buoyancy to the fluid and would thus assist the convection process. If the steam at the top of this plume can only dissipate itself through paths of high flow resistance, the pressures and temperatures in the plume zone may slightly exceed the values shown in Fig. 13.

When bores were first sunk at Wairakei, it was observed that the enthalpy of fluids collected by shallow bores of only a few hundred feet depth was about 575 btu/lb (320 cal/g) whereas that of deep bores of 2000 to 3000 ft depth was about 460 btu/lb (255 cal/g) – about 20% lower. This rather surprising observation was explained by the fact that the shallow bores were almost certainly collecting, with the hot water, a certain amount of steam that was boiling off from the hotter waters rising from below.

Actual underground temperatures in a wet field will seldom follow exactly the boiling-point/depth relationship, not only for the reasons given above but also because of changing patterns of ground-water flow and of convection currents, local differences of flow resistance, the presence of large masses of relatively impermeable rocks possessing 'thermal inertia' to temperature changes and other conceivable complexities. Measurements in static (non-flowing) bores sometimes even reveal temperature inversions.

If the temperatures in a non-discharging bore at all depth lie to the left of the curve of Fig. 13, the bore would be incapable of spontaneously 'blowing' because the vapour pressure would everywhere be less than the overlying hydrostatic pressure. To start such a bore discharging, it would be necessary to reduce the hydraulic pressure – either by using compressed air to blow out some of the 'top' water or by injecting some effervescent – so that the vapour pressure at depth could then take control. Once a bore starts to 'blow', it should, if the permeability of the aquifer at the base of the bore is adequate, continue to blow indefinitely owing to the reduced density of the bore fluid as it flashes off steam when rising up the bore, unless the temperature at the base of the bore is excessively low. However, if a bore is sunk to a point where the temperature lies to the right of the curve of Fig. 13, the mere opening of the valve at the top of the bore should set off spontaneous and sustained blowing. A recent proposal has been advanced [199] for stimulating the upward flow of bore water by injecting into the base of the bore a highly volatile 'secondary fluid' such as a freon or isobutane, and using that fluid to actuate a binary fluid generation cycle (see section 10.2.6).

Sometimes, if a bore diameter is rather too great and the aquifer permeability inadequate, the volume of flashing steam as the fluid rises up the bore will be insufficient to sustain buoyancy, and the well will suddenly quench itself.

5.10 Blowing of dry hyper-thermal bores

In a steam-dominated field a large volume of steam is trapped beneath the cap-rock under pressure owing to its inability to escape – by lateral migration into cooler surface waters, by leakage through surface manifestations or by artificial draw-off through bores – as rapidly as it is being generated at depth. The underlying water, deep down in the field, must boil at a temperature corresponding not to the hydrostatic pressure alone but to the sum of the hydrostatic pressure plus the steam pressure above its surface. A bore sunk into the steam zone will, owing to the prevailing pressure, automatically 'blow' without stimulation. This does not imply, however, that this state of affairs will last indefinitely, as will be shown below.

5.11 The phenomenon of supherheat in dry hyper-thermal fields

If, as is generally believed, a dry hyper-thermal field consists of a steam zone overlying a deep-seated water zone in the aquifer, it might have been expected that the steam drawn off from the upper zone would be dry saturated, since the two phases are in contact with one another. Both at the Geysers and at Larderello, however, the steam at the wellheads is moderately superheated – by about 12° C at the Geysers and about 70° C at Larderello. Moreover, the enthalpy of the steam at Larderello is about 703 cal/g (1265 btu/lb), which is greater than the maximum attainable enthalpy of dry saturated steam (669.8 cal/g or 1205.6 btu/lb). At the Geysers the enthalpy of the steam is just about the same as that maximum. At one time it was believed that the superheated condition was due to the admixture of very hot magmatic vapours rising from great depth with dry saturated steam at higher levels, but it is now known that the proportion of magmatic steam would be insufficient to account for the observed degree of superheat. Another theory entertained was that meteoric waters, after filtering down through a permeable zone, encountered very hot impermeable bedrock and immediately flashed into superheated steam. However, as it is now almost certain that there is in fact a water phase always present in the lower part of a steam-dominated field, this theory is no longer tenable.

The probable explanation of the superheat is two-fold. First it is known that steam fields, before exploitation, produced dry saturated steam. The process of exploitation has lowered the water/steam boundary level, because the inflow of recharge waters is insufficient to make good the mass of total drawn-off steam. It is believed that the water level has dropped several hundreds of metres since the Larderello field was first exploited [21]. In doing so it has left 'high and dry' large quantities of rock that have retained much of their original heat at a higher temperature than that at which the steam is boiling off at the water surface. The passage of dry saturated steam from the boiling water surface to the bores, through this hot rock, would cause superheating. Secondly, the mere throttling of dry saturated steam when flowing through a permeable or porous resistance to flow must also cause superheating; for throttling is an isenthalpic process which – as can be seen from a Mollier chart – both superheats and (paradoxically) cools it slightly owing to the Joule–Thomson effect.

5.12 Very deep, wet, hyper-thermal fields

Reverting to Fig. 13, it will be realized that at some very great hydrostatic pressure the boiling point must rise to the critical temperature for water – 374° C or 706° F. This temperature would only be reached at a water depth of about 3500 m, say 11 500 ft, and it is doubtful whether any formation could retain permeability, and thus act as an aquifer, at such depths at which the lithostatic pressure would be extremely high. However, if such a deep aquifer could in fact be found, the curious condition would arise of a very deep layer of boiling water floating on the top of a 'cushion' of very high pressure superheated steam, in the same manner as a globule of water dropped onto a very hot metal plate will float above the plate in the 'spheroidal state', carried on a very hot thin layer of superheated steam.

5.13 Blowing semi-thermal bores

Except where artesian or 'geopressurized' conditions prevail (see Section 5.14) a semi-thermal field will not 'blow' spontaneously without aid, as the vapour pressure is insufficient to lift the hot water to the surface. In some cases it may be necessary to raise the water by the continuous action of a submersible pump. In other cases, once a flow has been established by initial pumping from great depth, it may be sustained by thermo-syphon effect by virtue of the higher specific gravity of the less hot overlying water by comparison with the hotter water in the rising well. It all depends upon the depth below the ground surface at which the water table occurs, and upon the depth of the aquifer below that.

5.14 Geo-pressurized fields

Although not falling within the broad classification of geothermal fields given in Section 5.3 and Fig. 9, there is also a type of 'freak' field known as the 'geopressurized' field. Such fields are found along the North side of the Gulf of Mexico, mostly in the States of Louisiana and Texas [22], and also in Hungary [95]. An interesting feature of these fields is that they occur in *non-thermal areas*, where the temperature gradients range from about 27 to 30° C/km only. It is only on account of their great depth (up to about 6000 m, or about 20 000 ft, in places) that temperatures ranging from rather less than 93° C (or about 200° F) to rather more than 150° C (or about 300° F) are encountered. These fields are filled with pressurized hot water of 'connate' origin at pressures ranging from about 40% to about 90% *in excess* of the hydrostatic pressure corresponding to the depth. The explanation of these very high pressures is believed to be that gradual subsidence *via* growth faults has led to the ultimate isolation of trapped pockets of water contained in alternating layers of sandstone and shale. The trapped fluid supports a substantial share of the weight of the overburden and has prevented the full compaction of the formation. The pore pressures in these trapped pockets have values intermediate between the hydrostatic and lithostatic pressures. Thus one particular well in the Gulf of Mexico area, 4900m (about 16 000ft) deep and having a reservoir pressure of 871 atmospheres (12 800 psi) at the bottom, can give rise to a net *wellhead* pressure of about 439 atmospheres, or 6450 psi at no flow – an excess pressure of about 95% of the hydrostatic value. Geo-pressurized fields have the added advantage that they produce not only the heat energy of the pressurized hot water, but also *hydraulic* energy by virtue of the pressure in excess of the vapour pressure which may be used to drive a water turbine, and also the fluids may contain large quantities of dissolved natural gas which can account for further heat.

5.15 Reservoir capacity

When plans are being drawn up for exploiting a hydrothermal field, it is very useful to have at least a rough idea of the total amount of exploitable energy available, so that the rate of heat extraction may be related to commercial and economic considerations. Over-exploitation could result in the run-down of a field in a shorter time than the expected economic life of the exploitation plant and equipment; and that would be tantamount to a lamentable waste of capital expenditure and, in consequence, excessive production costs.

Estimates of the thermal capacity of a field can be made only very roughly, as they must depend upon the accuracy of the 'model' built up by the earth scientists. Whilst the methods commonly used for the making of such estimates are likely to err on the conservative side, it would be most unwise to tempt Providence by exploiting a field at anything approaching the maximum theoretical possible rate. For example, let it be supposed that the capacity of a field (in terms of power production) has been estimated at 10 000 MW-years. If a 25-year commercial life could be assigned to the power plant it could be argued that as much as 400 MW of plant could safely be installed (assuming that an adequate power market is available), so that the resources of the field would be fully used up by the time that the plant would have been written off. Quite apart from the philosophical aspects of whether it is wise to exploit a country's resources quickly on the 'short-life-and-a-gay-one' principle or to eke them out slowly on the 'thought-for-the-morrow' principle, it would be extremely unwise to install the full 400 MW at the outset. In the first place the estimate, though believed to be cautious, might in fact be over-optimistic. Secondly, the run-down of a field through excessive exploitation would be a gradual process involving, perhaps after a few years, a slow decline in steam temperatures and pressures; so that even though plenty of energy might still remain in the ground, it would not be possible to obtain full output from the installed plant. It would be far wiser to install a smaller plant at first so that if, by any chance, signs of declining steam conditions should be observed before the end of that plant's commercial life, then a second-generation plant could later be installed specially designed to operate at reduced pressure. In general, it is wise to develop a field rather cautiously, step by step, and to observe the behaviour of the field very closely during the years of exploitation. As more and more knowledge of the field is accumulated it will become steadily easier to plan the future rate of exploitation.

Banwell [23] has devised a well-argued method of estimating the thermal capacity of a wet field, based on the assumption of a permeable aquifer of known area and depth impregnated with water, everywhere at boiling point according to the temperature/depth curve of Fig. 13. He assumes reasonable values for porosities and specific heat of the rock and for energy conversion efficiencies and, taking into account the stored heat both in the water and in the rock, he deduces the family of curves shown in Fig. 14, which shows the energy stored in a field in terms of its horizontal extent in area, and of its depth. It will be noted that two horizontal scales are provided – one for

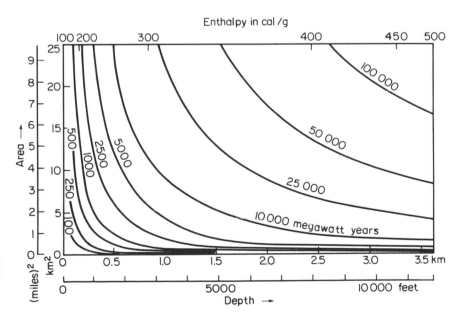

Figure 14 Approximate estimation of the power potential of a hyper-thermal field. (After Fig. 4 of [23].)

depth and the other for enthalpy at the lower limits of the field. The area and depth of the field have to be deduced from geophysical and other exploration data. The enthalpy, which follows from the depth and assumed temperature/depth relationship, should be confirmed by geochemical methods where possible. Clearly Banwell's premises cannot be precise, and the depth is likely to be uncertain; so his curves should be regarded as no more than a rough approximation. As they allow for stored heat alone, and make no allowance for continued heat infeed from depth, the curves are likely to err on the side of caution. The rate of heat extraction from a field during exploitation may, of course, greatly exceed the rate of heat in-feed at depth; but the latter could be quite considerable, and could even rise as the field temperature falls after long exploitation. As the exploitation of a wet field progresses, pressures and water levels in the aquifer will fall and the mass rate of fluid recharge will consequently increase owing to the greater head differential from outside the field. Most of the fluid in-flow would come from cooler ground waters beyond the confines of the field and, although these waters would gather stored heat, from the hot rocks of the aquifer there would be a slow but steady decline of temperatures within the field. Very probably the boundaries of the field will be ill-defined, and *warm* waters would intervene to some extent between the hot waters of the original field proper and the cold waters further afield. Moreover, very hot waters may well flow in from great depths to help replenish the depleted fluid stocks within the aquifer, so that the *heat* inflow could well increase on this account also. At Wairakei it has been established that heat is now flowing into the field, after more than 25 years of drilling and exploitation, at something like 70% of the rate at which heat is being extracted, and that the *fluid* inflow very nearly equals the fluid withdrawal. When the field was first exploited pressures fell rather rapidly, but later they approached fairly stable values. Obviously a day must come, if exploitation continues at a higher rate than the heat inflow, when the permissible rate of extraction must suffer. That point will not come suddenly: there will be forewarning in the form of falling temperatures as cool waters from outside gradually extract the stored heat in the aquifer rocks and cool them. Early in 1968 the draw-off at Wairakei was deliberately reduced to about one-third of the normal rate for about three months, in order to see whether the field would show signs of 'recovery'. There was an immediate, but slight, rise in field pressures.

Estimating the thermal capacity of a dry field is less easy than for a wet one, for it is necessary to know the level of the water/steam boundary that is believed to exist at depth. If a water level can clearly be detected, and if the temperature at its surface and the probable depth of water beneath that surface be known, then it is relatively simple to calculate the approximate amount of heat contined in the lower, water-saturated part of the field, including the heat stored in the rocks if reasonable assumptions are made as to their permeability and specific heat. In the upper steam-filled part of the field, the greater enthalpy of the steam is more than offset by its lower density in comparison with the same volume of water at the same temperature, so that the heat content of the steam could range only from about 1.8% at 6.8 atü (100 psig) to about 4.1% at 23.8 atü (350 psig) of that of the same volume of water at the same temperature. If the dimensions of the field be known, however, the heat contained in the steam and rock above the water boundary can also be calculated for different pressures and for different assumptions for the specific heat and permeability of the rock. Hence an estimate of the total heat stored within a steam-dominated field may be calculated; but the calculations cannot be expressed in the form of a family of curves as in Fig. 14 because everything will depend upon the depth at which the steam/water boundary occurs and upon the geometry of the aquifer.

5.16 Bore yield decline

The inevitable slow decline in the heat potential of a wisely exploited field can usually be concealed for many years (as at Larderello) by sinking new bores or deepening existing ones so as to tap new zones of the aquifer and thus maintain a fairly constant output from the field. This, however, is a process that cannot be prolonged indefinitely because the potential of a field is not infinite. But although the life of a whole field may be extended for a very long time by avoiding over-exploitation, the life of individual bores is definitely limited. With the passage of time the heat yield from an individual bore will usually decline fairly rapidly: in other words a bore will 'age'. This ageing process is in part due to chemical deposition, particularly silica, within the fissures and pores of the aquifer. In dry fields an additional cause of ageing is the increasing length of the path to be traversed by the steam from its point of generation (the steam/water interface) to the point

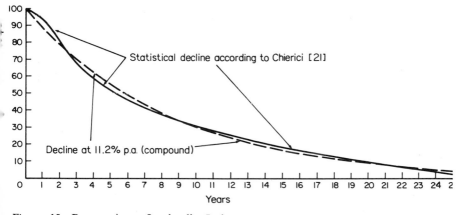

Figure 15   Bore ageing at Larderello, Italy.

of entry into the bore, as the water table falls in the course of exploitation. Chierici [21] has published a statistical relationship between the output and the age of the Larderello bores (Fig. 15). His original figure was plotted to a logarithmic horizontal axis, but, when plotted to linear axes, as in Fig. 15, the curious pattern of decline can perhaps be seen more clearly. The curve suggests an average ageing rate of only about 7 or 8% for the first $1\frac{1}{2}$ years, followed by a more sustained decline of about 12% p.a. over the next 20 years or so. A simplified approximation, which agrees with Chierici's curve at three points (after about 2, 9 and 20 years) would be to assume a steady compound ageing rate of 11.2% p.a. (also shown in Fig. 15). Similar ageing patterns have been observed for wells in other fields, both wet and dry, though the decline rates are individualistic to each field. Economically, a bore will have a finite life – usually about 10 years – after which its heat yield will become insufficient to justify the continued tying up of the capital invested in wellhead equipment and branch pipelines. After this life has been reached, it will pay either to deepen the bore or to scrap it and sink a new one elsewhere; in the later cause removing the wellhead equipment and whatever pipework can be salvaged to the new site.

 Exploration

6.1 General
Geothermal exploration is a subject so wide as to justify many separate books, each fully devoted in its own right to a single branch of the several exploratory sciences involved. To attempt to cover the whole subject in a single short chapter could well be regarded by any exploration specialist almost as an impertinence; but the reader is reminded that the purpose of this book is to introduce him in a rather general way to the geothermal sciences – not to instruct him fully in all of them. No more can therefore be attempted here than to describe very superficially some of the exploration techniques and their inter-relationships, and to refer the reader to the bibliography from which more detailed information may be acquired. These exploration techniques will here be treated separately under each of the several disciplines involved; but it is emphasized that all these disciplines are inter-dependent. Each specialist must rely to a large extent upon his technical colleagues, versed in other disciplines, to supplement his own observations, and to support or cast doubts upon his own deductions. In short, geothermal exploration is essentially a matter of *teamwork*, and the team members may aptly be likened to a squad of detectives and forensic experts in search of evidence.

6.2 The aims of exploration
The aims of the geothermal explorer, of whatever discipline, are as follows:

(i) To locate a geothermal field;
(ii) To decide whether such a field, if found, is semi-thermal or hyper-thermal;
(iii) To decide whether a hyper-thermal field, if located, is steam- or water-dominated;
(iv) To define as closely as possible the location, area, depth and probable range of temperatures of any located field

(v) From all this, to estimate the order of magnitude of the heat or power potential of any located field, and the grade of heat obtainable therefrom.

In the not far distant future the geothermal explorer is likely to become interested not merely in locating and evaluating geothermal *fields* (see Section 5.3) but also thermal *areas* (see Section 5.2), the exploitation of which is becoming increasingly probable. Even now, the discovery of thermal areas in the course of the search for thermal fields should be carefully noted for possible future use.

6.3 Likely locations of geothermal fields [24]

As already stated near the end of Section 5.4, the majority of hyper-thermal fields occur near places that exhibit surface manifestations of thermal activity, though this is by no means a *sine qua non*. Fields that occur in areas totally devoid of any surface evidence whatsoever must be sought in the light of the known geologic environment. Nevertheless, surface manifestations are a reasonably reliable sign of the probable presence of a hyper-thermal field within a moderate distance (though not necessarily in the immediate vicinity, as hot fluids can escape to the surface from below-ground along inclined faults and fissures from a source fairly distant from the visible surface phenomenon). In general, hyper-thermal fields are likely to be found within the seismic belt and in zones of recent vulcanism. Semi-thermal fields, not being restricted to such areas, can best be sought in areas where abnormally high temperature gradients are known to occur.

6.4 Analysis of records

Before undertaking any systematic exploration programme it is important that all available local physical and chemical data first be collected and carefully digested. Even in the most remote parts of the world there will usually be *some* relevant field data available, much of which will have been amassed for purposes other than geothermal exploration. All recorded data such as topography, meteorology, geology, hydrogeology, observations of hot springs, geysers and furmaroles, geochemistry and geophysical measurements should be carefully collected and reviewed. From all these data it may be possible to select promising regions, or even more narrowly defined areas, that are likely to repay closer investigation. It is important that data of interest should not be confined to areas that are obviously or apparently 'thermal'; for the more information that is available concerning non-thermal areas surrounding a thermal area, the more effectively can thermal anomalies be characterized.

6.5 The function of the geologist [24, 25]

The role of the geologist in geothermal exploration is particularly difficult to describe, and is made more arduous by his use of a jargon and vocabulary that is more or less unintelligible to the layman! McNitt [24] emphasizes that in geothermal exploration 'the geologist must be a generalist, sufficiently familiar with a great variety of specialized exploration, drilling and well-testing techniques to ensure their proper coordination in the overall exploration effort'. He goes on to say that 'geology is more of a subjective discipline in which conclusions are based on a minimum of information that is often internally inconsistent', and that the geologist's tasks include collaboration with the geophysicist and geochemist 'in choosing the specific exploration techniques which he believes will yield the best results' and to 'resolve the inevitable conflicting interpretations resulting from these techniques'. All this makes it sound rather as though the geothermal geologist should be a qualified witch doctor! This is not to suggest that the author disputes what Dr McNitt has said; but the quoted words do serve to show how difficult it is to expand in simple words upon the role of the geothermal geologist.

Reduced to its simplest terms, it could be said that the task of the geothermal geologist is to deduce, as accurately as possible and as funds and circumstances permit, a 3-dimensional 'model' of the geological structure of an allegedly thermal region to as great a depth as may be practicable, and to deduce promising drilling sites therefrom. By means of surface geological mapping; by the study of the tilt of outcrops; by the results of any cored soundings which may have been obtained by previous investigators (perhaps in pursuit of minerals or of ground water) or by his own exploratory drillings which he may consider to be justified; by the observation of faults and of surface thermal manifestations; by his deduction of the presence of 'cap-rock' formations that sometimes may serve to prevent the wholesale escape of steam from below-ground into the atmosphere; by all these means he should try to build up his hypothetical 'model', to suggest what zones may be permeable and contain hot fluids, and to pin-point promising drilling sites. He may further suggest from where such zones are recharged with water and whence the source of heat originates. Since all but a tiny fraction of his model will be concealed from the eye, he must rely to a great extent upon deduction rather than upon direct observation; and it is on this that his skill mainly rests. Supporting evidence from his exploration colleagues of other disciplines will be of great help to him in building up his model.

6.6 The function of the hydro-geologist

The task of the hydro-geologist is to collaborate closely with the geologist and to deduce the probable paths along which water will flow underground, through and between the various geological strata and boundaries of the geologist's model, and generally to assist the geologist to build up that model. The hydro-geologist (who may sometimes be the geologist himself, using a different facet of his knowledge, rather than another individual) should offer a plausible explanation as to how the thermal fluids reach the permeable zones of the alleged field, how they escape therefrom to the points where natural surface thermal phenomena are observed, how they are contained so as to prevent them from escaping elsewhere, and how they may be expected to behave when provided with artificial escape paths in the form of drilled bore-holes. He must study gradients, porosities and permeabilities of the constituent geological formations; and he will have to rely to a large extent upon the opinions of his exploration colleagues to confirm or to challenge his theories. He, in collaboration with the geochemist, should attempt to differentiate between magmatic and meteoric waters; and he may need to use isotopes to put to the test his theories of underground flow paths [26, 27].

6.7 The function of the geophysicist [28–30]

The science of geophysics is basically that of seeking and interpreting anomalies of all kinds. In, say, a very large flat prairie of homogeneous formation it would be expected that the intensity of gravity, the rate of heat flow from belowground, the electrical resistivity of the surface soil and subsoils, the magnetic qualities of the subterranean rocks and various other physical properties would be identical at all points in the prairie, except perhaps near the edges. On the other hand, in mountainous regions containing various mineral deposits and different geological formations, it is only to be expected that these physical properties will vary from place to place. The task of the geophysicist in geothermal exploration is to measure as accurately as possible several physical properties in many places, to detect anomalies, to plot them where possible in the from of isotherms, isogals, isoresistivity lines or other equipotential 'contours' so that the anomalies may clearly be recognized; and then to interpret them as plausible evidence of underground formations. The geophysicist, in his role of detective, has little more than 'finger-prints' to aid him in solving a mystery. But, as with his other exploration colleagues, he has *their* evidence and *their* interpretations to which he may refer for corroborative, or perhaps conflicting, evidence. It is only to be expected that a geothermal field – which in a general way may be described as a large volume of steam and/or hot water retained in permeable rocks – is likely to give rise to various anomalies by comparison with surrounding norms. Hence it is not surprising that geophysics can serve as a powerful tool in the detection of thermal fields.

Many techniques are available to the geophysicist, varying widely in reliability and cost. Fortunately, some of the most useful techniques are the cheapest; and it is on these that he should first concentrate before calling upon his resources of more expensive and sophisticated techniques to such an extent as he may consider to be justified.

One of the cheapest, simplest and most useful weapons in the armoury of the geophysicist is the thermometer, which may take many forms – the geothermograph, the Amerada gauge, thermo-couples, thermistors, platinum resistance thermometers or even mercury maximum thermometers: each has its own particular applications. By embedding thermometers of suitable type at various depths and at many points over a wide area and by observing their readings, the geophysicist can deduce temperature gradients and (with the help of the geologist) heat flow rates and can thus detect circumstantial evidence of more deeply seated local hot spots [31–33]. If he is a good detective he will be sceptical of *obvious* evidence derived from such temperature measurements, because they may perhaps be disguised by cross-flows of cool ground waters that will tend to displace the apparent position of hot spots away from their true positions. Deductions such as these form part of the skills of the geothermal geophysicist who, in this particular instance, would be aided by the hydrogeologist. Temperature measurements in close proximity to surface thermal manifestations should, as far as possible, be avoided as suspect. The geochemist, as will later be shown, can also supplement direct temperature measurements with *deduced* temperatures arrived at from chemical evidence; so that he too may aid the geophysicist in his temperature investigations.

The measurement of natural outward heat flow from a thermal field should by no means be confined to ground conduction measurements; for the amount of heat escaping to the atmosphere by conduction through the ground surface from a hot field beneath will usually form only a fraction of the total natural outward heat flow, the balance occurring at hot springs, geysers, etc. The amount of heat escaping by these latter courses will usually be disguised by dilution from local cool surface waters, but the degree of dilution can often be estimated fairly closely by chemical methods. To estimate the *total* natural heat flow from a thermal area it is necessary to take into account *all* heat and mass losses in and around the margins of the area, including possible seepages into rivers and lakes and ground water movements across the area. This can be a difficult and complicated task, but estimates with a reasonable degree of accuracy can usually be obtained by exercising great care and judgment and, after allowing for the degree of dilution, an approximate estimate may be made of the enthalpy of the escaping fluids by dividing the deduced heat loss by the estimated mass loss [28].

Total natural heat loss over an area is regarded as a measurement of some importance in the assessment of a geothermal field, but it would be fallacious to assume that the rate of natural heat loss is necessarily the same as the rate of heat *inflow* from below the field, for to do so would be to presuppose a state of equilibrium. It is always possible that a field is still in the process of forming and that the quantity of stored heat is still accumulating; or alternatively that it has passed its zenith and is beginning to decline by surrendering more heat than it is receiving. The difference between the heat inflow and the heat outflow is of course a measure of the changing quantity of *stored* heat in the field. It may take many years of careful observation to arrive at a reasonably reliable estimate of heat inflow.

In addition to thermometric studies, *electrical resistivity* measurements can also be performed fairly cheaply and quickly and can yield valuable information [34, 35]. They are achieved by injecting current into the ground through suitably spaced electrodes, and measuring the voltages between these electrodes. The techniques of resistivity methods can be adjusted so as to separate the horizontal from the vertical resistivity components within the area of survey. The interpretations of resistivity observations is a highly skilled art, for our 'detective' geophysicist may be beset by false clues. Not only can differences of resistance be caused by the physical differences in the rocks between the electrodes, but also by the presence of steam or of electrically conductive thermal waters held within the permeable rock zones. The chemical constituents of the thermal fluids will further complicate matters. The geophysicist, in conjunction with the geologist and geochemist, must learn to differentiate between the various causes of resistivity anomalies. In general, a hot water field will tend to produce a zone of *low* resistivity, mainly on account of the dissolved salts, whereas a steam field will tend to show up as a *high* resistivity zone. However, these broad tendencies, as already stated, can be masked by misleading factors that confuse the interpretation of the observed results. The presence of clay formations, for example, can give rise to low resistivity and can thus masquerade as hot water fields that may well be non-existent. Other geological factors can similarly ensnare the unwary.

Further techniques in the weaponry of the geophysicist include:

> gravity measurements [34];
> seismic measurements, reflective and refractive [35];
> micro-wave techniques;
> electro-magnetic techniques [36];
> radio frequency interference;
> ground noise measurements [37];
> micro-seismicity [38, 39];
> audio-magneto-tellurics;
> aerial scanning of infra-red radiation [40].

All these methods, some of which are costly to perform, may be regarded as *reserve* tools to be used only when the cheaper techniques of thermometry and resistivity produce apparently conflicting evidence or irreconcilable interpretations; but sometimes they can be justified for other reasons. It has been claimed, for example, that the presence of both vertical and horizontal fissures can be predicted by seismic measurements and could thus indicate promising drilling sites for penetrating highly permeable zones that might be capable of yielding large quantities of thermal fluids. Each technique has its potential value for different circumstances, and the judgment of the geophysicist is required as to which method is best suited to resolve any special problems encountered.

Scanning by artificial earth satellites is a new technique of growing importance, as very long range photography can reveal anomalies that are difficult to detect at closer quarters [41, 188].

6.8 The function of the geochemist [42]

Geochemistry affords one of the cheapest tools available to the geothermal explorer. It can even sometimes prevent the unnecessary wastage of exploration effort and expense at the very outset by demonstrating that some particular area is likely to prove to be geothermally worthless.

The geochemist's first task is to analyse the chemical constituents of natural thermal discharges, where such exist. The results of his analysis can 'serve as an important guide for decision making on subsurface exploration by drilling' [42]. After exploration bores have been sunk, the chemical analysis of deep geothermal fluids 'provides information on flow patterns of water and assists in selecting improved drilling sites' [42].

The chemistry of thermal fluids emitted from hot springs and fumaroles is determined by the interaction of underground fluids and rocks with which they come into contact. If the fluids are hot, chemical equilibrium will be achieved quite rapidly, and certain chemical features of fluids discharged at the surface are indicative of the temperature at which this equilibrium has been attained. Thus, the geochemist may be regarded as a highly sophisticated 'thermometrician' and can so aid the geophysicist and his other exploration colleagues in deducing the temperatures prevailing at depth in a thermal field. However, like his fellow 'detectives', the geochemist must always be on his guard to recognize falsified evidence resulting from dilution from cool surface waters and from chemical reactions between rising thermal fluids and rocks lying not far beneath the surface.

The three principal indicators of deep reservoir temperatures, to be sought in hot spring chemistry, are *silica, magnesium* and *sodium/potassium ratios*. Silica concentrations are more reliable for hot springs of high discharge than for those of low discharge or minor seepages. Magnesium is of limited value as a temperature indicator, but its total absence could be suggestive of economically useful reservoir temperatures (at least 200° C), as magnesium is retained in clay materials which are stable at high temperatures. The atomic ratio Na/K is a fairly reliable indicator of deep temperature – the lower the ratio the higher the temperature. The presence of free hydrogen in fumaroles is another indicator of subterranean temperatures exceeding 200° C [42].

The geochemist also has other useful analytical tools to help him deduce the quality and temperature of deep thermal fluids, both gaseous and aqueous, which are of the utmost interest to the specialist but which need not here be pursued. He is also able to detect the presence of valuable mineral constituents in thermal fluids that could be of interest to industrial geothermal developers, and he is able to form approximate estimates of the relative proportions of magmatic to meteoric waters present in the reservoir, thus aiding the conception of the field model. He must exercise the utmost care in his sampling methods.

Even after exploitation, the geochemist's art can be a valuable tool for detecting changes of temperature and water levels in a reservoir, and so in aiding an understanding of a field's behaviour.

6.9 Exploratory drilling

The final aim of *all* preliminary exploratory work is to select promising sites for exploratory drilling. Until such drilling has been undertaken, all the evidence collected by the team of earth scientists should be regarded more or less as 'circumstantial'. No detective can carry his investigations to their logical conclusion until he has obtained *proof* of his deductions. The object of exploratory drilling is to seek this proof.

From the combined evidence provided by all his team members, the project manager must sooner or later select promising sites for exploratory drilling which, he hopes, will provide him with the necessary supporting evidence to prove that his model of the field is broadly valid. Of course it is always just possible, if circumstances should be particularly unfortunate, that in the course of the preliminary exploration the considered collective opinion of the team will be that no commercially exploitable field exists at all, in which case the relatively high cost of drilling operations can be saved and the expenses of the abortive project may thus be limited to within a reasonable sum. However, assuming that the manager and his team are of the opinion that a useful field *does* exist, then, after making his best possible choice of site in the light of the evidence available, the manager will sink a bore to a depth at which he hopes to penetrate to a zone of permeable rock impregnated with hot thermal fluids. Expense and time may be saved if the first few exploratory holes are 'slim' – i.e. of small diameter: larger bores may follow after confidence in the field has been established. It may be necessary to sink the first hole to a greater depth than originally intended, to allow for inaccuracies of the model of the field, if the evidence collected during the drilling process – e.g. cores, downhole temperatures – is sufficiently encouraging. Quite possibly the first chosen site may yield a negative result, but if so, this should not necessarily be regarded as unduly discouraging. For every hole drilled, even non-productive ones, will provide additional evidence that will enable the project manager to modify his model so as to approach, step by step, more closely to the truth. Until a hole has been sunk, the explorer has only the evidence of his eyes and his instruments on which to construct his model. However, as soon as he has penetrated deeply with the drill – even if the bore should be totally unproductive – he can gain the direct evidence of cores, of deep temperature measurements and of chemical analyses at depth that will greatly reinforce his overall knowledge of the field, so that his next choice of drilling site should have a greater probability of success. His very first bore, regardless of its productivity, will have added a dimension to the explorer's knowledge.

Before many exploratory bores have been sunk, unless he has been particularly unfortunate, the project manager should succeed in striking a productive bore, and from then onwards the distinction between exploratory drilling and production drilling becomes steadily less clearly defined. After two or three good producing bores have been won, the main work of exploration has virtually been completed. A plausible model of the field will have been conceived and the pattern of future productive drilling will have been broadly determined, though both the model and the choice of drilling sites may well have to be modified to some extent as the evidence of each completed bore is added to the sum total knowledge of the field.

The work of exploration may, of course, still continue into areas beyond the presumed boundaries of the field until those boundaries have become clearly defined. Thereafter, exploration, as such, more or less ceases and 'field management' takes its place.

6.10 Extent of world-wide geothermal exploration as in 1976

Considering that, apart from Tuscany and a very few other places, almost no geothermal exploration had been undertaken anywhere until about 1950, it is impressive to contemplate the enormous amount of such exploration initiated in the third quarter of this century. By 1975, hyper-thermal and semi-thermal fields were already being *exploited* in the following countries (which exclude certain other countries where balneology was the only application of geothermal energy). In every case, of course, exploitation had been preceded by exploration:

El Salvador (P)
France (D)
Hungary (D, F)
Iceland (P, D, I, F)
Italy (P, F)
Japan (P, D, I, F, M)
México (P)
New Zealand (P, D, I, F)
Taiwan (M)
Turkey (P, F)
USA (P, D, M, F)
USSR (P, D, F)

Key

P, power generation
D, district heating
I, industry
F, farming
M, mineral extraction

In all these twelve countries geothermal exploration is still being vigorously pursued. Whereas in the USA geothermal exploitation has hitherto been mainly confined to California, exploration is now being extended to many other States, including Hawaii. In the USSR, where exploitation was first achieved in Kamchatka, investigations have now been extended to the Caucasus and many other parts of the Soviet Union.

By 1976 geothermal exploration programmes had either been proposed or, in the majority of cases, set in motion in the following 35 countries:

Algeria	Israel
Austria	Kenya
Chile	Montserrat
Colombia	Nicaragua
Costa Rica	Netherlands
Czechoslovakia	New Britain (Rabaul)
Djibouti	Panama
Ecuador	Philippines
Ethiopia	Poland
France	Portugal
French Antilles	Romania
Germany (West)	Santa Lucia
Greece	Spain
Guatemala	Switzerland
Honduras	Thailand
India	United Kingdom
Indonesia	Yugoslavia.
Iran	

This list is not necessarily complete, as certain countries have not publicized their activities, but even so the list is imposing and is indicative of the tremendous leap forward that has occurred within a quarter of a century in international concern for geothermal energy. Some of the listed countries hope to find high grade heat for power generation; others expect to win only low grade heat for other applications. It should be noted that the list contains such countries as France, Western Germany, the Netherlands and the United Kingdom, all of which until quite recently had been regarded as geothermally worthless. This is partly due to the highly successful exploitation of semi-thermal fields in Hungary, where no geothermal energy had even been suspected until fairly recently, partly to the economic development of nonthermal areas in France; and partly to the hopes that are now being felt of our ability to exploit hot dry rocks in semi-thermal areas within the fairly short term future.

7 Drilling

7.1 General

Drilling for steam and hot water, like exploration, is an important branch of technology in its own right, to which no more than very superficial treatment can be given in a single chapter. For more detailed descriptions, the reader is referred to references [43–57]. In hyper-thermal fields drilling depths generally range from about 500 m to 2000 m, although a few bores may lie outside these limits: depths of 600 to 900 m would cover the majority. In semi-thermal fields depths averaging about 1800 m are typical, while in geopressurized fields it may be necessary to penetrate as deeply as 6000 m or so.

7.2 Cellars

Before drilling operations are begun it is necessary to construct a concrete cellar (Figs. 16 and 39a) about 10 ft × 8 ft × 10 ft deep internally, to support the weight of the drilling rig and later to accommodate the wellhead valving and expansion spool. A concrete stairway normally provides access to this cellar, which is also served with means of drainage (by gravity wherever possible). Consolidation grouting should be injected into the surrounding ground, where this consists of pumice or other weak formations, over a diameter of about 60 ft and to depths ranging from about 50 ft at the peri-meter of the grouted area to about 100 ft beneath the cellar. This grouting not only gives support to the cellar but it can also serve to deflect away from the wellhead any steam that may accidentally ascend to the surface along the *outside* of the bore and its casings: such escapes of steam, though fortunately rare, have been known to occur, and the consolidation grouting provides a degree of safety by ensuring that the ensuing blow-out is diverted sufficiently far away from the cellar as not to endanger access thereto.

7.3 Optimum bore diameter

It is important that the diameter of a bore be neither too small nor too large.

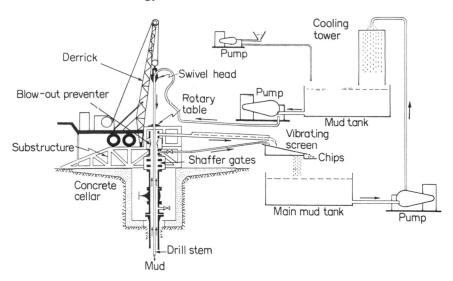

Figure 16 Typical rotary drilling rig and mud circulation arrangements. (After Fig. 2 of [43].)

Too small a bore will restrict fluid production by offering a high resistance to its upward flow. Too large a bore, on the other hand, will cost much money and time in sinking, thereby staking an excessive financial risk in view of the possibility (never entirely absent) that the bore may cease to be productive after a relatively short period of service, or even that it may be non-productive from the outset. Moreover, in a wet field, a large bore might strike a formation – permeable, but of high resistance to flow – incapable of producing a sustained output of geothermal fluid against the large mass of water that would tend to fill the bore, so that the well would become quenched. Theoretically there is an optimum bore diameter which is a function of the flow resistance within the bore itself, the flow resistance within the permeable formation from which the bore is fed with fluid, the cost of the bore, the probable success ratio in winning productive wells and the value attached to the geothermal fluid(s) when won. This optimum diameter more or less defies calculation and clearly cannot be determined even empirically at an early stage of development: it must be chosen in the light of the results from the first few production bores. At first it is advisable to drill bores of rather moderate diameter, and to increase the size if and when the results appear to justify doing so. Although the best choice of bore diameter must to some extent depend upon the 'quality', or wetness, of the fluid, the figures quoted in Table 4 may be taken as fairly typical of optimum bore sizes. The terms used in the 'classification' column will be explained in Section 7.5 below.

Table 4 Guide to optimum bore sizes (from p. 75, [44]).

Steam flow tons/hr	Open hole	Casing size o.d.	Casing classification
10 to 25	17-inch	13.375-inch	Surface
	12¼-inch	9.625-inch	Intermediate
	8.625-inch	7-inch	Production
	6¼-inch	4½-inch	Slotted liner
25 to 50	18-inch	16-inch	Surface
	14¾-inch	11¾-inch	Intermediate
	10.625-inch	8.625-inch	Production
	7.625-inch	6.625-inch	Slotted liner (o.d. of coupling skimmed by 1/16th inch)
50 to 80	22-inch	18-inch	Surface
	17-inch	13.375-inch	Intermediate
	12¼-inch	9.625-inch	Production
	8.625-inch	7-inch	Slotted liner

N.B. Inches are the usual standard. 1 inch = 2.54 cm.

7.4 Rotary drilling

The methods and equipment normally used for geothermal drilling do not basically differ very greatly from those used for winning petroleum or natural gas; but the harder rock formations and higher temperatures usually encountered in geothermal drilling, together with the possibility of striking highly corrosive fluids, call for certain improved techniques, materials and equipment. Penetration is normally effected by the chipping, spalling and abrasive action of special serrated tri-cone 'bits' (Plate 6) of very tough steel, the depth of the serrations being dependent upon the hardness of the rocks encountered. The bits are rotated, together with a hollow 'drill-stem' to which they are attached, by means of a mechanical drive at the surface – usually a diesel engine.

A drilling rig with all its appurtenances is quite an elaborate installation. First there will be a tall lattice steel tower, or derrick, containing a pulley system for positioning and withdrawing drill stems and casings so as to give access to them for examination and for joining sections together before they are lowered into the ground. Then there will be power units for rotating the drill, operating the derrick and driving the auxiliary pumps, air compressor, etc. There will be a rack for carrying a stock of casing pipes and drill stems ready for joining together and insertion into the ground; and there will be a circulating system for pumping, cooling, screening, settling and storing the cooling mud (Fig. 16 and Plate 7).

The purpose of the cooling mud [48 and 49] is to cool and lubricate the

Plate 6 Rotary drilling bits, as used at Wairakei, New Zealand [195].
Left: Tricone bit, with small heap of chippings.
Right: Coring bit, with piece of core.
(By courtesy of the New Zealand Ministry of Works and Development.)

bit and drill stem, to wash away rock cuttings from the hole, to prevent the bore walls from caving in and to cool the surrounding ground. The mud is forced downwards through the hollow drill stem and returns upwards through the annular space outside it. Various cooling muds are used, all having a higher specific gravity than water. For temperatures up to about 150° C bentonite or other clay-based muds are satisfactory, but at high temperatures such muds tend to 'gel' and the filtrate to increase in quantity. Chrome lignite/chrome ligno-sulphanate (CL/CLS) muds are then used, which are alkaline and contain a small amount of bentonite and traces of sodium hydroxide and of a de-foamer. Muds discharged from up a bore are screened to remove the rock chippings and are passed through a cooling tower (aided by forced draught if need be) before being pumped down the hole in re-circulation. Cooling through about 10 to 15° C is usually possible with a cooling tower using natural draught, though this depends upon the local climate: with the help of a fan the cooling range can be increased to about 20 or 25° C under typical conditions.

As an alternative to mud, compressed air is sometimes used [50]. Air drilling is cheaper and quicker than mud drilling and avoids possible damage to the production zone from circulating mud which may enter the zone when drilling. Nevertheless, this method cannot be used in formations that are very wet or which tend to slough. A common practice is to use mud until after the production casing (see Section 7.5 below) has been placed in position, and to use air drilling when penetrating the production zone.

Plate 7 Drilling rig in the Wairakei field, New Zealand. Note the mud cooling tower (right background) and the rack (foreground) for casings and drill stems. Several well-head silencers can be seen in the background. (By courtesy of the New Zealand Ministry of Works and Development.)

In the course of drilling, fissures and voids are liable to be encountered at various depths. If these are indicative of a useful producing 'horizon' at which hot thermal fluids can be won, they may be left undisturbed and a slotted casing inserted, but at shallower depths they will usually be un-productive, or even counter-productive, either by allowing cool ground waters to enter a bore or hot bore fluids from lower down to escape into higher formations. In such cases they should be sealed off by injecting cement, and later protected by means of an inserted casing. The presence of such fissures and voids will be evidenced by a loss of mud circulation.

7.5 Casings [51]

One of the most important factors in drilling is the provision of adequate steel casings to the correct depths. There will normally be as many as four

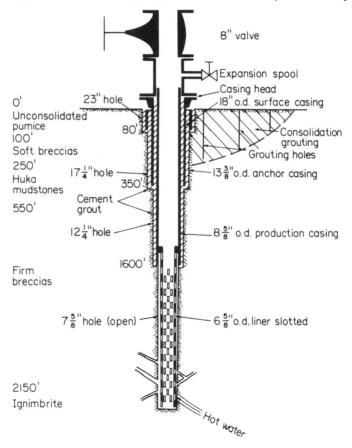

Figure 17 Casing lay-out and geological formations at Wairakei, New Zealand. (After Fig. 1 of [43].)

concentric casings in a single well, all made of high quality steel (J55 to API standards usually, but acid-proof casings should be used where thermal waters are highly corrosive). These casings are rigidly fixed by means of cement to the surrounding rock (except in the case of perforated linings) and are joined to one another by means of screwed couplings. The depth to which each casing is sunk will depend upon the nature of the geological strata through which the bore penetrates. Fig. 17 shows a typical arrangement of casings as adopted at Wairakei, New Zealand. The largest, or *surface* casing extends to about 60 to 80 ft depth. The upper end terminates at the wellhead cellar (Fig. 16) and the casing extends through the upper stratum of unconsolidated pumice, thereby providing a firm locating support to the system of inner concentric casings that are later inserted. The second, or *anchor* casing extends downwards to 300 or 400 ft depth through the zone of soft breccias into the upper layers of the Huka mudstone cap-rock which effectively anchors the lower end. The third, or *production* casing extends downwards through the firm breccias as deeply as may be necessary (say, 1000 to 2000 ft typically) to seal off non-productive or counter-productive fissures and to reach the productive zone of the aquifer. The lowest part of the bore may sometimes be left unsupported, but more generally it is provided with a slotted liner (Plate 8) extending to the bottom of the bore and standing at least ½-inch clear of the bore walls. The purpose of the slotted liner is to act as a strainer to hold back the larger pieces of rock which may here and there break off from the bore walls. The length of the innermost slotted liner may be anything, according to the levels at which productive horizons are struck, but only a few enter the top of the ignimbrite bedrock beneath the aquifer, as nothing is to be gained by deepening them further. Fig. 17 and the above description apply particularly to Wairakei: other fields will have different casing arrangements appropriate to the local geology and to the depths of the productive zones. For very deep holes it may be necessary to increase the number of concentric casings to more than the conventional three, as illustrated in Fig. 17, excluding the slotted liner.

In the early days of drilling at Wairakei, slotted liners were not used. When a newly drilled bore was first 'blown' it was not unusual for large quantities of rock and debris to be discharged up the bore and scattered over the countryside. After a while, the quantity of this discharge usually fell off to negligible amounts. The theory was held that the thermal fluids entering the base of the bore at high velocity broke down the bore walls, ejected the debris and formed cavernous chambers around the base of the bore. As these chambers grew in size, the yielding area increased, the fluid velocity at the chamber walls fell, and the disruption declined. Later it was found that bore walls continued to collapse spasmodically long after a bore had been deemed to have been cleared: so to prevent the discharge of debris into the wellhead equipment, slotted liners were then fixed as a

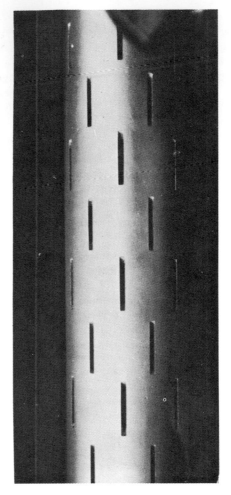

Plate 8 $6\frac{5}{8}$-in slotted casing, as used at Wairakei, New Zealand. 28 slots per foot of liner [195]. (By courtesy of the New Zealand Ministry of Works and Development.)

standard precautionary practice. This practice is now adopted in most other fields.

7.6 Cementing [52]
It is of the utmost importance that all casings, except the perforated liners, be firmly cemented into position. This is effected by injecting cement slurry down the casing by means of a plug which forces the slurry to flow around the bottom, or 'shoe' of the casing and to rise up the surrounding annular space until it reaches the surface. Ordinary Portland cement is suitable for

temperatures up to about 150° C but special cements containing various proportions of silica flour, perlite, fly-ash and other ingredients must be used for higher temperatures. Faulty cementing can give rise to subsequent troubles such as collapsed casings, and it is very important that the greatest care be exercised when cementing. Casings are held concentrically during the cementing process by means of spacers, so as to give uniform annular sections.

7.7 Blow-out prevention [53]
In case of a blow-out during drilling, two safety devices are provided. First are the Shaffer gates (Fig. 16) which can be closed almost instantaneously either manually or by means of compressed air. The gates have semi-circular rubber rams that can fit closely round the drill-stem, drill collar or casing, as may be appropriate, so as to seal off the surrounding annular space. Above the Shaffer gates is a hydraulically actuated rubber blow-out preventer which is sufficiently flexible to close tightly round a drill string or casing, as may be required.

7.8 Coring
By using special annular bits (Plate 6), core samples may be extracted from a bore so that a continuous geological record can be made available of all the penetrated strata. Coring is rather a slow process which should be performed only when the geologist is in need of the valuable information that it can contribute towards the knowledge of a field.

7.9 Drilling times and penetration rates
The rate of drilling is of course influenced by such factors as rock hardness, ease of access to the site, the power and efficiency of the rig, the quantities of casing required, the bore diameter, the lengths to be cored and the depths to be penetrated into the productive aquifer. The periods shown in Table 5 are not unreasonable expectations under favourable conditions [44].

Actual penetration rates are of course dependent upon the weight carried by the bit and on the speed of rotation, but if either is excessive too much heat will be generated and the circulating mud or air will be unable to cool

Table 5 Typical drilling times for different depths.

Depth (m)	Actual drilling (days)	Finishing and testing (days)
500	15–30	
1000	25–45	10
1500	35–55	
2000	50–70	

the bit effectively and a blow-out may occur. Under Wairakei conditions [43] typical bit loadings range from 150 to more than 1300 lb per inch of bit diameter, with increasing depth. Rotary speeds are of the order of 90 to 100 rev/min and penetration rates range from 23 to 40 ft/h (the lower limit applying to the larger diameter surface casing holes, and the upper limit to the deep small diameter holes in the production zone. In igneous rocks at fairly great depth penetration rates may be as low as 30 ft/day [138].

7.10 Corrosion and erosion

Some thermal fluids are highly corrosive and require the use of special resistant steel casings and wellhead equipment materials. If the fluid is too corrosive, a bore – or even perhaps a whole area – may have to be abandoned. Erosion of production casings and wellhead equipment can be caused by the ejection of sand and larger rock particles spalled from the bore walls, and this can lead to fractures and blow-outs. Slotted liners will normally reduce erosion to acceptable limits, though special tough quality steels may in rare instances have to be used.

7.11 Maintenance or enhancement of output

Sometimes the deposition of silica or of calcitic salts on the bore walls in a wet field can result in a marked decline in fluid yields. Chemical deposition commonly occurs just above the slotted casing in the lower part of the production casing, where flashing of the uprising hot water causes the concentration, and consequent precipitation, of the solubles, [54]. Usually, and fortunately, this is a slow process that can be remedied, at intervals of a year or two, by reaming out the deposit so as to restore the full internal diameter to the bore. In a few very persistent cases, however, the build-up of deposits has been so rapid that frequent reaming has proved to be uneconomic, and the bore has had to be abandoned.

In other cases, where the bore fluid yield has been disappointing, it has been possible to increase the flow by perforating the production casing by means of a bullet type gun [55] consisting of a special barrel, having four chambers per foot, each housing a hardened $\frac{3}{4}$-inch steel bullet (19mm) and an explosive charge. The barrel is lowered to a depth where the geologists believe there may be a sealed-off producing horizon, the charges are detonated and the bullets pierce the casing and its surrounding grout and penetrate into the rock formation. Appreciable yield increases have sometimes been achieved in this manner. Reaming will not of course influence the decline in bore yield due to chemical deposition within the fissures and pores of the aquifer, which can partly account for the decline illustrated in Fig. 15 and referred to in Section 5.16, but bullet performation *could* help to offset this phenomenon.

7.12 Directional drilling [43]

Standard techniques have been developed in the oil drilling industry, and are adaptable to geothermal drilling, for producing a *deviated* bore that does not descend vertically. The ability to aim a bore in any chosen direction can be extremely useful for either of two specific purposes:

(i) it can provide means of sealing off a 'rogue' bore that has broken out of control (see Section 7.13 below).

(ii) it offers great possibilities of economizing in surface pipework and in land acquisition costs, while at the same time giving access to parts of an aquifer that underly difficult or inacessible terrain.

Regarding point (ii) it may happen that parts of a field underly a built-up area, valuable cultivated land, a beauty spot, expensive land, or a terrain that is unstable, very steep or even precipitous, where the cost of levelling and stabilizing a suitable wellhead area would be excessive. In such cases it should be possible to tap the underlying aquifer by drilling deviated holes in the manner shown in Fig. 18. There could of course be legal problems which would have to be investigated, regarding the rights of the owner of the surface land overlying the tapped area, particularly if subsidence risks are foreseen; but there could be many occasions where deviated drilling would bring to a relatively small collection area the potentialities of a widely spread aquifer. An indirect advantage of doing this would be that the quantity of surface collection pipework could be greatly reduced, and it might even be possible to place the collection area close to the exploitation

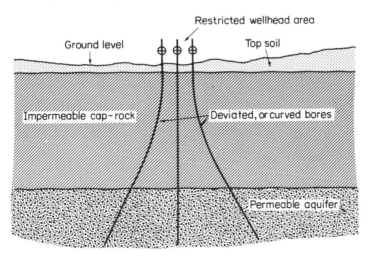

Figure 18 Suggested method of using deviated, or curved bores to tap an extended producing zone from a restricted wellhead area.

plant. The differential costs of the longer deviated holes, by comparison with shorter vertical holes, would have to be weighed against the saved costs of reduced surface pipework (see Section 15.10).

7.13 Mishaps [56]

Many kinds of mishap can occur during and after drilling, such as steam blow-outs, drill-pipe sticking, parting of a drill-pipe joint or the breaking off of a drill collar, casing collapse, parted casing couplings, casing rupture due to excessive erosion or thermal shock. To pursue all these potential troubles and to describe the appropriate remedial procedures would require too much detailed description for a book of this type, and the reader is referred to other more sophisticated papers and articles in the references [44, 53, 56 and 57]. In general, most mishaps can be prevented by the exercise of great care in the selection of bits, ensurance of adequate cooling during drilling, proper cementing of casings, consolidation grouting round cellars and the avoidance of thermal shock by heating or cooling bores very slowly when starting up or servicing.

At Wairakei a spectacular blow-out occurred in 1960 when a casing ruptured at about 600 ft below the surface and huge quantities of steam escaped to the surface and formed a miniature volcano close to the wellhead. This was followed by the collapse of a hillside, and the whole area soon resembled an inferno. By the taking of prompt action the equipment was rescued and no one was seriously hurt. The trouble was ultimately overcome by the very skillful drilling of a deviated, or curved, hole some 200 ft away from the original bore [43]. This new hole was so accurately aimed that it struck a point very close to the original bore at a depth of about 1500 ft just below its production casing (see Fig. 19). Cement and gravel were then injected down the new bore so as to seal off the 'rogue' bore at its base. The surface eruption immediately started to decline and it soon became possible to fill the whole of the defective bore with cement, thus completely blocking off the ruptured casing. The deviated hole was then cleaned out, cleared of its cement plug at depth, and deepened, thus becoming a new and useful production bore. This incident clearly illustrates the value of directional drilling for the first purpose mentioned in Section 7.12 above.

Another dramatic mishap occurred in Mexico in 1967 at the Cerro Prieto field when a wellhead pipe fractured below the shut-off valve. Vast quantities of steam and boiling water were escaping into the air, and it was dangerous to approach the well too closely. By the skillful cutting away of the jagged upper end of the fractured pipe, by means of a blow-lamp, the careful lowering, by means of a derrick, of a new open valve with attached pipe stem over the jet of escaping fluid until the pipe stem butted against the stub end of the faulty well; the welding together of the new pipe stem to the old stub end; the welding of a safety collar round the new welded joint; and by

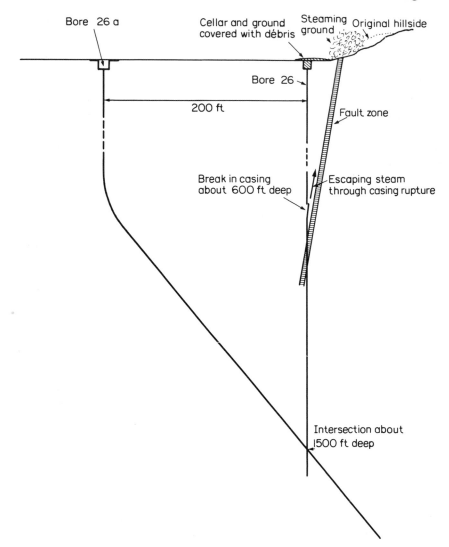

Bore 26 a   Cellar and ground   Steaming   Original hillside
             covered with débris   ground

Bore 26

200 ft

Fault zone

Break in casing
about 600 ft deep

Escaping steam
through casing rupture

Intersection about
1500 ft deep

Figure 19 'Rogue' bore (No. 26) at Wairakei, with deviated rescue bore (No. 26a).
(Not to scale.)
Simplified version of more detailed sketch appearing in the official Wairakei brochure
published by the New Zealand Government [195].

the final closing of the new valve; the well was ultimately brought under
control.

Other serious mishaps have occurred on rare occasions elsewhere; but
the experience gained from each has been put to good use, so that such
misadventures are now exceedingly rare.

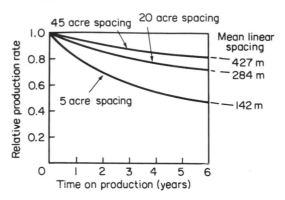

Figure 20 Effect of well density on production rate in terms of time (Geysers field). (After Fig. 5 of [58].)

7.14 Well spacing

No hard and fast rule can be laid down concerning well spacing. Clearly, the more closely the wells are spaced relatively to one another the less money will have to be spent upon collection pipework, but if they are too closely sited they may interfere with one another's performance. The mutual interference of wells is largely fortuitous, depending upon the random fissure patterns within the aquifer. Matsuo [44] suggests an empirical well spacing of 100 m for bores of 500 m depth to 300 m for bores of 2000 m depth. Budd [58] has established by means of a reservoir simulation model a relationship between well spacing and the rate of declining yield for the Geysers field, California, as illustrated in Fig. 20. The choice of optimum spacing then becomes a straight question of economics in terms of the prices of land, of drilling and of surface pipework. Budd qualifies the curves by stating, as would be expected, that they would be influenced by bore depth. A typical bore depth in the Geysers field is about 6000 ft (1800 m) according to p. 3 of Budd's paper, and if this depth be applied to Matsuo's very rough relationship between depth and spacing it would seem that the 20-acre spacing curve would be about appropriate. This curve shows an average annual decline over six years of about $5\frac{1}{3}\%$, which compares with about 11% p.a. for Larderello (Fig. 15); but as the Larderello bores are on average about half as deep as the Geysers bores, the decline would be expected to be more rapid in the former case, though the magnitude of the difference is perhaps rather surprising. It is only to be expected, however, that each field will have different permeability characteristics, and therefore interference characteristics. At Wairakei, the average bore spacing is about 100 m and the average bore depth about 600 m, which does not greatly differ from Matsuo's suggestions.

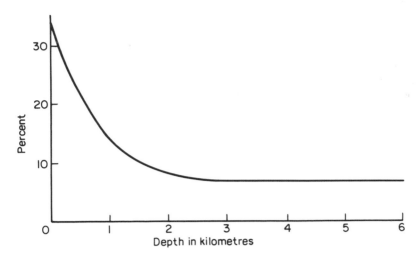

Figure 21   Cost/depth relationship for conventional drilling, showing approximate percentage cost increment for every additional 100 m drilled at various depths.

Note   This curve is based on statistical data for *petroleum* wells. The corresponding curve for geothermal wells could be appreciably different, though probably of similar general shape. In fact Garnish [138] quotes drilling costs that suggest a much less rapid quasi-exponential relationship than here implied.

7.15 The costs of conventional drilling

Although this chapter is not directly concerned with economic considerations, which will later be discussed in Chapter 15, the *pattern* of conventional drilling costs is of great importance. Fig. 21 illustrates this pattern: it is expressed in terms of the percentage *incremental* cost of drilling every additional 100 m at different attained depths. The absolute costs, which would vary widely according to the location and accessibility of the site, the hardness of the rock to be penetrated and the degree of monetary inflation, are for the present immaterial: it is the cost/depth *relationship* that is of importance. The figures used in the preparation of Fig. 21, which were derived from statistical data published by the American Petroleum Institute, clearly show a quasi-exponential relationship. Except for very shallow bores, where the costs are masked by the expenditure of bringing the rig to the site, setting it up and later removing it, the relative incremental drilling costs rise rapidly up to 2 or 3 km depth, after which they stabilise at about $6\frac{1}{2}$ to 7% cost increase (compound) for every additional 100 m of depth. Thus it could be expected that bores of 2000 and 4000 m depth would cost about $2\frac{3}{4}$ and 11 times (respectively) as much as a bore of 1000 m depth. Since in a wet field the temperature rises less rapidly than the depth increases (Fig. 13), and even in a dry hot rock thermal area the temperature rise is probably more or less proportional to depth, it follows that the drilling cost per million kgcal of heat won must rise rapidly with depth. This would be so even on the improbable assumption that the fluid yield from a bore remains constant regardless of depth. It is true that the *grade* of heat rises with increasing depth and that its potential range of applicability therefore increases, but the fact remains that conventional drilling is a process that yields diminishing returns, per unit of cost, with increasing depth. It is for this reason that geothermal drilling can seldom be justified at present for depths exceeding 2000 m or so, except perhaps in the case of geopressurized fields where the prize to be won is not only hot water and steam, but also hydraulic pressure and dissolved natural gas. Thus the recovery of very deep heat is likely to remain commercially unachievable *until* we can develop new methods of penetration for which the costs are far less sensitive to depth, or *unless* the quantity and grade of heat at great depth are so attractive as to offset the cost of reaching it. Even now we are *technically* capable of drilling to depths of the order of 15 km by conventional means, but the cost of doing so would be prohibitive. As long ago as 1971 it was estimated that a hole of this depth and about 25 cm diameter, even under ideal conditions, would take about $4\frac{3}{4}$ years to sink and that it would then have cost about US $20 m [185]. It is probable that by now (1977) the cost would have risen to something exceeding US $30m. To justify such a vast capital outlay a gigantic emission of heat would be necessary.

What are the reasons for this rapid rise in drilling costs with increasing depth? There are several. In the first place a very deep hole needs a larger and heavier rig to handle the great weight of long casing strings and drill stems, and more powerful driving units. Secondly, a very deep hole may need more than the conventional three concentric casings above the slotted liner (Fig. 17); and if the bottom diameter of the hole is to be retained at a required practical minimum, the upper casings will have to be larger and heavier to cater for the greater number of stepped enlargements. All this requires more steel, heavier loads more grouting cement, longer times to place and grout the casings in position, and slower penetration rates at the upper levels owing to the larger bits needed. Then, penetration rates, which may be as much as 200 to 300 ft/day in sedimentary rocks at fairly shallow depths will be far less in the harder igneous rocks liable to be encountered at great depths – perhaps as low as 30 ft/day. Yet another factor is that bits have a finite life, depending upon the hardness and temperature of the rock; so that in general the deeper the penetration the shorter the bit life. This not only raises the cost of the bits, but adds enormously to the time wasted in replacing them, for every time a bit is replaced it is necessary to raise and dismantle the entire drill stem, fit the new bit, reassemble the drill stem and lower it into the hole. The time required for this may be about 10 hours at 3000 m depth and about 24 hours at 10 000 m depth. As the life of a bit may sometimes be as short as 24 hours, the wasted replacement time virtually halves the effective penetration rate at this latter depth. Deeper holes require larger quantities of drilling mud to be stocked, and greater facilities for heat dissipation. All these factors contribute to the quasi-exponential cost/depth relationship of drilling.

7.16 Turbo-drilling

With conventional drilling the mechanical power drive is placed at the ground surface and the driving torque is transmitted to the bit through the long drill-stem. In very deep holes this driving torque may cause two or three complete rotations of the stem, so that the bit lags behind the drive by several hundred degrees of angle. Rotating drill stems are apt to whip, and thus to damage themselves, the casings and unlined bore walls. Various attempts have therefore been made to fix the driving unit at the base of the hole, close to the bit, so that the stem – though still twisted by the torque – need not rotate. Electric motors have been used for this purpose, but probably the most successful drill without a rotating stem is the Bristol–Siddeley turbo-drill, designed by Sir Frank Whittle. This incorporates a compact multi-stage turbine motivated by the drilling mud, an epicyclic speed reduction gearbox, and a device for automatically adjusting the pressure on the bit according to the cutting resistance so as to ensure a fairly constant speed of rotation and to reduce mechanical shocks. The drill claims improved penetration rates and easy adaptability to directional drilling. Although it has not yet seen a great deal of service there is no reason to doubt its reliability, and it is a finely engineered device. The cost pattern of using it at various depths has not yet been well established, but it seems likely to be comparable with that of conventional drilling. The turbo drill does not of course enable the drill stem to be dispensed with as the device has to be supported and the driving torque resisted; but the fact that the stem does not rotate eliminates the tendency to whip and avoids the consequential damage.

7.17 Other rock penetration methods

There can be no doubt that the quasi-exponential nature of the cost/depth relationship for conventional drilling has hitherto seriously limited geothermal development, for it virtually imposes a maximum economic depth below which it would not pay to penetrate. This economic limitation is reinforced by a *temperature* limitation, for at great depths – even in zones of moderate temperature gradient – the rock becomes too hot for the durability of the drilling bits. Many alternative methods of rock penetration have been devised, such as the percussion drill, the explosive drill, the laser drill and others, some of which may well show a marked economic advantage over conventional deep drilling. A jet piercing device, in which an oxygen-fuel flame is impinged upon the rock so as to cause spalling by thermal stress has had limited success with certain rock types, but not with most of the common igneous rocks which do not spall easily. By far the most promising device for the future deep penetration into the crustal rocks is the *subterrene*, which will be described in some detail in Chapter 19.

Bore characteristics and their measurement

8.1 General

To succeed in bringing to the surface hot water and/or steam from below-ground is, of course, only the first step in geothermal exploitation. Before we know to what practical purpose and on what scale these fluids can be applied we must first study their available quantities and their qualities, both physical and chemical. Several important measurements must be taken as soon as possible after a new bore has been 'blown' (or, in the case of certain semi-thermal fields, 'pumped') in order to ascertain the physical properties and energy potential of the discharged fluids. It will also be necessary to determine the *chemical* characteristics of the fluids, but these will be dealt with in Chapter 16. The most important variable physical features of a productive bore are as follows; and as some of them will be interdependent, much of the gathered information will take the form of a series of curves rather than fixed figures:

  (i) the wellhead pressure,
 (ii) the wellhead temperature,
(iii) the yield, or mass flow, of the steam (if any),
(iv) the yield, or mass flow, of the hot water (if any),
 (v) the fluid enthalpy,
(vi) the fluid 'quality' – i.e. the dryness or wetness factor.

Although loose references will later be made to 'typical' bores, it is important to recognize at the outset that well characteristics will vary widely from field to field and from bore to bore. Strictly speaking there is no such thing as a 'typical' bore any more than there is a 'typical' man or woman, but a certain broad range of features can be discerned from the study of many fields, so that orders of magnitude at least may roughly be determined.

8.2 Wet hyper-thermal fields

With bores yielding water/steam mixtures, saturated conditions must of course prevail, so the variables (i)/(ii), and (v)/(vi) must be interdependent. The relationships between these four variables can be obtained accurately from the steam tables, or approximately from Figs. 22 and 23. The fluid yields, variables (iii) and (iv), will differ widely from bore to bore and from field to field, and must be determined by test measurements: they will vary with the wellhead pressure. These yields will be dependent upon the tempera-

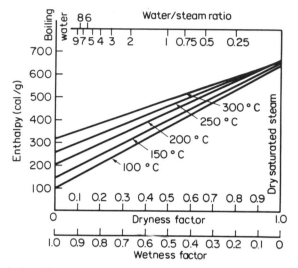

Figure 22 Relationship between enthalpy, temperature and quality for saturated water/steam mixtures.

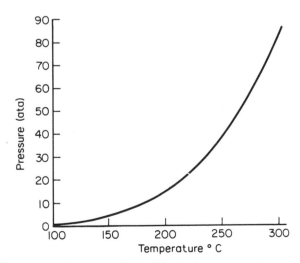

Figure 23 Temperature/pressure relationship for dry saturated steam.

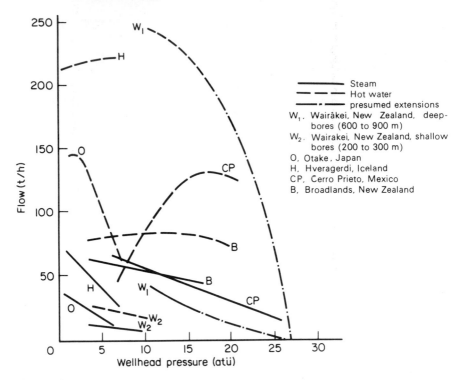

Figure 24   'Typical' bore characteristics for 'wet' geothermal fields.

ture at depth and upon the flow resistances through the fissures of the aquifer and up the bores; and since these factors will differ in every case, the flow characteristics of every individual bore will be unique to that bore. Fig. 24 shows some typical bore characteristics at various exploited wet fields; but the word 'typical' requires further qualification. First, for each field a random number of bores has been taken according to the available published data, and *average* yields have been deduced therefrom. The characteristics of any one of the selected bores may well differ appreciably from the average. Secondly, the average figures do not all lie exactly upon the smoothed approximate curves here shown. Thirdly, the characteristics of a bore are not altogether constant. As already explained in Section 5.16, bore yields tend to decline in the course of time. In wet fields they often become drier (i.e. the enthalpy increases) as well as less productive. The curves shown in Fig. 24 are all more or less representative of bores tested fairly early in the development of each field.

When a bore is closed, the flow of both water and steam must of course fall to zero. The pressure instantaneously registered at the moment of closure is known as the 'shut-in' pressure, which may later decline as the fluid

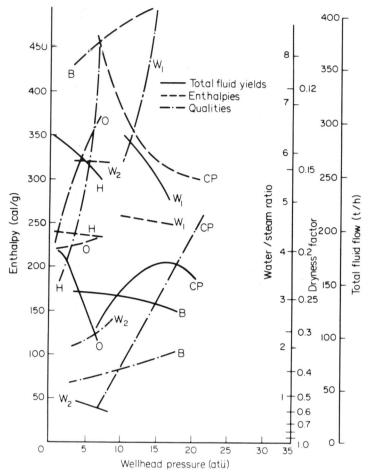

W₁ Wairakei, New Zealand. deep bores (600 to 900m)
W₂, Wairakei, New Zealand. Shallow bores (200 to 300m)
O, Otaxe, Japan
H, Hveragerdi, Iceland
CP, Cerro Prieto, Mexico
B, Broadlands, New Zealand

Figure 25 Typical total bore yields, enthalpies and fluid qualities for 'wet' geothermal fields.

in the bore cools off. In the case of the Wairakei deep bores (W_1 in Fig. 24) presumed extensions of the water and steam characteristics would suggest a shut-in pressure of about 27 atü, though curvatures sometimes fall off rather abruptly when a bore is gradually quenched by closure. It will be noted that steam yields tend to approximate to straight lines over the pressure range of greatest interest, while hot water characteristics are usually curved.

From the steam and hot water characteristics it is a simple matter to deduce the total fluid flow* (by adding the two), the quality (by comparing the two) and the enthalpy (from the steam tables or from Fig. 22). These further characteristics for the same fields as in Fig. 24 are shown in Fig. 25. They reveal the following features:

(i) the very great variety of bore characteristics found from field to field.

(ii) the relative constancy of the fluid enthalpy over a fair range of pressures at Wairakei (both shallow and deep bores), Hveragerdi and Otake.

(iii) The *rising* enthalpy at Broadlands and the *falling* enthalpy at Cerro Prieto, with increasing wellhead pressure.

(iv) In every case rising wellhead pressure causes increasing wetness.

(v) The enthalpy of the shallow Wairakei bores is greater than that of the deep bores in the same field.

If a bore were to tap a zone of boiling water, the stagnant enthalpy of the flashing mixture (i.e. the enthalpy after allowing for the kinetic energy of the bore effluent) emitted by the bore would be expected to be constant, at the value of the enthalpy of the hot water tapped at depth. Thus the two classes of bore at Wairakei and also the bores at Hveragerdi and, to a lesser extent, Otake approximately conform with this expectation. The steeply rising and falling enthalpy curves for Broadlands and Cerro Prieto respectively suggest that the bores are fed from more than one 'horizon' in each case, each horizon contributing different proportions of the total heat with different pressure distributions down the bores. Such phenomena as these show that the true nature of fields must in fact be considerably more complicated than the simple stylized model of Fig. 10 would suggest. Increased wetness with rising pressure is only to be expected, especially if the enthalpy does not vary very greatly, since the higher pressure will tend to suppress flashing. The phenomenon of the higher enthalpy of the Wairakei shallow bores by comparison with that of the deep bores has already been commented upon in Section 5.9.

8.3 Dry hyper-thermal fields

As the steam discharged from dry hyper-thermal fields is usually somewhat superheated, there can clearly be no direct relationship between the variables (i) and (ii) of Section 8.1, while variable (v) will follow from these two and variable (vi) will have no meaning. At Larderello, in Italy, the degree of superheat ranges from about 83° C at a wellhead pressure of 6 ata to about 56° C at 16 ata. As low wellhead pressures are associated with high steam flows, there would be increased throttling action at low pressures as the

---

*The *total* fluid flow should of course include incondensible gases, but in the present context only the 'H_2O' fluids are being considered.

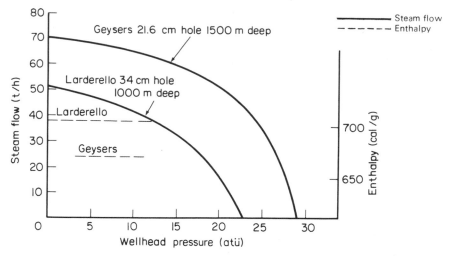

Figure 26   Typical bore characteristics in 'dry' geothermal fields.

steam passes at high velocity through the permeable formation, and this could account for the higher degree of superheat, as would be expected with isenthalpic expansion. At the Geysers field in California too the degree of superheat declines with rising well-head pressure, ranging from about 28° C at 4 ata to about 10° C at 10 ata. The enthalpy at the Geysers varies somewhat from bore to bore, but is generally rather close to the maximum attainable enthalpy of rather less than 670 cal/g for saturated stream.

The steam flow and enthalpy characteristics, against a wellhead pressure base, are shown for Larderello and the Geysers – the two most famous 'dry' fields in the world – in Fig. 26.

8.4 Hot water fields

The characteristics of low enthalpy hot water fields are too idiosyncratic for generalizations to be made about them. Such fields range from deep water reservoirs in semi-thermal areas, initially at pressure equilibrium and requiring either pumping or thermo-symphonic action to lift the water from depth, to geopressurized fields that gush under partial lithostatic pressure [22].

8.5 Measurements

The measurements of wellhead temperatures and pressures offer no problems: they can simply be made by means of thermometers and pressure gauges. For wet bores, in fact, it is only necessary to measure one of these two, since they are inter-related in accordance with the curve of Fig. 23 or, for greater accuracy, in terms of the steam tables. It is the measurement of mass flows that requires particular care. Once these measurements have also been obtained – for each phase in a wet mixture – the derivation of enthalpies and of dryness or wetness factors can be quite simply deduced, either from the steam tables or, in the case of wet fields, from Fig. 22 approximately.

Mass flow measurements of dry saturated steam or of superheated steam can be made quite simply by means of sharp-edged orifice meters, after making sure that sufficient lengths of straight piping are placed upstream and downstream of the orifice to ensure non-turbulent flow (at least 25 pipe diameters upstream and at least 10 diameters downstream). The mass flow will then be a function of the manometric pressure drop across the orifice, of the upstream pressure and temperature and of the orifice restriction ratio: a calibration chart will be provided with each orifice flow-meter.

An alternative method of measuring dry steam is to discharge it at sonic velocity (so that the precise downstream pressure is immaterial) to the atmosphere through a cone of partly standardized dimensions, as shown in Fig. 27. The method is not very accurate owing to the number of rival formulae (quoted on Fig. 27) that can be used to interpret the results, but for approximate measurements its simplicity has something to commend it when the upstream pressure is at least 2 ata. Although the upstream pipe diameter and the cone length are standardized, the cone taper – and therefore the discharge area – may be varied so that suitably sized cones may be chosen for testing at convenient pressures. The accuracy of the various formulae will depend to some extent upon the degree of cone taper, but the BSS 752 formula is probably as reliable as any.

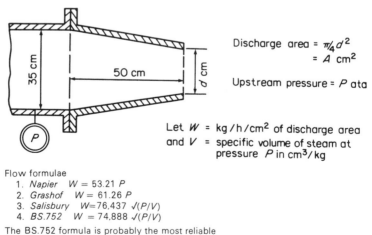

Discharge area = $\frac{\pi}{4}d^2$
= A cm^2

Upstream pressure = P ata

Let W = kg/h/cm^2 of discharge area
and V = specific volume of steam at pressure P in cm^3/kg

Flow formulae
1. *Napier*　$W = 53.21\ P$
2. *Grashof*　$W = 61.26\ P$
3. *Salisbury*　$W = 76,437\ \sqrt{(P/V)}$
4. *BS.752*　$W = 74,888\ \sqrt{(P/V)}$

The BS.752 formula is probably the most reliable

Figure 27　Measurement of dry steam discharge to atmosphere through cones at sonic velocity.
Correction for superheat: multiply W by 1.001 17 S, where S is the number of degrees of superheat in °C.

The measurement of water/steam mixtures presents greater difficulties. The following are some of the methods used:

8.5.1 *Calorimetry* [59]. The whole of the water/steam output of a bore is discharged for a measured time into a tank containing a known mass of cold water at a known temperature. When the well discharge has ceased, the additional weight of water in the tank is a measure of the mass flow during the period of the test; (the water content is simply added while the steam content is condensed). From the rise in temperature of the water it is a simple matter to deduce the added heat, and by dividing this by the added mass the enthalpy of the bore fluid can be deduced and the dryness factor will be a derivative of this. Despite its simplicity, this method has its limitations. Unless a very large tank is used, the method is suited to bores of small output only; otherwise the duration of the test would have to be very short and the accuracy of the test would be sensitive to the speed of operating the test valves. A special rocking arm has been devised in New Zealand [59] to reduce this difficulty; but if valves are used it is necessary to *start opening* them at the beginning of the measured time and to *start closing* them at the end of the measured time, so that the periods of valve operation will tend to cancel one another out. The method can be used for bores of larger output by splitting the discharge in two at a T-joint, measuring the discharge from one branch only and doubling it. If a valve on the *un*measured branch is adjusted to give the same back-pressure as on the measured branch, it is not

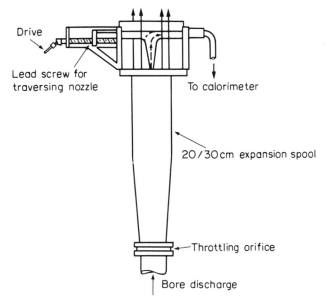

Figure 28   Bore enthalpy measurement by sampler. (After Fig. 2 of [59].)

unreasonable to suppose that the flow resistance in each branch is the same, so that each accepts one half of the total.

8.5.2 *Sampler measurement* [59]. By means of a travelling sampler tube, the flows at different points on the diameter of a discharge pipe can be measured by calorimetry and integrated, so as to arrive at the total discharge across the whole pipe. This is a very cheap method, as it requires only a small calorimeter and a light portable traversing device (Fig. 28), but it is not very accurate. The speed of traverse can be arranged so that the time in each position is proportional to the area of the annular ring being sampled. Ideally, the velocity of the fluid drawn off should be the same as the velocity of fluid in the main discharge pipe, so that the sampling tube causes a minimum of disturbance to the flow pattern.

8.5.3 *Phase separation and the measurement of each phase separately.* This is the most accurate, but the most expensive method. Its reliability also depends upon the efficiency of the separator used, but as efficiencies of the order of 99.9% can usually be achieved this limitation should be insignificant. The well fluid is passed through a cyclone separator, the steam flow is measured by calibrated orifice and manometer, and the hot water is measured in the same way. It is important that the water first be cooled sufficiently to

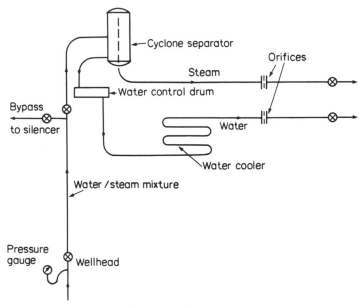

Figure 29   Well yield measurement by phase separation and orifices. (After Fig. 1 of [59].)

ensure that boiling does not occur when passing through the orifice, otherwise the water flow readings will be quite unreliable. The arrangement is
shown in Fig. 29,

8.5.4 *Critical lip pressure method* [60]. If hot water or a water/steam
mixture is discharged at sonic velocity through an open-ended pipe, and if
the pressure is measured at the discharge lip of the pipe, this pressure is a
measure of the *heat flow* of the discharged fluid. James [60] has deduced the
following formula, expressed in f.p.s. units.

$$G = \frac{11{,}400 \; P^{0 \cdot 96}}{h^{1 \cdot 102}},$$

where G = flow in lb/sec/ft² of discharge area.
P = critical lip pressure in psia.
h = fluid enthalpy in btu/lb

Where the two phases of a water/steam mixture are segregated by means of a
cyclone separator the water can be measured either by a cooled orifice
(as in 8.6.3.) or, after allowing for the flashed steam, by means of a V-notch
or weir. Alternatively, however, the water may simply be discharged at
sonic velocity through an open-ended pipe and measured by applying the
above formula. By measuring the temperature of the water leaving the
separator its enthalpy is made known, and G can be deduced from P. It may
be mentioned that the very existence of a positive gauge pressure for P is
proof that the velocity is sonic or supersonic. If no such lip pressure is registered, then a smaller pipe must be used for the discharge until a lip pressure can
be clearly measured. With water/steam mixtures the enthalpy of the mixture
is *not* known; but the combination of an upstream orifice and a lip pressure
gauge can provide all the information required for deducing the enthalpy and
the mass flow in the manner shown in Fig. 30. The orifice factor K must first
be determined by measurement, and the enthalpy when deduced from the
lower graph after substituting the value K in the ordinate and observing the
upstream pressure and the manometer pressure, as well as the critical lip
pressure. This is a cheap method of measuring bore output characteristics,
and is reasonably accurate within the idiosyncratic variations of a bore's
performance from time to time.

8.5.5 *Method of cones.* This is a relatively simple method of measuring
the mass flow, quality and enthalpy characteristics of a wet bore by using a
cyclone separator and a series of measuring cones of the type illustrated
in Fig. 27. The method avoids the separate measurement of the water
(though a lip pressure discharge pipe could provide useful corroborative
evidence) and enables the dryness and enthalpy of the bore mixture to be

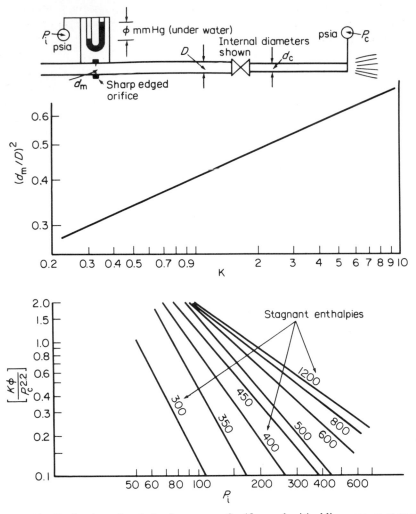

Figure 30   Derivation of enthalpy by means of orifice and critical lip pressure at sonic velocity discharge. (After Fig. 1 of [60].)

Notes: (i) d_c/D assumed to be 0.75, which gives convenient values of P_c. For other values of d_c/D use the correction formula $P_{c_1}/P_{c_2} = (d_{c_2}/d_{c_1})^{2.082}$.

(ii)   Recommended straight pipe lengths:

upstream of orifice $\not< 25\,D$

orifice to valve $\not< 10\,D$

valve to discharge $\not< 25\,d_c$.

(iii)   James' original chart, giving pressures and enthalpies in British units is reproduced here. For conversions:

1 cal/g $= 1.8$ btu/lb

1 psi $= 0.068\,03$ atm.

deduced indirectly. It also dispenses with the need for long straight pipes required when a sharp-edged orifice is used for the steam measurement. The small sketch of Fig 31(a) shows a wet bore discharging directly into a cyclone separator after passing through a throttling valve V_1 which enables the wellhead pressure to be varied. The water is rejected (perhaps after corroborative measurement) and the steam discharge is connected to a measuring cone placed as closely as possible to the separator. Pressure measurements are taken just below the wellhead valve V_1 and also upstream of the cone. With various settings of the valve V_1 and with various sizes of cone, readings of both pressure-gauges are taken over a wide range of wellhead pressures. By using one of the cone formulae given in Fig. 27 – preferably BS. 752 – the steam S_c discharged from the cone is deduced from the pressure P_c, and these steam flows are plotted against wellhead pressure in the manner shown in Fig. 31(b). However, none of these curves is the well steam yield characteristic because, owing to the pressure drop from P_w to P_c a fraction of the water phase will be flashed into steam, so that

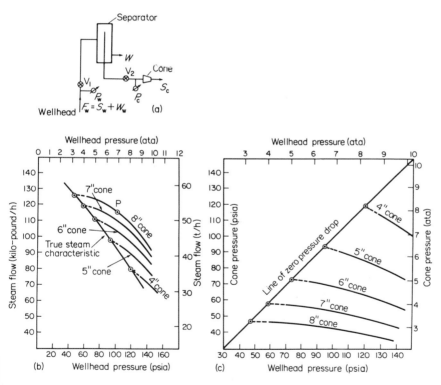

Figure 31 Measurement of well characteristics by means of separator and cones. (a) Test connections (b) Cone flow characteristics (c) Deduction by extrapolation of points on true steam characteristic

S_c will exceed the mass of steam issuing from the bore at the wellhead. The first task is to plot a graph of the pressure readings of both gauges (Fig. 31(c)) and to extrapolate the curves to meet the line of equal pressures. On that line, where there is no pressure drop from the wellhead to the cone, there would be no flashing of the water. The points where the cone curves meet the line of equal pressures will therefore truly represent the well steam yield characteristic. These points are then transferred to Fig. 31(b) and joined up to form the steam characteristic of the well. To deduce the water characteristic a single example will illustrate the method. At a wellhead pressure of 100 psia the cone steam discharge (Point P) is 116 500 lb/h, while the bore steam is 93 000 lb/h. The difference of 23 500 lb/h represents the steam flashed off from the water phase in dropping its pressure from 100 psia to 43 psia – the pressure at which an 8-in cone would discharge 116 500 lb/h of dry steam according to the BS. 752 formula. The enthalpy of saturated hot water is 298.7 btu/lb at 100 psia and 240.7 btu/lb at 43 psia, while the latent heat at 43 psia is 931.2 btu/lb. The proportion of water flashed will therefore be (298.7–240.7)/931.2, or 0.06228. Hence the water issuing from the wellhead will be 23 500/0.062 28, or 377 810 lb/h. Hence the mixture enthalpy at the wellhead can be deduced so:

$$
\begin{array}{lll}
\text{water heat} & 377\,810 \times \;\;298.7 = 112\,850\,000\;\text{btu/h} \\
\text{steam heat} & \underline{\;\;93\,000 \times 1188.2 = 110\,500\,000\;\text{btu/h}} \\
& \overline{470\,810\,\text{lb/h}} \qquad\quad 223\,350\,000\;\text{btu/h}
\end{array}
$$

$$
\text{Enthalpy} = \frac{223\,350\,000}{470\,810} = 474.4\;\text{btu/lb.}
$$

From this figure, with the help of steam tables, it is a simple matter to deduce the water content of the bore fluid at a wellhead pressure of 100 psia. The exercise may be repeated for several other points, thus enabling the enthalpy and water characteristics to be added to the steam characteristics. The accuracy of this method is limited by that of the cone formula used and by the efficiency of the separator, but the results should lie well within the vagaries of the bore itself. The method also ignores the very small pressure drop from the separator chamber to the cone entry, and if the separator were 100% efficient there could be a very small degree of superheat due to this small pressure drop if the pipe between the separator and the cone were lagged. The separator itself and the entry pipe should be lagged. If the water discharge is separately measured for corroborative evidence it would of course be necessary to add the quantity flashed, in order to arrive at the water quantity at the wellhead. As many cones may be used as are considered necessary to obtain consistent results, but if very small cones are used, care must be taken to avoid building up pressures in the separator higher than its design rating.

Cones can also be used for determining the characteristics of dry bores, without the intermediary of a separator, but thermometers must be used to ascertain the degree of superheat, so that a suitable correction may be applied to the cone discharge in accordance with the formula quoted on Fig. 27.

8.5.6 *Beta-rays and isotopes.* Various suggestions have been made of how to determine the quality of fluid emitted from a wet bore by passing a stream of beta-rays through it and measuring the absorption rate, which would be a function of the fluid density. Unfortunately, as will be shown in Section 8.11, the velocities of the two fluids will not be the same, so the apparent and true densities will differ unless the velocities of both fluids are the same. Even then, the result would be affected by the distribution of the water across the pipe cross-section. These problems are not easily solved. Alternatively, selective isotopes could be injected – one to be carried by the water and the other by the steam – so that selective measurement of radiation from each isotope at a chosen point over a fixed period of time would provide a means of determining the density which, in conjunction with the steam tables, would enable the enthalpy of the fluid mixture to be deduced. Some such method would enable a portable instrument to be taken from bore to bore for the analysis of flows without taking the bores out of service. Whether the obstacles are economic or technical, it would seem that no simple device has yet been produced, but the solution of the difficulties offers a challenge.

8.6 Heat capacity of a bore

Once the characteristics of a bore are known, it is a simple matter to deduce its heat capacity, provided that its application is known. An example will illustrate this point. Fig. 32 shows the flow characteristics of a hypothetical wet bore producing fluid at a constant enthalpy of 278 cal/g (500 btu/lb). The figure is arbitrary, but not improbable. It will be seen that the total heat emitted by the bore is fairly constant at low pressures but declines rapidly at higher pressures. This feature follows from the shape of the total fluid flow curve and the constancy of the enthalpy. With rising pressure and temperature, the steam contributes less, and the hot water more, of the total heat. Fig. 32 is only illustrative: other bores would behave differently.

It will be noted that there are three heat-yield curves shown in Fig. 32. The upper one represents the heat yield at the wellhead, reckoned above $0°C$, as taken from the steam tables. The middle curve also refers to the heat yield at the wellhead, but reckoned above an assumed local ambient temperature of river or lake water at about $19°C$, or $67°F$. It is, after all, only in the fact that the bore fluid exceeds some such temperature that it can be regarded as 'hot'. The lowest curve assumes a heat loss of 5% in transmission from the bore to the exploitation plant, and therefore represents

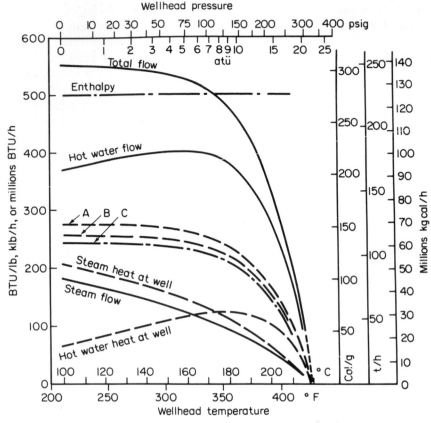

A. Total heat at well
B. Net heat at well reckoned above 67° F (19.4° C) assumed ambient
C. Usable heat assuming 5% transmission loss

Figure 32 Typical heat flow pattern for a 'wet' bore, assuming a constant enthalpy of 500 btu/1b or 278 cal/g.

Note: It is assumed that the heat both of the steam *and* of the water is used.

a probable estimation of the *usable* heat. Now let it be assumed that this particular bore is to be used for a timber-drying plant which, according to Fig. 2, needs heat at about 160° C; and let it further be assumed that a temperature loss of 3° C is incurred between the well and the plant. Fig. 32 shows that the particular well represented could supply about 59 million kgcal/h (net) at the required temperature. A paper factory, on the other hand, requiring heat at 180° C plus a further 3° C temperature drop from wellhead to plant, could obtain from the same well only about 52 Mkgcal/h at the required temperature.

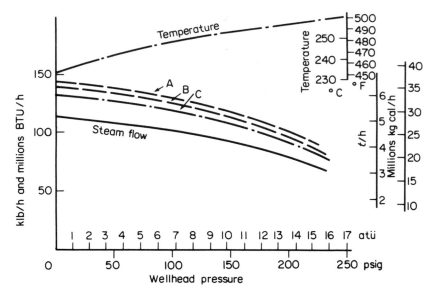

A. Total heat at well
B. Net heat at well reckoned above 67° F (19.4°C) assumed ambient
C. Usable heat assuming 5% transmission loss

Figure 33   Typical heat flow pattern for a Larderello type 'dry' bore.

If a typical Larderello type of dry bore were to be used, as illustrated in Fig. 33, it will be seen that the grade of available heat is higher at all wellhead pressures than required by either of the industries assumed in the above example for a wet bore. It would thus pay to operate the dry bore at the lowest practicable pressure – perhaps even sub-atmospheric – in order to obtain the maximum heat yield, and sometimes even to degrade that heat by dilution to the temperature required for the particular industrial application contemplated. With a 'Geysers' type of dry bore the temperatures are rather lower, though the steam flows are rather higher, than with a 'Larderello' type bore. In either case, except for processes requiring the highest grades of heat, such as power generation, it would usually pay to operate the wells at the lowest practicable pressure (consistent with plant design requirements) at which heat yields are still rising with falling pressure. As with Fig. 32, three heat-yield curves are shown.

8.7 Power capacity of a bore
When it comes to the generation of electric power, however, the situation is quite different from that of other heat-consuming industries, for the efficiency of a power plant declines with temperature owing to inescapable thermodynamic constraints. Thus the wellhead pressure of a bore at which maximum

power potential is attained will not coincide with that at which maximum *heat* potential occurs. Rising wellhead pressure has two opposing effects: it *reduces* the steam yield but *raises* the extractable energy per kilogram of steam. The net result of these two opposing effects is to give a steam power

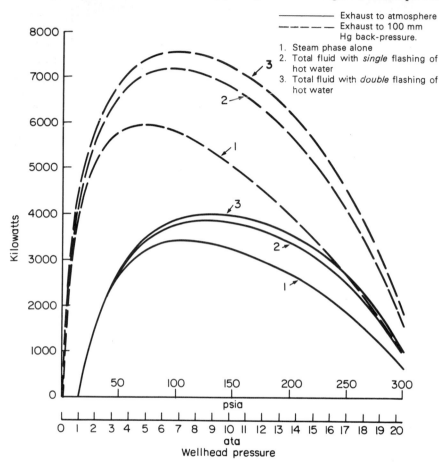

Figure 34 Power potential from typical 'wet' bore.

Note: The powers shown are at the generator terminals, without deduction for the power absorbed by gas exhausters or other auxiliaries.

Assumptions: (i) Same bore as for Fig. 32.

(ii) *Condensing* turbines sited at central power station fairly remote from bore. 25% of the absolute wellhead pressure and 6% of steam condensation assumed to be lost in steam transmission. In the case of flash steam a $2\frac{1}{2}\%$ heat loss has been assumed in hot water transmission.

(iii) *Non-condensing* turbines sited fairly close to the bores. 10% of the absolute wellhead pressure and 1% of steam condensation assumed to be lost in steam transmission.

potential curve that attains a maximum value at one particular wellhead pressure, falling off with either increasing or decreasing pressure on either side of that optimum If consideration is also taken of the power extractable from the hot water phase of wet bores, a similar, though rather different, optimum pressure can be determined – despite the fact that over certain pressure ranges the hot water yield may sometimes *rise* with increased wellhead pressure (as in Fig. 32). Much will depend upon the cycle adopted, the fluid enthalpy and upon the turbine back-pressure. A typical pattern of power potential for a *wet* bore, supplying straight condensing and non-condensing turbines, is illustrated in Fig. 34. The use of better vacua would improve the power potential of the bore when supplying condensing turbines, but the gain would not necessarily be economically worthwhile because of the greater power required for gas exhaustion and the need for larger cooling facilities. (The power potentials of Fig. 34 are *gross*, and exclude deduction for auxiliary power consumption.) The question of single and double flashing will be covered in Chapter 10, but it may here simply be noted that by using the hot water and adopting two stages of flashing after the wellhead, in the example chosen for Fig. 34, a power gain of about 26% could be won by comparison with the use of steam alone in condensing turbines. With lower enthalpy fluids (i.e. having a higher proportion of hot water to steam) the proportional power gain could be substantially more.

For a Larderello type of dry bore the pattern of power potential would be as illustrated in Fig. 35. The smaller steam yield, by comparison with the chosen example of wet bore, tends to be offset to a greater or lesser extent by the superheat, with the result that the steam phase alone of the chosen wet bore yields about 14% *less* power with non-condensing and about 5% *more* power with condensing plants at 100 mmHg back-pressure. If the hot water from the wet bore were also exploited for power by double flashing, both wells would yield about the same amount of maximum power when used with non-condensing plant. A Geysers type of dry bore (Fig. 26) could produce more power than either of the two bores considered above.

It is emphasized that Figs. 34 and 35 are examples only: every individual bore will have a different power potential curve, depending upon its mass flow and enthalpy characteristics and upon how the bore is to be used. But in every case there will be an optimum pressure at which the power potential will attain a maximum value. This optimum pressure, however, will not necessarily be the same as the *overall economic optimum pressure*, (see next Section).

8.8 Economic optimum wellhead pressure for power production.
Operating a bore at a pressure at which the power yield is a maximum will clearly save capital expenditure in drilling costs and in wellhead equipment and branch pipe-lines, in that fewer bores will be needed to produce a desired

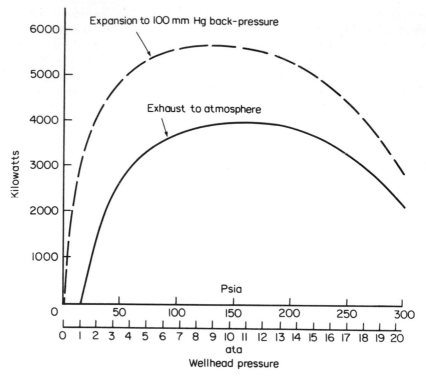

Figure 35   Power potential from Larderello type 'dry' bore.

   Note: The powers shown are at the generator terminals, without deduction for the power absorbed by gas exhausters or other auxilliaries.

   Assumptions: (i) Same bore as for Fig. 33.

   (ii) *Condensing* turbines sited at central power station fairly remote from bore. 25% of the absolute wellhead pressure and 5% of the heat content assumed to be lost in steam transmission.

   (iii) *Non-condensing* turbines sited fairly close to bores. 10% of the absolute wellhead pressure and 1% of the heat content assumed to be lost in steam transmission.

quantity of power. On the other hand there are other economic criteria to be considered. Low pressures imply high specific steam volumes and therefore bulky pipework and vessels. Low pressures will also tend to shorten the life of a finite reservoir of heat by drawing down the stored fluid too quickly. High pressures, on the other hand, involve the use of short turbine blades which are more conducive to blade gap losses than long blades and more vulnerable to the build-up of chemical deposits. High pressure valves and fittings tend to be costlier, and the associated higher temperatures require thicker thermal insulation. The interplay of all these considerations has been carefully studied by James[61], who has concluded that for fluids of enthalpies exceeding 400 btu/lb (about 220 cal/g) – and this covers most hyperthermal fields – the overall economic wellhead pressure both for dry *and* wet fields seems to lie in the neighbourhood of 6 ata. This comparatively 'precise' figure is somewhat surprising in view of the fact that the optimum power yield pressures range from about 5 to 11 ata in Figs. 34 and 35. However, the curves in these figures are relatively 'flat-topped', and by choosing a wellhead pressure of about 6 ata the sacrifice in power from the theoretical optima would be within 3% for all condensing conditions. Only with atmospheric exhaust – rarely adopted for geothermal power generation – would the power sacrifice amount to about $11\frac{1}{2}\%$ with a Larderello type bore, and much less in other cases. Hence there is plenty of scope for the other economic considerations.

8.9 The characteristics of artificially created 'dry rock' bores

As bores of this type are still in the experimental stage, it would be inappropriate to speculate too much upon their expected characteristics. The power absorbed by the circulation pumps, except where thermo-syphon conditions can be established (see Chapter 19), would form an additional variable of importance, so that the characteristics would probably consist of curves of mass flow and temperature (and hence enthalpy) against a base of pumping power. Stabilization times after each change of flow rate could also be of importance.

8.10 Open discharge pipes

A certain amount of rough qualitative knowledge can be gained from observing a well freely discharging into the atmosphere through an open-ended pipe. Superheated steam will tend to emerge from the pipe in a parallel jet with a fairly long transparent gap between the pipe end and the point at which the steam starts to condense into an opaque stream of water droplets. After some distance the air friction dissipates the kinetic energy of the jet and the condensed steam forms a rising billowing cloud (Fig. 36a). Dry saturated steam produces a similar formation, but the transparent zone is quite short as the steam starts to condense very soon after leaving the pipe (Fig. 36b). Wet steam produces a characteristic 'tulip' discharge (Fig. 36c and Plate 9) due to the evaporative 'explosion' of the water particles as they flash as soon as the pressure fals to atmospheric, thus laterally distending the jet: there is no transparent zone. When a test cone is used, a slightly tulip-shaped formation is apt to occur, even with dry steam, owing to the directional inertia of the fluid sliding along the tapered cone surface (Fig. 36d). Too much should not be read into these discharge patterns, as much will depend upon the humidity of the atmosphere and upon the wind; but very wet or highly superheated steam will usually betray their quality by the distinctiveness of the discharge jet pattern.

Plate 9 Typical 'tulip' discharge of wet bore fluid from open pipe at Cerro Prieto, Mexico [189]. (By courtesy of the Hydrothermal Power Company Ltd.)

8.11 Fluid velocities at wellheads

In a dry field, if the steam enters the lower uncased part of a bore at several different levels, the steam velocity will accelerate as it ascends from the bottom, partly because of the increasing number of feeding points and partly because of the falling pressure (and consequently the increasing specific volume) of the steam as it rises. At the point where the uncased zone

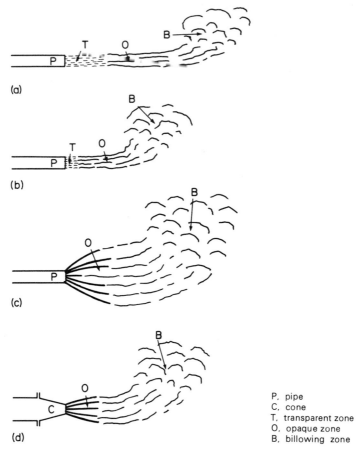

Figure 36 Typical steam discharge patterns. (a) Superheated steam (b) Dry saturated steam (c) Wet steam (d) Dry steam from cone

ends and the steam enters the production casing, there may be an increase in the sectional flow area (see Fig. 17) which will cause a momentary check to the velocity; but after that the steam will again accelerate as it approaches the top of the bore. At the wellhead, it is a simple matter to calculate the average steam velocity if the mass flow, pressure and temperature are all known. From the last two the specific volume may be deduced and the mass flow converted to volumetric flow which, when divided by the cross-sectional area of the bore, will give the average steam velocity. Owing to skin friction, the velocity close to the bore walls will be somewhat less than in the middle of the bore. Obviously, for a given mass flow, the smaller the bore diameter the higher will be the steam velocity; and if this should be allowed to attain sonic velocity a maximum bore output will be imposed. The fact that the two bore characteristics shown in Fig. 26 are still rising at atmospheric wellhead pressure, with further pressure reduction, shows that subsonic flow still prevails (for the unspecified bore size used).

At Larderello most production casings are of 34 cm diameter – i.e. 0.0908 m² section. If a bore is normally operated at a wellhead pressure of, say 9 ata (8 atü), the bore yield according to Fig. 26 would be about 44 t/h. At this pressure, at which the temperature is about 250° C, the specific volume of superheated steam would be 0.256 m³/kg. Hence the mean steam velocity at the wellhead would be $(44\,000 \times 0.256)/(3600 \times 0.0908)$, or 34.46 m/sec (113.1 ft/sec).

In the case of a wet bore conditions are less simple. In the lower part of the bore the fluid would probably be hot water, possibly fed at different levels, and the upward velocity would increase as the fluid rises, with a check at the point of entry into the production casing. Owing to the high density of water, the velocities at depth would tend to be far lower than in a dry steam bore. But at some point in the bore, probably near the lower end of the production casing, flashing would develop at an increasing rate and thus reduce the specific gravity of the water/steam mixture, thereby imparting buoyancy to the fluid to assist the upward flow. The proportion of steam to water would steadily increase as the fluid rises, and the velocity would accelerate rapidly. The pressure drop over the last few metres before the wellhead valve will be considerable. At the wellhead the water phase will tend to cling to the bore walls while the steam phase will stay in the middle of the stream. Skin friction will thus ensure that the water velocity will always be appreciably less than the steam velocity. To take as an example the hypothetical wet bore of Fig. 32, and again assuming a working wellhead pressure of 9 ata, the bore yield would be 223 t/h, of which 177 t/h would be hot water and 46 t/h steam. At this pressure under saturated conditions the fluid specific volumes are 0.2124 and 0.00112 m³/kg for steam and water respectively, so that the *mean* fluid specific volume would be:

$$\frac{(177 \times 0.001\,12) + (46 \times 0.2124)}{223}, \text{ or } 0.044\,69\,\text{m}^3/\text{kg}.$$

With the same sized bore as assumed for Larderello – 34 cm diameter, 0.0908 m² – the *mean* fluid velocity at the wellhead would be:

$$\frac{223\,000 \times 0.044\,69}{3600 \times 0.0908} = 30.43\,\text{m/sec (99.83 ft/sec)}.$$

The water velocity, however, is likely to be considerably lower, and the steam velocity higher than this mean value. There will even be velocity variations within each phase. The greater the mean velocity difference between the two fluids, the greater will be the proportional area of the bore section occupied by the water. The relationship between the mean velocities of each phase, for the chosen example, would be as shown in Fig. 37.

As already mentioned, if the fluid yield is sufficiently high and the bore diameter sufficiently small, sonic velocity will be attained at the wellhead and this will impose a lower limit to the wellhead pressure. If a wellhead valve be wide open so that the bore discharges freely to the atmosphere, the wellhead pressure will register atmospheric so long as the discharge is subsonic; but as soon as sonic value is reached a positive gauge pressure will start to build up at the wellhead. This will be equivalent to the 'lip pressure' referred to in Section 8.5.4. This means that with certain very powerful wells with slightly undersized bore diameters, as are to be found in the Cerro Prieto field in Mexico, it is impossible to reduce the wellhead pressure to atmospheric. Lower pressures above the wellhead valve could only be obtained by throttl-

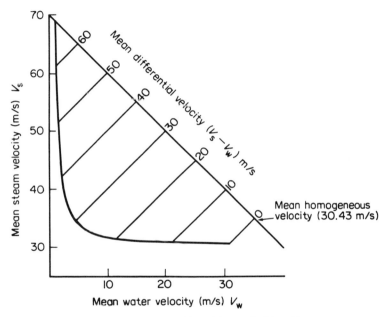

Figure 37   Relative fluid velocities at wellhead for typical 'wet' bore.
Assumptions: bore diameter = 34 cm (internal)
wellhead pressure = 9 ata
fluid flows {steam = 46 t/h
water = 177 t/h
total = 223 t/h
dryness factor = 20.63%
wetness factor = 79.37%
water/steam ratio = 3.85
enthalpy = 278 cal/g (500 btu/1b)

ing, and increasing the pressure below the valve. Only the substitution of a larger diameter bore could enable atmospheric pressure to be attained at the wellhead when freely discharging with the valve fully open, but this would be unnecessary.

The total flow characteristic of the hypothetical wet bore of Fig. 32 is almost horizontal at atmospheric wellhead pressure, which shows that this bore, with the (unspecified) diameter in use, would just about attain critical flow – i.e. sonic velocity – when discharging freely to atmosphere.

The sonic velocity of a fluid is proportional to the square root of the elasticity/density ratio. With a perfect gas the elasticity is proportional to the absolute pressure. With wet steam mixtures the elasticity mainly depends upon that of the steam, while the density is very sensitive to the wetness. The sonic velocity of wet mixtures is therefore usually lower than that of dry steam at the same temperature, except at very low dryness factors when the high elasticity of water prevails and the sonic velocity rises rapidly towards that of water, which is very high.

Fluid collection and transmission

9.1 Wellhead gear

Except where pressurized or two-phase fluid transmission is adopted (see Sections 9.11 and 9.12) it is necessary for a quantity of equipment to be assembled at every wellhead in a wet hyper-thermal field, to provide the means of controlling the fluids that emerge from the bore, of separating the water and grit from the steam, of disposing of unwanted fluids, of silencing the bore fluids when discharged to the atmosphere and of dissipating their energy, and of protecting the equipment and pipelines against excessive

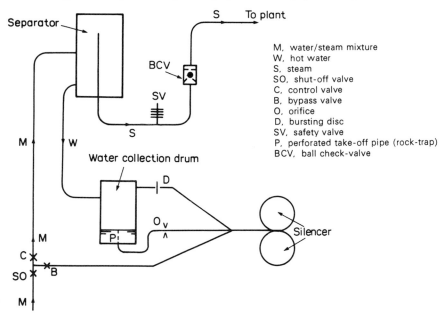

M, water/steam mixture
W, hot water
S, steam
SO, shut-off valve
C, control valve
B, bypass valve
O, orifice
D, bursting disc
SV, safety valve
P, perforated take-off pipe (rock-trap)
BCV, ball check-valve

Figure 38 Diagrammatic arrangement of wellhead equipment and safety devices at a 'wet' bore. (After Fig. 5 of [63].)

Plate 10 Typical wellhead gear assembly, Wairakei, New Zealand. Note the pipe expansion looping, the tall water/steam separator (centre), the water-collection drum (left centre), the grille-covered access steps into the cellar (left foreground) and the twin silencers (left background). Note also some trees (right) damaged by silicated spray (see Section 17.9). (By courtesy of the New Zealand Ministry of Works and Development.)

pressures. A typical assembly of wellhead equipment at such a bore is shown in Fig. 38 and Plate 10. The various components are connected together by means of pipework with conventional provision for thermal expansion – i.e. loops or bellows pieces – and most of them and their associated pipework will be thermally lagged. Where pressurized or two-phase transmission is adopted the wellhead equipment can be greatly simplified, as no separator, ball-check valve or water collection drum is required: some simplification is also possible at the wellheads in dry hyper-thermal fields.

9.2 Wellhead valving
Fig. 39a shows a typical wellhead valving arrangement as used at Wairakei.

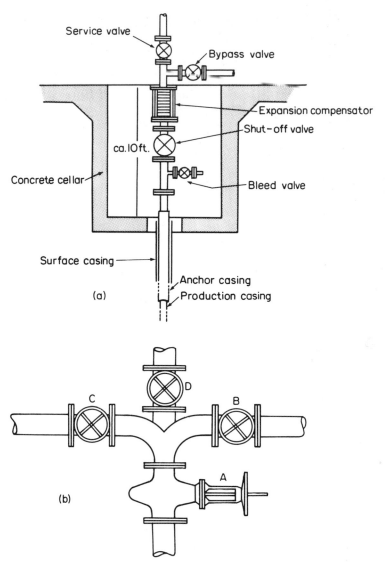

Figure 39   Wellhead valving arrangements. (a) Typical wellhead valving arrangement at Wairakei. (After Fig. 3 of [68].) (b) 'Christmas-tree' valve assembly, commonly used in Italy, California and elsewhere. (After Fig. 2 of [200].)

The valves are accommodated either in or just above the concrete cellar referred to in Section 7.2 and Fig. 16. The service valve is used to regulate the flow and pressure of the emergent fluids during testing, while the shut-off valve enables the well to be isolated for maintenance purposes. A bleed valve permits the removal of non-condensible gases in a quenched bore, and a bypass valve permits the bore output to be discharged to waste. When a bore is taken out of service it is advisable always to bypass at least some of the fluid so as to keep the casings hot and thus avoid thermal shock due to alternating heating and cooling of the bore; and when a bore is in service the control and shut-off valves should normally be wide open so as to reduce wear and to eliminate the inefficient process of throttling: the control of pressure and flow is then exercised by the *system control* (see Chapter 14). An alternative valving arrangement, favoured in the dry fields of Larderello and the Geysers, is the 'Christmas-tree' assembly (Fig. 39b), where A is the shut-off valve, B the service valve, C a bypass control valve for discharging the steam to waste through a silencer when necessary, and D is to admit instruments or reamers to be lowered vertically into a bore.

9.3 Separators

The earliest crude separator tried out in a wet field was a simple 180° U-bend with a steam branch leading off from the inner wall of the downward discharging branch (Fig. 40). The water/steam mixture, in passing round the 180° bend, is subjected to very high centrifugal forces which throw the water against the outer wall so that (theoretically) only dry steam can be withdrawn from the inner wall. If a mean water velocity of 20 m/sec is

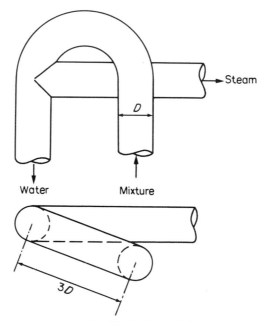

N.B. D is the internal diameter

Figure 40   Crude U-bend separator. (After Fig. 1 of [62].)

taken (a not unreasonable figure according to Fig. 37), and a U-bend radius of 50 cm be assumed at the outer internal wall, the water acceleration would attain a value exceeding 80g. Under such conditions it might be thought that all the water would cling to the outer wall and that dry steam only would be drawn off from the inner wall, but owing to turbulence the device would remove only from 80 to 90% of the water. Some of the Wairakei bores

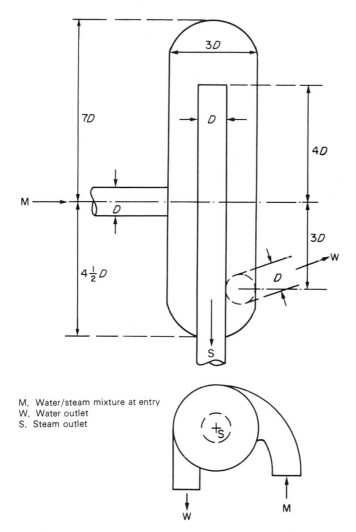

M, Water/steam mixture at entry
W, Water outlet
S, Steam outlet

Figure 41   Schematic arrangement and recommended proportions of Webre type steam/water separator with spiral inlet. (After Figs. 8 and 10 of [62].)

Note: Internal diameters range from 76 to 137 cm according to the mass output of the bore.

yielded fluid containing as much as 8 parts of water to 1 of steam, so that with a U-bend efficiency of, say 85%, the steam lead-off pipe would still contain fluid of more than 50% wetness factor – a totally unacceptable figure. Various sophisticated separators have been devised, but one of the cheapest and most efficient is the Webre type [62], schematically illustrated in Fig. 41 and pictured in Plate 10 and capable of attaining an efficiency of 99.9% or more provided it is not over-loaded. The spiral inlet shown gives an improved performance over that of a simple tangential inlet that was tried earlier. At first, these cyclones were preceded (at Wairakei) by a U-bend so as to reduce the burden on the separator, but later it was found that this was not really necessary, so the water/steam mixture is now generally led through an expansion loop directly from the wellhead into the separator, thus greatly simplifying and cheapening the pipework.

In dry fields the steam may contain a quantity of grit or dust, which is usually removed by means of a coaxial 'swirl' separator in the steam take-off pipe, the dust being discharged continuously through a 6 or 7 mm orifice permanently open to the atmosphere. The loss of steam through this orifice is negligible. In wet fields fine pieces of rock and dust are washed away with the water discharge from the separators, but the occasional coarser fragments are trapped in the water collection drum by the perforated water take-off pipe (P in Fig. 38).

When considering separator performance it is essential to distinguish between mass wetness and volumetric wetness. Although this is obvious, the quantitative differences between the two figures can be surprising. For example, at a wellhead pressure of 8 ata, and with a water/steam mass ratio of 7 : 1, the *mass wetness* would be 87.5% while the *volumetric wetness* would be only 3.15%. This, of course, is due to the enormously differing specific volumes of the two fluids. At lower pressures the difference between the two figures would be greater, and at higher pressures it would be less.

Although the efficiency of the Webre type separator is very high under properly designed conditions, it deteriorates rapidly if the fluid inlet velocity is allowed to exceed about 50 m/sec. The pressure drop absorbed by the separator is more or less proportional to the *volumetric* wetness at the inlet: except with very wet fluids it should be less than half an atmosphere.

9.4 Hot water discharge

Often the hot water from a wet field is (regrettably) rejected to waste because the economics of extracting its inherent energy are not regarded as sufficiently attractive. Although the water/steam ratio is fairly steady at a constant pressure, it is nevertheless subject to short term fluctuations (gulping) and even to gradual long term variations. The capacity of the water collection drum can take care of moderate short term variations, but some cheap means of disposing of the hot water from this drum at the same average rate as that at which it is collected is necessary, if the water is to be discharged to waste. The use of float-controlled discharge valves would be costly and could sometimes be mechanically troublesome. So the problem has been solved in various wet fields simply by discharging the hot water through suitably sized bell mouthed orifices. Such orifices, when passing boiling water, possess the convenient property of enabling a wide range of water flows (about 3 : 1 typically) to be discharged without upstream flooding on the one hand and without loss of seal on the other hand. The explanation of this phenomenon has been given in simple terms by Armstead and Shaw [63] and in more sophisticated terms in some of the bibliography references given by them. If the hot water is to be exploited for some useful purpose, it should either be pumped, or removed by gravity if sufficient fall is available, from the water collection drum to the exploitation plant.

9.5 Silencers

When a bore is bypassed to waste, the noise can be ear-splitting and can even cause deafness to people subjected to it for too long without adequate protection. At Wairakei, even when the steam phase of a bore is being transmitted to the power plant, the hot water is normally rejected to waste, and the steam that flashes off from this hot water may sometimes be as much as, or more than, the useful steam (in mass). When the whole of a bore is being bypassed to waste, the volume of discharged fluids can be enormous (see Plate 11). Thus there is need at Wairakei for silencers at all times, whether the bores be in service or bypassed. The same applies to other wet fields where the hot water is rejected to waste. Even where the hot water from a wet field is usefully exploited, and also for dry bores, it is sometimes necessary to bypass to waste the full fluid output during emergencies or maintenance periods. Hence it is essential to provide silencers at *all* geo-thermal bores, even though in some cases they may be needed only occasionally.

A simple design of silencer, devised some years ago in New Zealand, is used at Wairakei and has found favour in other fields also [64]. It is diagramatically illustrated in Fig. 42. The fluid impinges upon a cusped dividing plate of toughened steel at the tangential junction of twin cylindrical concrete towers. The water swirls round the cylinders, losing its kinetic energy in friction, and is passed out over a measuring weir to a waste channel. (When using this weir for measurement purposes it is of course necessary to make due allowance for flashed steam in order to arrive at the quantity of water issuing from a wet bore). The steam escapes to the air from the top of the cylinders. This type of silencer deflects most of the noise skywards and, more important, lowers the pitch of the noise from a high frequency scream to a tolerable deep-noted roar. Silencers of this type can be seen in Plates 10–15.

Plate 11 Wairakei steam field, New Zealand. Discharge of superheated water through twin silencers. Despite the dramatic impact of this picture it represents waste and pollution. The steam is flashed off from high pressure hot water on reduction to atmos-

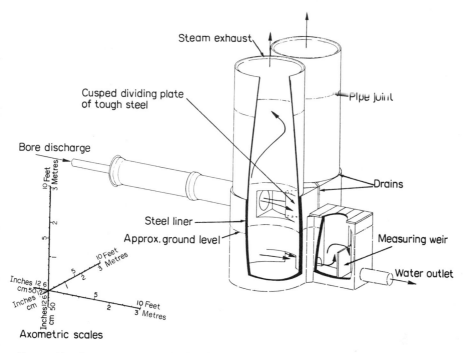

Figure 42   Cut-away axometric sketch of twin cyclone silencer as used at Wairakei, New Zealand. (After Fig. 6 of [64].)

9.6 Safety devices

The working pressure of a steam collection system will usually be a fraction of the shut-in pressure of the bore, and although the bores and wellhead valves will be designed to withstand that shut-in pressure, all equipment beyond will be designed for a lower pressure than the bore and its valving. A safety-valve should therefore by provided at each wellhead as a statutary requirement. But as the lifting of a safety-valve is an occurrence to be avoided, except as a last resort, (because of the erosion damage to seatings and because of the difficulty in getting them to re-seat without appreciably lowering the pressure) a cheaper safety device is also often provided at each wellhead in the form of a bursting disc. Such discs can be easily replaced and are relatively cheap, but as they are not usually accepted by Boiler Inspectors they have to be backed up by conventional spring- or weight-loaded safety-valves (see Fig. 38). Conventional safety-valves have the advantages that the lifting pressure can be varied at will and set precisely, and that they will re-seat when the pressure has fallen sufficiently; but they have the disadvantages of high cost and of the difficulty of maintaining an absolutely tight seal. In geothermal conditions this may result in the precipitation of dissolved solids in the area

Plate 12 Typical branch steam line, Wairakei, New Zealand [195]. Note the stiff right-angled expansion loops, the concrete 'apron' to prevent erosion of the pumice sloping ground surface, and the twin silencers from a bore at the top of the slope. (By courtesy of the New Zealand Ministry of Works and Development.)

Plate 13 General view of bore field, Wairakei, New Zealand. Note the three lagged 30-in steam pipes on the left and the five 20-in steam pipes to the right. On the extreme right is the 19-in experimental hot water transmission pipe, also lagged. To the far right can be seen one of the control vent-valve houses discharging a moderate amount of steam, while in the background can be seen several wellhead silencers, steam pipe loops and branch steam lines [195]. (By courtesy of the New Zealand Ministry of Works and Development.)

Plate 14 Vertical steam pipe expansion loop, allowing vehicle access beneath, Wairakei, New Zealand [195]. See Section 9.8 and Fig. 43a. Note the twin silencers in the background. (By courtesy of the New Zealand Ministry of Works and Development.)

around the seatings, which could necessitate increased maintenance and sometimes even prevent the valve from discharging its full rated output. A further disadvantage of the safety-valve is that its capacity when passing a steam/water mixture cannot be precisely known. Bursting discs, on the other hand, are cheap, simple and easily maintained. They have the disadvantages that they cannot be set precisely, that their bursting pressure depends upon temperature, that once burst the well cannot be used until the maintenance staff have replaced the broken disc with a new one, and that the material of the disc can be weakened by age, vibration and chemical attack, so that a disc may burst at a lower pressure than intended. If this should occur during normal operation, the system may become disturbed. A further safety device is sometimes provided at wet wells in the form of a steel ball-float which lifts to shut off the steam supply to the steam collection system in the event of a separator flooding or failing to perform its duty properly. The entry

Plate 15 Aerial view of the Wairakei geothermal development, New Zealand. Note the bore field in the middle distance, the steam transmission pipelines, the A and B power stations, the cooling water pump-house, the Waikato River in the foreground, the cooling water outfall (right), the step-up substation and high voltage transmission lines (behind the B station) and the Wairakei Hotel (left). The equipment between the A and B buildings is the flash vessels and scrubbers used for the experimental hot water transmission scheme (see Sections 9.9 and 14.3; see also Fig. 64). (By courtesy of the New Zealand Ministry of Works and Development.)

of a gulp of water into a steam main, and perhaps ultimately reaching a turbine, could be disastrous; hence the need for the ball check-valve. The lifting of the ball against its upper seating would tend to induce the shut-in well pressure in all the wellhead equipment; hence the bursting disc and safety-valve should be placed between the top of the bore and the ball-valve at the entry to the steam collection system, as in Fig. 38. Normally, the disc will burst on the lifting of the ball and the entire bore fluid will be discharged to waste through the silencer. But if it should fail to act, the safety-valve will lift and relieve the pressure in the wellhead equipment.

9.7 Steam branch pipes (see Plate 12)

From each well a steam branch pipe, sized to the output of the well and to allow for the pressure-drop over its length, will conduct the almost dry steam from the wellhead to the nearest convenient point in the steam mains that carry the steam from the field to the power station or other utilization plant. These branch pipes must be anchored at certain points by means of solid ground supports, and provision must be made for thermal expansion and contraction between anchor points either by 'zigzagging' or 'snaking' the pipes – usually in a horizontal plane, though sometimes by means of vertical loops where road access must be provided from one side of the branch pipe to the other. Pipe supports between anchors may take the form of simple sliding pads on concrete blocks, of flexible upright metal rods capable of bending one way or another to take up pipe movement, or of suspension rods hanging from light portal or 'gallows' frames.

9.8 Steam mains (Plate 13)

The carrying capacity of the steam mains will normally increase as they progress from the remoter parts of the bore field towards the exploitation plant, gathering additional steam in the process. The increased capacity may simply be provided by enlarging the diameter of the pipe(s), or alternatively by adding more pipes in parallel. Wherever there is a change in the number of pipes it is convenient to provide a valved manifold to enable individual pipes to be isolated for servicing. All such manifolds must be at anchor points. Junctions with well branch-pipes should also be at anchor points, where the pipe movement is zero; and anchors must of course also be provided wherever the pipeline changes direction, so as to take up the large lateral components of the expansion forces. Compensation for thermal expansion and contraction can be provided in a number of ways:

9.8.1 *Zig-zagging or lyre bends.* These methods are suitable for pipes of moderate diameter. As with branch pipes adequate anchoring is necessary, and suitable sliding or suspension supports must be provided between anchors.

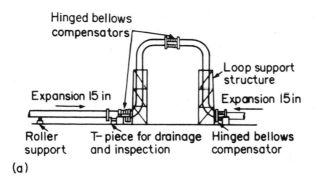

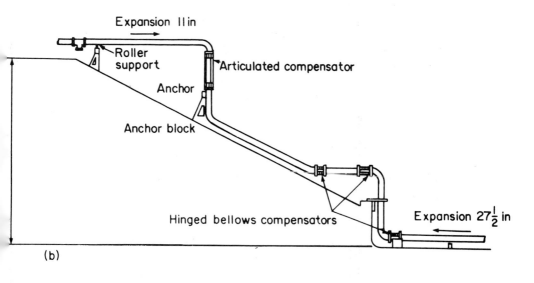

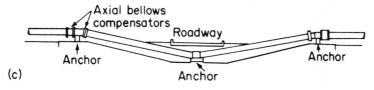

Figure 43 Use of bellows expansion compensators of various types for steam and hot water transmission at Wairakei, New Zealand. (a) Steam line at loop (b) Steam line at bluff (c) Hot water line under road. (After Fig. 7 of [68].)

9.8.2 *Axial bellows pieces.* These, of the type shown on a wellhead in Fig. 39a, transmit high compressive or tensile forces to the anchors, which must in consequence be sufficiently robust to resist these forces. They provide a rather expensive means of compensation, and are therefore used only where space restrictions prohibit the use of any form of looping – e.g. within a power house building, or in a hot water transmission pipe where vertical looping (for reasons that will be explained in Section 9.9) cannot be adopted. Internal metal sleeves greatly reduce steam friction in corrugated axial bellows.

9.8.3 *Hinged bellows pieces.* These are capable of angular flexure without net axial yield. They transmit only minimal axial thrusts when used in three-pin arch loops, as shown in Fig. 43a. Loops of this kind are commonly adopted at Wairakei (Plate 14), where they provide the additional asset of road access to motor vehicles beneath. The loops are supported by means of light steel frames. Each loop is capable of absorbing at least ¾m of pipe movement, and at Wairakei they are spaced at intervals of about 300 m or so. The relatively high cost of hinged bellows pieces tends to be offset by the lightness and wide spacing of anchors. The straight pipe lengths between loops and anchors are normally supported on rollers carried on concrete pads let into the ground. Welded onto the pipes are short steel contact pieces long enough to penetrate the lagging so as to present a metal surface to the rollers and so prevent damage to the lagging when pipe movement occurs.

9.8.4 *Articulated bellows pieces.* These are fairly long pieces of pipe provided with flexible corrugations and with flanges at either end. The flanges are interconnected by means of a pair of long hinged rods which permit the assembly to flex sideways while keeping the flanges parallel to one another. Such units, in conjunction with hinged bellows pieces can absorb complicated expansion movements such as occur where a pipeline passes down a steep hillside (see Fig. 43b and Plate 16).

9.8.5 *Slip joints.* These, though extremely simple in conception are not often entirely satisfactory, as they need careful maintenance and are vulnerable to chemical deposition. They consist simply of one tube sliding concentrically within another and provided with a stuffing box and gland packing.

Different designers of geothermal installations favour different solutions to the problem of pipeline expansion. In New Zealand, zigzagging is generally used for branch lines while hinged and articulated bellows pieces are commonly used for steam mains. In the dry fields of Larderello and the Geysers, bellows pieces are also used to a limited degree, but greater reliance is placed upon zigzagging and upon vertical loops. With all forms of expansion compensation it is usual to take up half of the pipe expansion movement when the system is cold, so that the initial pipe stresses are more

Plate 16 Steam transmission over difficult terrain, Wairakei, New Zealand. Note the variety of expansion arrangements where the steam pipes descend a steep bluff (see Section 9.8 and Fig. 43b). (By courtesy of the New Zealand Ministry of Works and Development.)

or less equal (but reversed) to the final stresses when the system is hot. Where necessary, intermediate supports may be provided with lateral or overhead steel guides to prevent possible buckling of pipes under stress; this can apply both to branch lines and to mains. A most erudite analysis of pipeline construction problems and of stressing under thermal changes has been prepared by Pollastri [65].

As already mentioned, mild steel – fortunately a cheap material – is adequate for pipelines and pressure vessels filled with geothermal fluids. This is because of the virtual immunity of this metal against H_2S attack in the absence of oxygen, – which is never present in bore fluids except in combination with other elements (as in CO_2). The immunity derives from a self-protective skin of iron sulphide which rapidly coats the inside walls (see Section 16.3.4). In the early days at Wairakei, in order to minimize the amount of welding (which could possibly be a seat of chemical or physical weakness) *seamless* Mannesmann tubes were used for the main pipelines. These, however, were expensive; and after it had been clearly shown that welding, if very carefully performed, constituted no danger, it was decided thereafter to use *seam-welded* tubes for the steam mains, though seamless tubes continued to be used on branch lines. Tubes of this type are now almost universally used for steam mains in all geothermal installations throughout the world. Seam-welded tubes are normally machine-bent to a cylindrical shape, and the edges are machined to a suitable profile and held securely together after bending. The joint is then welded longitudinally along the the pipe. The seam welding may most cheaply be done at the site, as this enables the steel to be shipped in compact plate form instead of in cylinders. The longitudinal seam welds may be done manually or automatically, and after tubes of convenient length have been produced they are mounted in position and joined together by means of manual circumferential welds to form long sections of pipeline. Another form of pipe sometimes used is the *spiral-seamed* pipe, formed by winding long strips of fairly narrow steel plate helically, so that one edge abuts the other. The strip edges have been previously shaped, and the spiral butt joint is then welded to form a solid tube. Carefully prepared welds can possess almost the same strength as the plate itself. Welding should be undertaken only by highly skilled and qualified welders who are periodically subjected to exacting tests to ensure that their skill has not deteriorated. Welds should be X-ray tested (50% of longitudinal and 100% of circumferential were so tested at Wairakei), and sample test pieces should be out off for physical testing. With thick plates of more than $\frac{1}{2}$ to $\frac{3}{4}$-inch thickness (say $12\frac{1}{2}$ to 19 mm) it is necessary to relieve the welding stresses by heat treatment. Bolted flanges, screwed and subsequently welded circumferentially, are obligatory wherever dismantling may be needed for maintenance purposes; elsewhere consecutive pipe lengths are circumferentially welded together – this process being known as 'stove-pipe' welding. At Wairakei, the steel plate used for the welded steam mains is to BS1501-151B, and the finished tube is to BS806 standard. Other countries may call for different standards.

9.9 Hot water transmission

Although the transmission of moderately hot water for district heating purposes (see Chapter 11) at temperatures near or below atmospheric boiling point is commonly practised in Iceland and elsewhere, only once (hitherto) has the transmission of very hot water at temperatures of 200° C or more been attempted over a distance of at least 1½ km. This was for an experimental installation at Wairakei for the extraction of power from hot water that would otherwise have been wasted. The theoretical aspects of extracting power from hot water will be dealt with in Chapter 10, but the *transmission* of this water raises special problems that fall within the scope of the present chapter.

The primary problem of transmitting very hot water is to prevent boiling during transmission, as the formation of steam pockets and their subsequent collapse could give rise to very high and unpredictable stresses which might cause a burst pipe. It is necessary to recognize that superheated water (i.e. above 100° C and pressurized) is an explosive fluid many times as dangerous as steam at the same temperature. For example, although hot water at 200° C has an isentropic heat drop to atmospheric pressure of only about 11% of that of steam at the same temperature, its density is 110 times as great. Hence a pipe or vessel containing water at 200° C has about 12 times the potential for doing mechanical work of the same pipe or vessel containing steam at 200° C; and this may be taken as a measure of its explosive capacity. A burst pipe transmitting very hot water could have disastrous consequences. The only way of preventing boiling in a hot water pipe is to ensure that at all points the *hydraulic* pressure exceeds the *vapour* pressure. The hydraulic pressure is a complex function of the initial vapour pressure at the point of collection, pipe friction, pipe gradients and dynamic heads arising from changes of water velocity due to valve movements. In the Wairakei scheme a safe margin between hydraulic and vapour pressure was assured by a combination of pumping, attemperation (by injecting lower temperature water from second grade wells) and of carefully controlled speeds of valve operation. A full description of this scheme may be found in the bibliography [67]: technically it was entirely successful, but unfortunately it had to be abandoned after about one year of operation because the wells selected for the experiment, which for obvious reasons were those closest to the power plant, proved to be near the edge of the field – a fact not suspected when the project was launched – and their output of hot water dried up. The hot water pipeline was subsequently converted into a subsidiary steam line. Due to a fear that the whole field might ultimately yield dry steam, the hot water transmission scheme was not extended further into the heart of the bore field where there was (and still is) a large quantity of hot water available, with the result that several tens of megawatts of potential power from that hot water has been going to waste ever since the Wairakei installation was first initiated late in the 1950s.

Thermal expansion of the hot water mains in the experimental scheme was taken up by means of internally sleeved axial bellows pieces so as to reduce pipe friction to a minimum and so as to avoid the unsparable loss of hydraulic head that would result from the adoption of vertical expansion loops (of about 6 m in height) as with the steam mains. At road crossings, therefore, the pipes were led beneath the road (Fig. 43c). Fairly heavy anchors had to be provided to take the high axial thrusts.

It has been mentioned that attemperation formed a feature of the experimental hot water transmission scheme at Warakei. This was achieved by taking advantage of the fact that in that field there were two bore systems

Plate 17 Head tank for experimental hot water transmission scheme, Wairakei, New Zealand. See Sections 9.9 and 14.3; see also Fig. 64. (By courtesy of the New Zealand Ministry of Works and Development.)

operating at different pressures and temperatures [67 and 68]. The hot water from the higher pressure system was lifted by low head pumps into a head tank (Plate 17) about 12 or 13 m above ground level so as to provide the necessary surplus pressure. This head of water 'floated' on the upper end of the hot water transmission pipe, (see Fig. 64 later). Hot water from the lower pressure bores was injected into the pipeline by means of high head pumps. The injected water reduced the mean temperature of the transmitted water and thus lowered the vapour pressure, at the same time making it possible to recover the energy of some of the lower temperature water, the separate transmission of which would have proved uneconomic. As an alternative to the use of pumps, with their attendant problems of maintenance and of using special materials, the author devised a system of vapour pumping, or, to use a more graphic if less scientific term, 'pumpless pumping', which would have worked on the principle illustrated in Fig. 44 and described more fully in the bibliography [67]. Although the system nominally relied upon the dual pressure system of Wairakei, it could equally be introduced into a single pressure system at reduced efficiency simply by using the atmosphere as the lower pressure. As with the actual Wairakei system a collection tank, A, is provided for each (high pressure) well and a common head tank, H, would serve a whole group of wells. Associated with each collection tank is a lift cyclinder, L, connected as shown in Fig. 44. The water in A is allowed to rise and fall alternately between pre-designed maximum and minimum

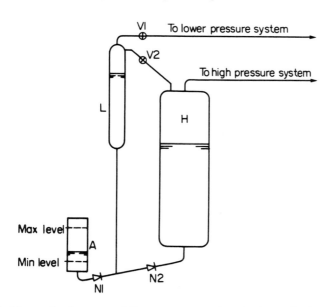

Figure 44   The author's proposal for vapour-pumping (or 'pumpless' pumping) for imparting static head to a hot water transmission system. (After Fig. 5 of [67].)

levels. When it rises to its maximum height, a level-sensitive device opens a valve, V_1, thus connecting the associated lift cylinder to the lower pressure system (valve V_2 at this time being closed). The pressure difference between the two systems impels the water from A into L, passing on its way through a non-return valve, N_1. As soon as the water in A falls to the prescribed minimum level, a signal causes valve V_1 to close and V_2 to open. This equalizes the pressure between L and H (being that of the high pressure system) and the water drains from L into H. Meanwhile A starts to refill and the cycle is repeated. The net effect of each cycle is to raise into the head tank, without the use of pumps, a quantity of high pressure water equal to the capacity of A between maximum and minimum levels. A similar cycle is meanwhile operating with other high pressure collection tanks and their associated lift cylinders. Since the operation of the various control valves is actuated solely by the water levels in the collection tanks, there is nothing to prevent a group of wells thus being connected to the same lift system even if they are at different ground levels. (With pumps, differences of ground level would require different pump characteristics for each well.) Standard collection tanks would be provided for each well: those wells having a high water yield would operate at a higher cycle frequency than those of low yield. Although there are certain secondary problems which need not here be described, 'pumpless pumping' would have the advantage of extreme mechanical simplicity. The lifting of water in this way would be effected at the cost of some high pressure steam and of some condensation on the water surfaces of the lift cylinders when the equalizing valves V_2, are opened. These losses provide the necessary lifting energy, but the temperature reduction in the lifting cylinder would provide useful attemperation. Although the efficiency of the process is only about one-third of that obtainable by mechanical pumping, the energy is still quite small – about $1\frac{1}{2}\%$ of the transmitted energy as against $\frac{1}{2}\%$ at Wairakei for lifting high pressure water.

The method of controlling the rate of movement of the valve admitting the hot water to the plant – the third ingredient principle of hot water transmission – will be described in Chapter 14.

It is possible that some future projects in wet fields may require the transmission of very hot water over fairly long distances, and that the principles briefly described here may be put into practice. However, it is also possible that the adoption of two-phase transmission or of submersible pumps (see Sections 9.11 and 9.12) may relegate hot water transmission, as such, to the level of historic and academic interest only, unless perhaps for non-power uses.

9.10 Thermal lagging, condensation in steam pipes and heat losses

During its passage from the wellheads to the exploitation plant, both steam and hot water will lose a certain amount of heat. All vessels and pipes, other than waste pipes and silencers should therefore be lagged with any of the thermal insulating materials (e.g. magnesia) commonly used in power station practice, or with other materials such as vermiculite which may be locally available, cheap and effective. The cost of lagging can be very high, and it is necessary to strike an economic balance between its cost and the value of the heat saved. In dry fields, which usually deliver superheated steam at the wells, it is advisable to ensure that the temperature does not fall during transmission to condensation point. With hot water transmission, quite a modest amount of lagging can ensure a temperature drop of about 1° C per km. With dry saturated steam, to which the steam leaving the wellhead separators in wet fields approximates, it would probably not be economic to reduce condensation on the pipe walls to less than about 5–7% during transmission, depending upon the length of the pipelines. Moreover, a certain amount of condensation is useful as a 'scrubbing device' for purifying the steam in transit. Even the most efficient of wellhead separators will let through small quantities of bore water with the steam, and these waters are likely to contain chlorides which, even in very small proportions, in the presence of hydrogen sulphide (almost invariably a constituent of geothermal steam) can do great damage to turbine blades. The scrubbing is effected by providing downward facing T-pieces at intervals of perhaps 150 m with drainage facilities (see Fig. 43a). Condensed steam, forming on the pipe walls, will mostly tend to collect along the bottom of the pipe, so that at each T-piece much of the water will fall into the downward pointing branch, from which it will be drained away. Originally, mechanical traps were provided to open intermittently whenever sufficient water had collected. More recently, at some geothermal installations, it has been decided that a simple open orifice is cheaper, despite the higher fluid loss entailed. The scrubbing process, which may be likened to rinsing, results from the repeated dilution of the condensate by further condensation, and its repeated partial removal through the drains. The degree of purification attainable can be astonishing, as will be illustrated by a hypothetical example.

Let it be assumed that the fluid issuing from a wellhead separator consists of 20 t/h of steam and 100 kg/h of hot water. This implies a wetness factor of nearly 0.5% – a very cautious assumption. Let it further be assumed that drains and traps are spaced at intervals of 150 m and that each is 75% efficient – i.e. that three-quarters of the accumulated water in the pipeline is removed at each drain and that one-quarter remains behind to be carried along with the steam. Finally let it be assumed that 100 kg/h (or 0.5% of the initial dry steam flow) is condensed in each 150 m section of pipeline between traps. The figures of Table 6 clearly reveal the enormous degree of purification that results from the scrubbing effect of repeated dilution and partial draining. It will be seen that even after the fourth drain

Table 6 Scrubbing effect of imperfectly lagged pipeline carrying slightly wet saturated steam.

Location		Steam kg/h	Water kg/h	Total salts kg/h × 10^{-6}	Salt concentration ppm
At exit from wellhead separator		20 000	100	100.S	S
1st trap	entry	19 900	200	100.S	0.5.S
	exit	19 900	50	25.S	0.5.S
2nd trap	entry	19 800	150	25.S	0.166 6667.S
	exit	19 800	37.5	6.25.S	0.166 6667.S
3rd trap	entry	19 700	137.5	6.25.S	0.045 4545.S
	exit	19 700	34.375	1.5625.S	0.045 4545.S
4th trap	entry	19 600	134.375	1.5625.S	0.011 628.S
	exit	19 600	33.594	0.3906.S	0.011 628.S
5th trap	entry	19 500	133.594	0.3906.S	0.002 924.S
	exit	19 500	33.398	0.0977.S	0.002 924.S
6th trap	entry	19 400	133.398	0.0977.S	0.000 732.S
	exit	19 400	33.349	0.0244.S	0.000 732.S
7th trap	entry	19 300	133.349	0.0244.S	0.000 183.S
	exit	19 300	33.337	0.0061.S	0.000 183.S
8th trap	entry	19 200	133.337	0.0061.S	0.000 046.S
	exit	19 200	33.334	0.001 53.S	0.000 046.S
9th trap	entry	19 100	133.334	0.001 53.S	0.000 011.S
	exit	19 100	33.333	0.000 38.S	0.000 011.S
10th trap	entry	19 000	133.333	0.000 38.S	0.000 003.S
	exit	19 000	33.333	0.000 09.S	0.000 003.S

Total condensation $= \dfrac{20\,000 - 19\,000}{20\,000} = 5\%$

Initial wetness $\dfrac{100}{20\,100} = 0.498\%$. Final wetness $= \dfrac{33.333}{19\,033} = 0.175\%$

Salt reduction $\dfrac{100}{0.000\,09} = 1.11 \times 10^6$ times

Salt concentration reduction $\dfrac{1}{0.000\,003} = 333\,333$ times

(600 m) 99.61% (100 − 0.39) of the solubles have been removed, while after the tenth drain (1500 m) less than one-millionth part of the original solubles remain in the pipeline. At Wairakei, the chloride content of the bore water may sometimes be as high as 3500 ppm, so that after the fifth trap the chloride concentration would only slightly exceed 10 ppm, while after the tenth trap it would have fallen to 0.0105 ppm only. Even the first figure would just be about acceptable to turbines, while the second figure virtually implies distilled water. It will also be seen from Table 6 that the condensation loss in 1500 m would be 5% − a fairly realistic figure − and that the steam enters the plant in a *drier* condition than it leaves the wellhead separator. (This latter feature would not be true, however, if the separator were much more efficient than has been assumed in this example). The figures in Table 6 are, of course, illustrative only; but the following points may be noted:

(i)   The total residual salts are a function only of the number and efficiency of the traps: they are independent of the rate of condensation in the pipelines.

(ii) The salt concentration in the water phase *does* depend on the rate of condensation in the pipeline.

(iii) the residual wetness depends upon the rate of condensation in the pipeline.

9.11 Pressurized water transmission

It has long been realized that fluid transmission could be enormously simplified and cheapened if it were possible to install submersible pumps at the bottom of wet bores, capable of raising to the surface and transmitting from the wellhead to the exploitation plant very hot pressurized water extracted from depth in a water-dominated field. The advantages of so doing would be as follows:

(i)   Hot water is economically transmittable over longer distances than steam.

(ii) No wellhead equipment other than a simple shut-off valve, and perhaps an emergency or testing bypass, would be needed.

(iii) By the suppression of boiling, no steam − with its inconveniently large specific volume − would be formed until after the pressurized water has reached the exploitation plant.

(iv) Little or no chemical precipitation would occur in transit if the heat losses are kept low by means of good lagging. For example, with silica present in the form that permits supersaturation, the water would probably be clear of the plant and rejected, having yielded up its energy, before any precipitation occurs. For the exploitation of geothermal brines this could be of immense value.

(v) Transmission pipes would be far smaller and cheaper than for steam.
(vi) Transmission velocities could be kept fairly low without undue extravagence in pipe diameters.
(vii) Heat losses in transmission would be much less than for steam.
(viii) On arrival at the exploitation plant the pressurized water could be flashed in any convenient number of stages, and steam could then be made available at pressures only slightly above convenient plant admission pressures, with virtually no steam transmission problems (owing to the short distances involved) except for scrubbing – see Section 10.7.

Up till now, no one has succeeded in producing a pump on a commercial basis capable of standing up to the arduous conditions involved, although hopes have been expressed that such pumps will be available by 1980 or even earlier. The availability of such pumps would revolutionize both hot fluid transmission and geothermal power generation (see Chapter 10). This method of fluid transmission may soon become a reality, rather than a dream.

9.12 Two-phase fluid transmission

An alternative method of simplifying wellhead equipment and pipework has not only been considered but also put to the test with, it is claimed, considerable success [66, 69]. This is two-phase fluid transmission – water and steam transmitted together in a single pipe, without the need for wellhead separators. Separation, and perhaps subsequent further flashing, would be done at the utilization plant. For some time, engineers were blinded by the undesirability of steam formation in hot water mains and of hot water in steam mains; the common bogey being water hammer. Certain investigators, notably James, [66] argued that as two-phase fluids rose up the bore casings without any disastrous effects, a pipeline forming a continuation of the bore might equally survive the rough treatment of transmitting flashing, boiling water/steam mixtures over the ground surface. Test rigs in New Zealand and Japan have shown that with two-phase fluid transmission no water hammer has been experienced, even with fairly rapid valve movements; no noticeable corrosion or erosion has occurred in the pipelines; and that although pressure drops are higher than when transmitting steam alone, they are not unacceptably high. Pressure drops have been found to conform approximately to the following formula [66]:

$$\Delta P_{\mathrm{m}} = \frac{\Delta P_{\mathrm{s}}}{d^{0.5}}$$

where ΔP_{m} = pressure drop through a given pipe passing water/steam mixture;
ΔP_{s} = pressure drop through the same pipeline passing dry saturated steam;
d = gravimetric dryness fraction of the mixture.

(The initial steam pressure is assumed to be the same for both fluids.)

Thus with a water/steam mixture of, say 4 : 1, d would be 0.2 and the pressure drop for the mixture would be 2.236 times as great as for dry steam. But as the pressure drop is also inversely proportional to the fifth power of the diameter (see Section 9.13), this could be compensated by increasing the diameter by about $17\frac{1}{2}\%$. In any case, pressure drop is not without a modest 'dividend' in the form of some additional flash steam formed in transit.

At the Hatchobaru power plant, now under construction in Japan (and which will probably be in service by 1978), two-phase fluid transmission is to be adopted as a design feature and is fully expected by the designers to be successful [70]. In future geothermal plants, two-phase fluid transmission could perhaps become common practice in wet fields, unless the submersible pump is first developed satisfactorily. If so, the apparent theoretical advantages could be great.

(i) The savings in wellhead gear could outweigh the extra cost of the rather larger pipework.

(ii) The adoption of very large separators close to the utilization plant would have a large economic 'scale effect' advantage by comparison with the use of many smaller individual wellhead separators, and their maintenance could be more easily and cheaply handled.

(iii) Significant amounts of additional power could be extracted from the hot water by multiple flashing at the plant without the use of a costly separate hot water collection and transmission scheme with its attendant controls (see Chapter 14). Under Wairakei conditions the power capacity could have been raised by about 50% if power were extracted from all the hot water available.

(iv) The wasteful rejection of hot bore water at the wellheads would be avoided, thereby not only conserving energy and avoiding heat pollution but also saving the very costly drainage channels that would otherwise be necessary, and would eliminate the emission of vast quantities of flash steam into the air from the wellhead silencers. With multi-stage flashing at the plant, the water may be rejected at about atmospheric pressure, with very little consequent flashing, and the effluent water would be discharged from a single point only – near the exploitation plant.

(v) Two-phase flow pipes can negotiate moderate uphill gradients.

(vi) The relative levels of different bores would be of no importance, as would be the case with separate hot water transmission where, unless 'pumpless pumping' is adopted, each pump must be designed to suit the level of the wellhead (above datum).

(vii) Fluids discharged from many bores may simply be merged together by joining wellhead branches to a single main pipeline (or more, if required).

(viii) Great aesthetic gains would result from the elimination of many elaborate wellhead equipment assemblies, and from the large reduction in the amount of escaping steam.

In short, the system would appear to be both economically and aesthetically advantageous. Certain precautions, however, would have to be taken in the interests of safety: it would first be necessary to heat and pressurize a pipeline with steam before admitting the entry of mixed fluids, so as to avoid water hammer. The reasons why water hammer, which can be so troublesome if water is present in a steam line or if steam is present in a hot water line, is apparently absent from pipes transmitting very wet mixtures is rather obscure: it may have something to do with the annular flow of the water, clinging to the pipe walls, rather than with 'rivulet' flow when smaller quantities of water are present in a steam line.

Despite the apparent advantages, it is well to point out that two-phase fluid transmission, as with hot water transmission, would require the use of scrubbers (see Section 10.7) for the removal of chlorides – normally effected in steam transmission pipelines by repeated condensation and drainage (Section 9.10). Scrubbers are large and costly pieces of equipment that are not altogether free from operating troubles. Moreover, although water hammer as such does not occur, unexplained pipe vibrations have sometimes been observed on two-phase transmission pipes. For these reasons the New Zealand authorities do *not* intend to adopt this form of transmission for their projected Broadlands installation. Japanese experience at Hatchobaru is awaited with great interest. Meanwhile it would be wise to suspend judgment on two-phase transmission.

9.13 Pressure drops in steam pipes

The well known 'Babcock' or 'Gutermuth and Fischer' formula, usually expressed in f.p.s. unit is

$$\Delta p = 0.4716 \left(1 + \frac{3.6}{d} \right) \frac{LVw^2}{d^5}$$

where Δp = pressure drop (psi),
L = pipe length (ft),
V = specific volume of steam (ft^3/lb),
d = internal diameter of pipe (inches),
w = mass flow (lb/sec).

Thus, for example, the pressure drop in 20-inch internal diameter straight pipe carrying dry saturated steam at a pressure of 8 ata and at a flow rate of 100 tons/h would be 2.56 psi per 1000 ft. To be accurate, since the specific volume continuously increases as the pressure falls, it would be necessary to assume a series of short lengths and adjust for the changing value of V each time, integrating the whole; and even to allow for condensation. A simpler, and sufficiently accurate method is to do the sum for the initial condition, deduce V at the delivery end of the pipe, and then to repeat the calculation using the mean value of V at the two ends. Actual pressure drops are in variably complicated by the presence of bends, fittings, elbows etc., each of which is equivalent to larger lengths of pipe than that which they occupy. For example a bend of radius equal to four times the pipe radius is equivalent in flow resistance to a straight length of pipe about four times the actual length(along the curve) of the bend. In a long and complicated pipeline with bends, loops, valves, bellows-pieces, etc., the effective pressure drop may be a considerable percentage higher than that calculated for a plain straight pipe. Various handbooks are available for providing the necessary data for calculation.

The presence of small quantities of water in the steam does not greatly affect the formula, so long as corrections are made for V and w, but with very wet water/steam mixtures the formula gives too high a result – which is fortunate for two-phase flow transmission.

9.14 Limiting fluid transmission velocities

With steam and two-phase fluid transmission, the limiting velocity of the fluids in pipelines is imposed by the following conditions:

(i) The permissible pressure drop between the assumed economic wellhead pressure and the designed inlet pressure to the exploitation plant must not be exceeded. Where a wellhead separator is installed, it would be wise to allow a pressure drop of about 10% of the absolute wellhead pressure to be absorbed in the separator and its associated pipework. Pressure drops incurred in the branch lines and mains should be separately calculated.

(ii) Owing to the risks of erosion and vibration in fittings such as expansion compensators and valves, it is inadvisable to allow steam fraction velocities to exceed about 60 m/sec or mean velocities of wet mixtures to exceed about 50 m/sec.

Electric power generation from geothermal energy

10

10.1. Evolution and situation in 1976

In Fig. 1 (p. xxv) the pattern of development of world geothermal power, between 1900 and 1960 is shown, while in Table 7 the approximate situation in 1976 with regard to geothermal power is summarized. Information was lacking from one or two countries that did not respond to enquiries, but any inaccuracies in Table 7 were then believed to be minor. Attention, however, is drawn to the note on Fig. 1.

Despite the inefficiency inherent in the process of converting heat into power, imposed by inescapable thermodynamic constraints, and despite the desirability (previously stressed) of encouraging other applications for geothermal heat, electric power generation is likely to remain for some time the commonest form of geothermal development. This is partly for the reasons given in the early part of Chapter 2; and partly because hyperthermal fields, though limited in number and location, are likely to be exploited more rapidly than semi-thermal fields (and probably than very deep artificial fields, should they materialize), and this fact could commercially offset the inherent inefficiency of the power generation process. Electric power generation, although it may gradually yield its present pre-eminence to other applications of geothermal energy, will for long retain great importance.

10.2 Cycles

The various practical cycles used for geothermal power generation are illustrated diagramatically in Fig. 45.

10.2.1 *Indirect condensing cycle* (Fig. 45a). In the early days of power development at Larderello in Italy, the corrosive nature of the steam was though to be too aggressive for its direct admission to the steam turbines. Accordingly an 'indirect system' was at first adopted, which involved the use of a heat-exchanger by means of which 'clean'

steam was raised from contaminated natural steam. The system included the means of disposal of non-condensible gases accompanying the
steam, and also enabled the (then) valuable boron and ammonia to be
recovered from the concentrated liquor in the heat-exchanger and from the
gas discharges. Before the metallurgy of turbine materials had advanced
very far, and when the recoverable minerals and gases commanded good
prices, this cycle was economic despite the fact that from 15 to 20% of

Table 7   World geothermal power generation: position in 1976.

Note: Most of the figures herein quoted have been directly communicated to the
author or have been extracted from recent publications. With a few countries, either no
response has been received to enquiries, or plans are known to be under consideration
but have not yet been positively declared. In these cases no figures have been shown in the
tables, which may therefore give a slight underestimate of the true position. But since
the figures for all those countries at present in the lead in geothermal power development
are closely known, the total megawatt capacities here shown should give a reasonably
close estimate of the position of world geothermal power generation as in 1976, both of
plants already in service and of those then definitely or tentatively planned. Nevertheless,
the note appearing on Fig. 1 shows that the situation may have materially changed
since 1976. Megawatt capacities shown are *gross* – i.e. as at the generator terminals
without deduction for auxiliary consumption.

Table 7a   Summary.

Country	Gross capacities (MW)		
	In service	Definitely planned	Contemplated
USA	522	1621	
Italy	420.6	41	60
New Zealand	202.2	165	
Mexico	78.5	325	
Japan	70	100	
El Salvador	60	30	
USSR	5.8		
Iceland	2.8	70	
Turkey	0.5	15.5	
Philippines		443	
Nicaragua		65	
Indonesia			30
Chile			15
Santa Lucia		1.5	
Kenya			30
Totals	1362.4	2877	135

In service	1362 MW
In service and definitely planned	4239 MW
In service, planned and contemplated	4374 MW

Table 7b   Details

Country	Location	Unit numbers	Plant type	Number and size of units MW	In service (MW)	Planned Definite (MW)	Tentative (MW)
Italy	*Larderello area (Total condensing plants: 362.7 MW)*						
	Larderello 2 {		C	4 × 14.5 }	69		
			C	1 × 11			
	Larderello 3 {		C	3 × 26 }	120		
			C	1 × 24			
			C	2 × 9			
	Gabbro		C	1 × 15	15		
	Castelnuovo {		C	1 × 26 }	50		
			C	2 × 11			
			C	1 × 2			
	Serrazzano {		C	2 × 12.5 }	47		
			C	2 × 3.5			
			C	1 × 15			
	Lago 2 {		C	1 × 14.5 }	33.5		
			C	1 × 12.5			
			C	1 × 6.5			
	Monterotondo		C	1 × 12.5	12.5		
	Sasso Pisano 2 {		C	1 × 12.5 }	15.7		
			C	1 × 3.2			
	(Total non-condensing plants: 57.9 MW)						
	Sasso Pisano 1		N–C	2 × 3.5	7		
	Lagoni Rossi 1		N–C	1 × 3.5	3.5		
	Lagoni Rossi 2		N–C	1 × 3	3		
	Valonsordo		N–C	1 × 0.9	0.9		
	Molinetto		N–C	1 × 3.5	3.5		
	Travale–Radicondoli {		N–C	1 × 15 }	18		
			N–C	1 × 3			
	Monte Amiata area						
	Bagnore 1		N–C	1 × 3.5	3.5		
	Bagnore 2		N–C	1 × 3.5	3.5		
	Piancastagnaio		N–C	1 × 15	15		
	Monterotondo (1978)					10	
	Travale (1978)					15	
	Location unspecified (1979)			2 × 8		16	
	Torre Alfina						15
	Efficiency improvements to existing plants						45
	Total for Italy				420.6	41	60

Table 7b (Contd.)

Country	Location	Unit numbers	Plant type	Number and size of units (MW)	In service (MW)	Planned Definite (MW)	Planned Tentative (MW)
USA	*The Geysers field*						
		1	C	1 × 12	12		
		2	C	1 × 14	14		
		3 & 4	C	2 × 28	56		
		5, 6, 7, 8, 9 & 10	C	6 × 55	330		
		11	C	1 × 110	110		
		12 (1978)	C	1 × 110		110	
		13 (1979)	C	1 × 140		140	
		14 (1978)	C	1 × 114		114	
		15 (1978)	C	1 × 57		57	
		16 & 17 (1980)	C	2 × 114		228	
		18 (1981)	C	1 × 114		114	
		19 & 20 (1982)	C	2 × 114		228	
		21 (1983)	C	1 × 114		114	
		22 & 23 (1984)	C	2 × 114		228	
		24 (1985)	C	1 × 114		114	
		25 (1986)	C	1 × 114		114	
	Salton Sea & Brady Hot Spring	(1978)				60	
Total for USA					522	1621	—
El Salvador	*Ahuachapan*	1 & 2	C	2 × 30	60		
		3 (1977)	C	1 × 30		30	
Total for El Salvador					60	30	
Chile	*El Tatio*						15
Total for Chile							15

Table 7b (Contd.)

Country	Location	Unit numbers	Plant type	Number and size of units (MW)	In service (MW)	Planned Definite (MW)	Planned Tentative (MW)
New Zealand							
	Wairakei	1, 4, 5 & 6	B–P	4 × 11.15	44.6		
		7, 8, 9 & 10	C	4 × 11.15	44.6		
		2 & 3	B–P	2 × 6.5	13		
		11, 12 & 13	MFC	3 × 30	90		
					192.2		
	Kawerau		B–P		10		
	Broadlands (1981)					165	
Total for New Zealand					202.2	165	
Japan							
	Matsukawa		C	1 × 22	22		
	Otake		C	1 × 13	13		
	Onuma		C	1 × 10	10		
	Onikobe		C	1 × 25	25		
	Hatchobaru (1977)		MFC	1 × 50		50	
	Kakkonda (1977)		C	1 × 50		50	
Total for Japan					70	100	
Mexico							
	Pathé		N–C	1 × 3.5	3.5		
	Cerro Prieto	1 & 2	C	2 × 37.5	75		
		3 & 4 (1978)	C	2 × 37.5		75	
		5 (flash)	C	1 × 30		30	
		6 (1980)	C	1 × 55		55	
		7 (1981)	C	1 × 55		55	
		8 (1982)	C	1 × 55		55	
		9 (1983)	C	1 × 55		55	
Total for Mexico					78.5	325	
Nicaragua							
	Momotombo	(1977)				10	
		(1978/79)				55	
Total for Nicaragua						65	

Table 7b (contd.)

Country	Location	Unit numbers	Plant type	Number and size of units (MW)	In service (MW)	Planned Definite (MW)	Tentative (MW)
Iceland	Namafjall		N–C	1 × 2.8	2.8		
	Krafla						
	(1977)		C	2 × 35		70	
Total for Iceland					2.8	70	
Turkey	Kizildere		N–C	1 × 0.5	0.5		
	(1980)					15.5	
Total for Turkey					0.5	15.5	
USSR	Pauzhetka		C	2 × 2.5	5.0		
	Paratunka		B	1 × 0.75	0.75		
Total for the USSR					5.75		
Philippines	Tiwi		N–C	1 × 3		3	
			C	4 × 55		220	
	Los Baños		C	4 × 55		220	
Total for the Philippines						443	
Indonesia	Kamojang						
	(West Java)					30	
Total for Indonesia						30	
Santa Lucia (Antilles)			N–C	1 × 1½		1.5	
Total for Santa Lucia						1.5	
Kenya						30	
Total for Kenya						30	

Note: The dates in brackets are the tentative years of commissioning the definitely planned plants. The key to 'plant types' is as follows:

C,	Condensing
N–C,	Non-condensing
B–P,	Back-pressure
MFC,	Mixed flow condensing
B,	Binary cycle

the power potential of the steam had to be sacrificed in the heat-exchanger. Examples of this cycle were still in operation in Italy until about the middle of the twentieth century, but improved metallurgy on the one hand and the declining economic attractions of mineral extraction (by comparison with other production methods) on the other hand have now rendered this cycle

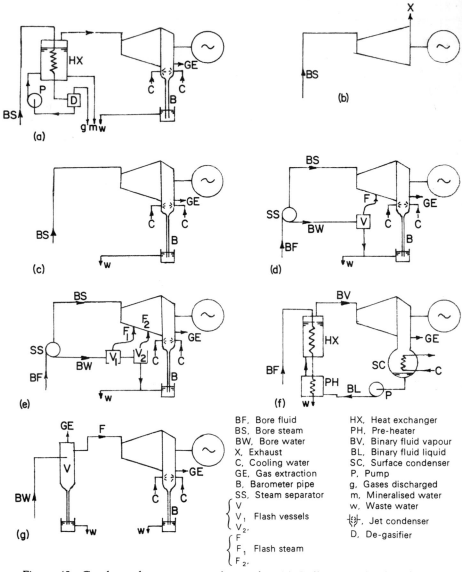

Figure 45 Geothermal power generation cycles. (a) Indirect condensing (b) Non-condensing (c) Straight condensing (d) Single flash (e) Double flash (f) Binary fluid cycle (g) Hot water sub-atmospheric

virtually obsolete, though it could still serve a purpose if exceptionally valuable minerals in high concentrations were present in the fluid emitted from a field. It could also perhaps be used where the geothermal steam contains from 10 to 50% by weight of non-condensible gases [71]. Such steam is often emitted from new bores in the Monte Amiata district in Italy, though the gas content declines considerably in the course of time.

10.2.2 *Direct non-condensing cycle* (Fig. 45b). This is the simplest and, in capital cost, the cheapest of all geothermal cycles. Bore steam, either direct from dry bores or after separation from wet bores, is simply passed through a turbine and exhausted to atmosphere. Such machines may consume about twice as much steam (for the same inlet pressure) as condensing plants per kilowatt of output and are therefore wasteful of energy and costly in bores. Nevertheless, they have their uses as pilot plants, as standby plants, for small local supplies from isolated bores, and even perhaps (though to a very limited extent) for supplying peak loads (see Section 10.15 (iv)). Non-condensing machines must also be used if the content of non-condensible gases in the steam is very high (greater than 50%), and will usually be used in preference to the condensing cycles for gas contents exceeding about 10%, because of the high power required to extract these gases from a condenser.

10.2.3 *Straight condensing cycle* (Fig. 45c). Since there is no need to recover the condensate for 'feed' purposes, as in a conventional thermal plant, direct contact jet condensers with barometric discharge pipes can be used in place of costlier surface condensers.

10.2.4 *Single flash cycle* (Fig. 45d)*. In wet fields it may be possible to extract quite a lot of supplementary power from the hot water phase by passing it to a flash vessel operating at a lower pressure than that at which the main steam is admitted to the turbine(s). The flash steam may then be used to pass through the lower pressure stages of the prime mover(s). Ideally, the maximum power yield from the hot water can be obtained if the flash vessel operates at a temperature midway between that of the collected hot water and that of the condenser. The unflashed fraction of the hot water is then rejected to waste or put to some industrial or other use.

Notes concerning flash cycles: First, it is emphasized that although in Figs. 45d and e a single turbine has been shown admitting pass-in steam, it would be equally possible for the plant to take the form of separate turbines connected in cascade, with the exhaust from an upstream turbine mingling with flash steam and passing on to the entry of a downstream turbine. Secondly, it is well to point out an ambiguity of *nomenclature* with flash cycles. Some people prefer to call Fig. 45d 'double flash', arguing that the first stage of flashing occurs in the bore itself. If the boiling water deep in the aquifer is regarded as the original source of energy, there is logic in this argument. The same people would regard Fig. 45e as 'treble flash'. To the author it seems more reasonable to treat the *wellhead fluid* as the starting point of the thermal cycle from the engineering viewpoint: otherwise Figs. 45b and c would have to be called 'single flash' in wet fields although no surface flash vessel is involved.

10.2.5 *Double flash cycle* (Fig. 45e). Ideally, the maximum power would be extracted from the hot water phase in a wet field by using an infinite number of flash vessels connected in cascade – or at any rate by using one flash vessel for every turbine stage; but such a procedure would be an economic absurdity. It *can*, however, sometimes pay to provide two flash vessels as shown in Fig. 45e, operating as closely as possible at temperatures of one-third and two-thirds way between the collected hot water temperature and the condenser temperature. This was the cycle adopted for the experimental hot water transmission scheme at Wairakei, described in Section 9.9. Approximately 68 to 70% of the supplementary power from the hot water, under Wairakei conditions, was supplied by the first flash, F_1 : this shows that there would be diminishing returns from further stages of multiple flash which would render yet another stage uneconomic.

10.2.6 *Binary fluid cycle.* Much thought has been, and is being, given to the use of refrigerant fluids of very low boiling point, such as freons and isobutane, in a closed turbine-feed-boiler cycle as shown in Fig. 45f. The theoretical advantages of the binary cycle are:

(i) It enables more heat to be extracted from geothermal fluids by rejecting them at a lower temperature.
(ii) It can make use of geothermal fluids that occur at much lower temperatures than would be economic for flash utilizations.
(iii) It uses higher vapour pressures that enable a very compact self-starting turbine to be used, and avoids the occurrence of sub-atmospheric pressures at any point in the cycle.
(iv) It confines chemical problems to the heat exchanger alone.
(v) It enables use to be made of geothermal fluids that are chemically hostile or that contain high proportions of non-condensible gases.
(vi) it can accept water/steam mixtures without separation.

There are, however, the following disadvantages:

(i) It necessitates the use of heat exchangers which are costly, wasteful in temperature drop and can be the focus of scaling.
(ii) It requires costly surface condensers instead of the cheaper jet type condenser that can be used when steam is the working fluid.
(iii) It needs a feed pump, which costs money and absorbs a substantial amount of the generated power.
(iv) Binary fluids are volatile and sometimes toxic, and must be very carefully contained by sealing.
(v) Makers are generally inexperienced, and high development costs are likely to be reflected in high plant prices – at any rate unless and until the cycle becomes commonly adopted in practice.

The only known operational example of a binary cycle geothermal plant, as distinct from laboratory prototypes, is the small 750 kW (nominal) unit at Paratunka in Kamchatka, USSR. This unit uses geothermal hot water at 81°C as the primary heat source, Freon 12 as the binary fluid and cooling water at 6 to 8°C. High parasitic losses reduce the net output to 450 kW only [72].

The binary cycle has its enthusiasts, and several units are being built in the USA. The case for it cannot yet be regarded as commercially proven for general use, though there could undoubtedly be conditions where it would be economic. Moreover, if the art of re-injection (see Chapter 17) should ever become so perfected as to ensure that discharged hot geothermal waters can usefully serve as 'boiler feed' to the underground heating cycle, the force of advantage (i) would be weakened.

10.2.7 *Hot water sub-atmospheric cycle* (Fig. 45g). Only one case of this cycle is known to the author, namely at Kiabukwa, in the province of Katanga, Zaire [73] where a small British-made 220 kW power plant was put into service as long ago as 1953. It is believed to be no longer in use. This plant used hot spring water at 91°C which was sprayed into a chamber at a low pressure of 0.3 ata, and the resulting flash steam passed through a 3-stage low pressure turbine exhausting to a condenser supplied with cooling water at 24°C. The plant was not self-starting, and the vacua in the flash chamber and condenser had first to be established by means of an ejector supplied with steam from an auxiliary wood-fired boiler. The natural occurrence of a hot spring enabled drilling costs to be avoided, but it is difficult to understand how such a plant could have been economic. Nevertheless, the plant was of great technical interest and is believed to have given good service for several years. It is doubtful whether this cycle will ever be repeated. An interesting phenomenon occurs at Kiabukwa, in that the discharge of hot water from the hot spring fluctuates in a tidal cycle of $12\frac{1}{2}$ hours, although the site is about 1300 km from the sea. This is probably due to the action of 'land tides' opening and closing horizontal subterranean fissures by very small movements which are nevertheless significant in relation to the width of what are probably very narrow fissures.

10.3 Power potential of geothermal fluids
The fact that geothermal fluids occur in nature at relatively low temperatures by comparison with those adopted in conventional thermal power plants means that the cycle efficiency of a geothermal power plant must always, as a matter of thermodynamical necessity, be low. Such plants are economic solely by virtue of the extreme cheapness of the *heat* (see Chapter 15) by comparison with fuels: they are also desirable from the aspects of energy conservation and minimal pollution.

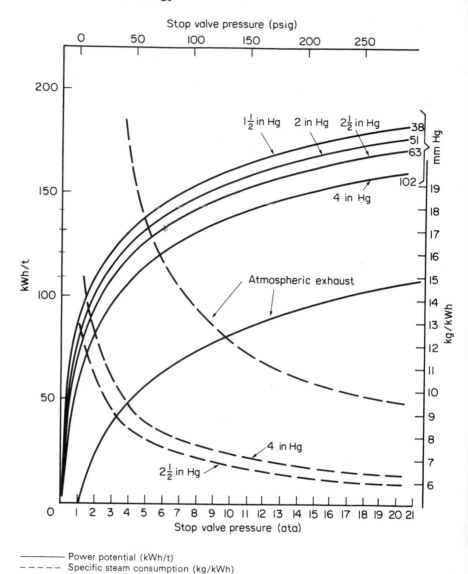

Power potential (kWh/t)
Specific steam consumption (kg/kWh)

Figure 46   Typical power potentials of dry saturated steam at various turbine admission pressures, exhausting to atmosphere or to different vacua.

Note: The broken lines are, of course, a multiple of the reciprocals of the full lines.

The actual power potential of steam delivered to a power plant will depend not only upon its pressure, temperature and quality, and upon the back-pressure at the turbine exhaust, but also upon the configurations of pipework, separators and valving within the power station, in all of which will be induced losses of pressure and energy. It will also depend upon the quality of the generating plant. Considerable differences in power potential will therefore be experienced from plant to plant, but the curves shown in Fig. 46 may be regarded as fairly representative for dry saturated steam. They allow for reasonable turbine and alternator efficiencies and are expressed in terms of energy delivered at the alternator terminals; no allowance being made for the power consumed by gas-exhausting equipment, cooling water pumping and miscellaneous station auxiliaries. These parasitic demands can vary widely according to the quantity of non-condensible gases present in the steam and the vacuum maintained in the condenser and method of cooling. They will vary from virtually zero for a non-condensing turbo-generator unit to as much as 47 or 48% for a condensing plant where the non-condensible gas content of the steam is as high as about 20%.

Where dry superheated steam is available, its power potential cannot be represented by a simple series of curves as in Fig. 46 because much will depend upon the degree of superheat. To calculate the power potential in such cases, the available heat drop must be deduced from a Mollier chart and allowance made for the combined efficiencies of the turbine, the alternator and the station pipework, etc. These efficiencies would probably range from about 70 to 76% according to the admission pressure and the size and arrangement of the plant units.

The power potential of boiling water, as recoverable by the use of single or multiple flashing is approximately as shown in Fig. 47a and b. (The expressions 'single flash' and 'double flash' here relate to the number of flashing operations *above-ground* and exclude the bores – see foot note p. 149.) Although this potential, when expressed per kilogram of fluid, is only a few percent of that of dry saturated steam at the same pressure, it should be remembered that the water/steam ratio is sometimes very high – perhaps 6 or 7 to 1 – so that the *total* power potential of the hot water may form a very substantial fraction of that of the steam in a wet field. It has already been mentioned in Section 9.12(iii) that the power generated at Wairakei when the field was first exploited could have been raised by about 50% if full use had been made of the hot water. With the passing of the years the fluid at Wairakei has become less wet, so this proportion would now have dropped somewhat, but it would still be impressive. The generation of power from the hot water of a wet field could effect a large saving in the cost of bores, wellhead equipment and steam pipework (per kilowatt); and this could well outweigh the cost of transmitting and flashing the hot water. Although the flash potential for non-condensing turbines has been shown in Fig. 47, this is really of academic interest only: it would almost certainly never be economically worthwhile. The curves in Fig. 47 are, of course, based upon the adoption of the economic flashing temperatures mentioned in Sections 10.2.4 and 5.

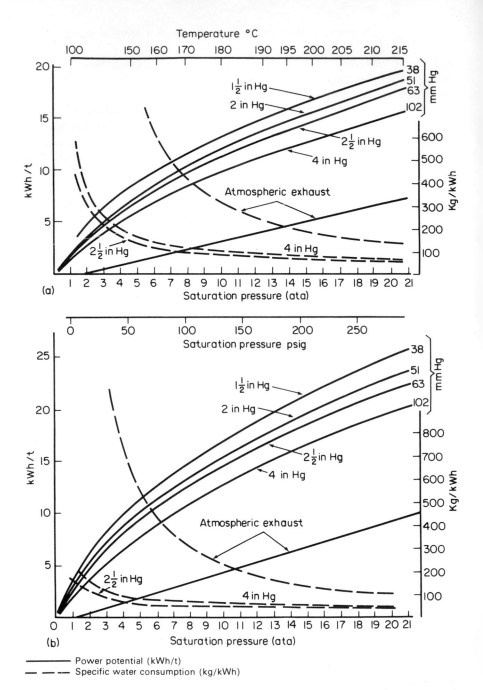

Figure 47 Typical potentials of boiling water at various pressures (and temperatures –see top scale) flashing at optimum pressures and exhausting to atmosphere or to different vacua. (a) Single flash (b) Double flash.

Note: The broken lines are, or course, multiple of the reciprocals of the full lines.

It is emphasized that when using Fig. 46 the chosen point on the base line should be the pressure at the power plant – *not* at the wells, owing to the inevitable pressure and condensation losses incurred in transmission. When using Fig. 47, however, (which has a temperature scale), the chosen point on the horizontal scale should be only a degree or two below the mean temperature prevailing at the wellheads, since hot water can be transmitted in insulated pipes over a kilometre or two with a very small loss of temperature, and since the hydraulic pressure (so long as it exceeds the vapour pressure) is immaterial. This of course applies to the separate transmission of hot water and steam. (Figs. 34 and 35, which attempted to show the typical power potentials of *wells*, made reasonable assumptions as to steam losses in transmission while embodying the characteristics of Figs. 46 and 47.)

10.4 Efficiencies of geothermal power plants

The poor thermal efficiencies of geothermal power plants, mainly due to the low fluid temperatures and pressures, are not helped by the absence of a closed feed cycle; though the cheapness of direct contact jet condensers, normally used with geothermal plants, by comparison with the cost of surface condensers, would largely offset any economic gain attainable from feed-heating, with its attendant heat-exchangers. The thermal efficiencies of geothermal power plants corresponding with the *empirical* power potential curves of Figs. 46 and 47, and reckoned above an assumed natural ambient water temperature of 15° C, are shown in Fig. 48. They are not impressive. If James's theory (Section 8.8) of an optimum wellhead pressure of about 6 ata is correct, with perhaps $4\frac{1}{2}$ ata at turbine entry, a thermal efficiency of only about 15% could be expected with a 4 inch (102 mm) Hg back-pressure, if steam only were used. When it is remembered that a modern conventional thermal plant should have an efficiency of well over 30%, this shows how extremely cheap earth heat must be for geothermal power to be economically competitive (as it undoubtedly can be) with fuel-fired conventional thermal power.

If a wet field is being exploited, and use made of the hot water by double flashing above ground, the generating efficiency would be reduced although the amount of power generated would be increased. For example, let it be assumed that a 4:1 water/steam mixture be delivered (separately) at a power plant with steam at $4\frac{1}{2}$ ata and hot water at 158° C (1.4° C below the wellhead temperature, assuming a 25% absolute steam pressure drop in transmission), and that the hot water is double-flashed at the power plant and that the back-pressure is 4 inch Hg (102 mm Hg). Then it can be shown that the mean efficiency would be only 10.05%. At first sight this reduction of efficiency from almost 15% to about 10% despite the increased use of heat may seem surprising. But the explanation lies in the fact that low though

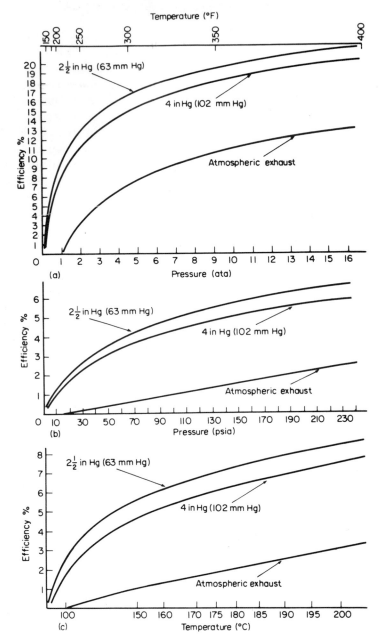

Figure 48 Typical geothermal generation efficiencies. (a) Dry saturated steam
(b) Boiling water (single flash) (c) Boiling water (double flash).

Notes: (i) Pressures and temperatures are at turbine entry

(ii) All horizontal scales apply to all three figures

(iii) No allowance is made for gas exhausters or other auxilliaries.

the efficiencies of Fig. 48 may be, they are not nearly so low as the *overall* efficiencies if reckoned on the total fluid heat yielded at the wells. In the example cited above, the efficiency of using the steam phase alone, though 14.9% intrinsically, is only 7.27% when compared with the total heat yielded up by the wells. This large difference is explained partly by the rejection of the hot water and partly by the energy losses in transmission. Countless other examples could be cited to show the very low overall generation efficiencies in terms of total well heat yields in wet fields. Although the use of the hot water will raise this overall efficiency, it will still be considerably less than the efficiency of power generation by steam alone owing to the very low efficiency at which power can be generated from flashed hot water. This, however, does not imply that the use of the hot water for power generation cannot be economically justified. There is more to the economics of power generation than the efficiency of just one part of the process.

With dry bores efficiences should be slightly better than indicated by Fig. 48a owing to the higher average dryness of the steam during its passage through the turbine resulting from the use of superheated steam. But as the degree of superheat is usually moderate by the time the steam reaches the plant, the differences are likely to be small and may well be swamped by the quality differences from plant to plant. For example, at the Geysers the makers of one of the 55 MW (gross) units specify a steam consumption of 907 530 lb/h at 100 psig/355° F entry with a 4 inch Hg back-pressure. This is equivalent to an efficiency of 17.3%, which lies almost exactly on the appropriate curve of Fig. 48a. As there would be no condensation losses in transmission, owing to the superheated condition of the steam, and only very small dry drain losses; and as there is no hot water to be rejected, the overall plant efficiency reckoned on the total bore heat yield could probably be arrived at approximately by allowing about 5% heat loss in transmission – i.e. 17.3 × 0.95 or 16.43%.

If two-phase fluid transmission is adopted, use is sure to be made of the hot water. The absence of rejected condensate in transmission, and the gain in steam quantity from flashing as a result of the pressure drop should help to make generation rather more efficient (as well as cheaper in capital outlay) than with separate fluid transmission.

The lowest curves in Figs. 48b and c (non-condensing) explain the remark in Section 10.3 that flashing would not be economic in conjunction with non-condensing sets.

It is emphasized that *all* the empirical curves of Figs. 46, 47 and 48 must be viewed in the light of the fact that the non-condensible gas content in geothermal fluids vary widely from field to field, and that this variation will affect overall plant performances and efficiencies. Hence the figures, for this reason also, can be no more than indicative of conditions that have broadly been experienced in exploited fields.

Although low efficiencies are undoubtedly a feature of geothermal power generation at present, there is no reason why this should always be so. On the one hand there is the possibility of raising the efficiency of the geothermal power cycle by the use of binary fluids which would enable the geothermal fluids to be rejected at far lower temperatures (though the wisdom of doing this – as shown in Section 10.2.6 – is still in some doubt). If melt-drilling (Section 19.4) should ever become economically attractive as a means of overcoming the 'depth barrier' imposed by the quasi-exponential cost characteristics of conventional drilling (see Section 7.15 and Fig. 21) it may become possible to attain far higher fluid temperatures and pressures before long, which, as can be seen from the trends of the curves in Fig. 48, and as must also follow from thermodynamic considerations, will certainly greatly raise power generating efficiencies. Again, there is the possibility of improving the efficiency of field use as such, by means of re-injection, (see Section 17.4). Although a primary purpose of re-injection is to avoid pollution, an advantageous side-effect could be the virtual provision of 'hot feed', to mitigate or delay the infeed of cold waters from outside the confines of a field as the hot fluids are withdrawn in the process of exploiting the latter. It is true that re-injection *in the wrong place* could *reduce* the heat output of a field, but the judicious re-introduction of hot bore fluids near the confines of a field could almost certainly delay the decline of a field's potential and thus prolong its life by acting as the equivalent of hot 'boiler feed'. This does not imply that the actual generation efficiency in terms of bore fluid would be improved, but it *could* mean that the efficiency of exploiting a field could be raised, and could offer an alternative – and possibly a cheaper one – to the direct use of the hot water issuing from a wet field by adding to the field life rather than by extracting more heat from the bore fluids. The whole question of the efficiency of geothermal exploitation is undoubtedly very complex.

10.5 Use of hot water: general
Since wet hyperthermal fields are more common than dry, very careful thought should be given to the various ways in which energy can be extracted from the hot water phase, bearing in mind what has been said in the preceding Section. Much of this problem has been discussed and analysed elsewhere [67]. It is instructive to list the various methods:

  (i) *Re-injection*. See Section 10.4.
 (ii) *Use of binary fluids*. See Section 10.2.6. The use of binary fluids need not of course be limited only to the water phase of wet fluids: it could equally be applied to both the steam and hot water phases simultaneously.
(iii) *Use for some non-power purpose* in a dual or multi-purpose project. (See Chapters 2 and 13.)

(iv) *Use of hydraulic pressure.* Except in the case of geopressurized fields (See Section 5.14) this is seldom of much value, as it is usually insignificant in comparison with the *thermal* pressure resulting from the temperature of the water.

(v) *Thermal use in conjunction with conventional power plant.* In 1961 Hansen [74] suggested that hot geothermal waters could be used as a means of feed-heating in a conventional fuel-fired plant in place of the usual bled steam feed-heating system, thus allowing *all* the steam to pass right through the turbine into the condenser and so generate more power. He claimed that the performance of a 150 MW conventional steam station (100 kg/cm²/538° C/3.825 mm Hg) could be improved by a 7% reduction in heat rate if about 450 kg/h of geothermal hot water at 197°C were used in place of bled steam in all but the highest pressure feed-heater. The objection to this interesting proposal is that it presupposes just the right combination of fuel requirements and geothermal availability, and a nice balance of costs. Nevertheless, the broad idea was again raised in 1975 by Harrowell [75] when consideration was first being given to the possibility of exploiting the fairly high thermal gradients found in Cornwall in south-west England. Harrowell suggested that any heat which might be won could be used to endow outdated thermal plants with a second lease of life in this way.

(vi) *Isentropic expansion,* by converting the heat energy of the hot water into kinetic energy to be harnessed in a turbine. This proposal is fraught with difficulties, but it will be discussed in Section 10.14.

(vii) *Isenthalpic expansion,* or flashing. This is the conventional method of using the heat energy of geothermal hot water, as proposed in Sections 10.2.4 and 5, and illustrated in Figs. 45d and e. Flashing is sufficiently important to warrant a separate section (Section 10.6).

The fact that the intrinsic energy of hot water per kilogram is far less than that of steam at the same temperature is sometimes offset by the far higher gravimetric proportion of water to steam, which may mean that the energy content of the water phase may be comparable with that of the steam phase of a mixed fluid issuing from a geothermal bore. Indeed, the *heat* content of the water may exceed that of the steam, though its lower grade may make it less valuable than the steam as an exploitable form of energy.

10.6 Flashing

The quantity of flash steam emitted from boiling water when its pressure is reduced to a value below saturation level, and when expansion takes place isenthalpically – i.e. when all kinetic energy temporarily generated is reconverted by friction into heat – can be deduced from the formula:

$$f = \frac{H_w - h_w}{L},$$

where f is the fraction of the water flashed into steam;
    H_w is the enthalpy of the hot water before flashing;
    h_w is the enthalpy of the residual water after flashing; and
    L is the latent heat of evaporation at the lower pressure.

The interpretation of this formula can be effected from the steam tables, but for approximate ready reckoning Fig. 49 may be useful. The flash steam quantities, when multiplied by the power potentials of dry saturated steam deduced from Fig. 46 will give the power potential of the hot water as deduced from Figs. 47a or b, provided that the flashing temperatures are taken at the optimum values between the initial and 'sink' temperatures of the cycle.

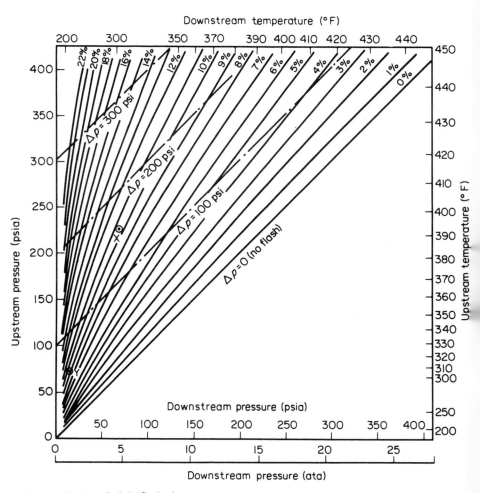

Figure 49   Isenthalpic flash chart.

Example. 1 t/h of boiling water at 200° C (15.34 ata) is to be double-flashed into a turbine, the condenser of which is maintained at a temperature of 52° C (102 mm Hg). The optimum flashing temperatures will be:

1st state $200 - \frac{1}{3}(200 - 52) = 150.7°$ C (equivalent to 4.77 ata)
2nd stage $200 - \frac{2}{3}(200 - 52) = 101.3°$ C (equivalent to 1.04 ata)
First state flash $H_w = 203.6$ cal/g $h_w = 151.6$ cal/g $L = 504.9$ cal/g

Hence $f = \dfrac{203.6 - 151.6}{504.9} = 0.103$ t/h

Residual water $= 1 - 0.103$, or 0.897 t/h
Second stage flash $H_w = 151.6$ cal/g $h_w = 101.2$ cal/g $L = 538.5$ cal/g

Hence $f = \dfrac{151.6 - 101.2}{538.5} \times 0.897 = 0.0839$ t/h.

Power from 1st stage flash at 4.77 ata $= 0.103$ t $\times$ 114 kWh/t (Fig. 46)
Power from 2nd stage flash at 1.04 ata $= 0.084$ t $\times$ 58 kWh/t (Fig. 46)
        i.e. $11.74 + 4.87$, or *16.61 kW.*
Check. From Fig. 47b, power potential of boiling water at 15.34 ata
        $= 16.6$ kWh/t.

According to these figures, the first and second stage flash proportions amounted to 10.3% and 9.36% respectively, and are represented by points X and Y on Fig. 49.

Despite the optimum flashing temperatures determined by dividing the total temperature range by an integral number of flashing stages (2 for single- and 3 for double-flashing), it may often be convenient to choose rather different flashing temperatures (and pressures) in order to take advantage of existing pressure systems that have been established for other reasons. For example, in the experimental hot water scheme at Wairakei, where the temperature of the hot water arriving at the plant was normally about 193° C and the vacuum temperature was 33.3° C, the optimum flashing temperatures would have been 140° C and 86.5° C for two-stage flashing. Apart from the undesirability of adopting a sub-atmospheric pressure for the second stage, it was convenient to flash into two established steam systems at the power station operating at about 148° C and 102° C. This involved a sacrifice of only about 4.3% of the ideal power obtainable from the hot water, because the curves of power potential against various intermediate temperatures are fairly 'flat'.

If flashing is allowed to take place by simply admitting boiling water into a flash vessel at a lower pressure, the operation passes momentarily through an *isentropic* stage in which the heat energy of the boiling water is at first converted into kinetic energy. Although this kinetic energy is subsequently destroyed by friction, so that the overall process becomes *isenthalpic*, high water jet velocities are temporarily achieved. For pure isentropic expansion, boiling water can attain a jet velocity of $91.4\sqrt{\Delta H}$ m/sec, where ΔH is the heat drop in cal/g. This figure will be modified in practice by imperfect conversion of heat into kinetic energy, but the actual jet velocities can nevertheless be substantial. Fears were felt at Wairakei, before the hot water flashing experiment was undertaken, that the wear and tear on flash vessels might be excessive. For example, in expanding from 193° C to 148° C the isentropic heat drop is 2.683 cal/g and the theoretical jet velocity would approach 150 m/sec. Various ways were considered for minimizing the isentropic factor so that isenthalpic expansion could be effected as gently as possible. These included a gang-operated group of valves in series which would reduce the heat drop (and therefore the velocity) across each valve, a re-boiler, a resistance pack approximating to an ideal 'porous plug' and a thermo-compressor arrangement. However, it was found that by admitting the flashing water tangentially into a cylindrical vessel – i.e. by adopting in effect a simple cyclone separator – the wear and tear from the flashing process was negligible.

Although the experimental hot water scheme at Wairakei (see Section 9.9) had to be abandoned through lack of bore water, moderate use continued to be made of some of the hot water by what became known as 'inter-flashing'. Owing to the adoption of two separate steam systems operating at different pressures it was possible to pass the discarded hot water from some of the higher pressure wellhead separators through other (flashing) separators situated close to the bores which discharged their flash steam into the lower pressure steam pipework system, discarding their hot water at a lower temperature than would have been possible had the water been rejected from the higher pressure separators alone [76]. Although this enabled about two-thirds of the power potential of some of the hot water from the higher pressure wells to be recovered, and avoided the expansive hot water transmission system, it involved the transmission of low pressure steam from the field to the plant, and this is the least efficient method of heat transmission (by comparison with high pressure steam and hot water). Nevertheless, the arrangement possessed overall economic attraction.

10.7 Flash steam scrubbing
As already mentioned in Section 9.10, the presence of even very small quantities of chlorides can do severe damage to turbine blades in the presence of bore gases. In the case of remote bore steam it has been shown (again in Section 9.10) how the scrubbing effect of long pipelines can remove virtually all chlorides that may be carried into the pipelines as a result of imperfect separation at the wellheads. Interflashed steam would also be purified in this way, but when flash vessels are located close to the plant it is vitally necessary to 'scrub' the flash steam before admitting it into the turbines, because with flash vessels a small amount of highly concentrated chloridic water is unavoidably carried over.

Fig. 50 illustrates the type of scrubber used at Wairakei for the experimental hot water scheme. The flash steam with its small entrainment of chloridic water droplets enters at the side near the bottom of the cylindrical scrubber and passes vertically upwards through three mesh separator packs and two impingement 'bubble-plates', leaving at the top for admission to the turbines. Stainless steel wire meshes were first used at Wairakei for the separator packs, but these were rapidly eaten away; which proved the need for scrubbing, especially as the H_2S content of flash steam is very small by comparison with that of the bore steam with which the former would mix when passing through the turbines. Inert plastic meshes were successfully substituted for stainless steel. Chloride-free wash water is flushed over the impingement plates, falling by gravity to a discharge point lower down the scrubber, as shown. The turbine makers had stipulated a maximum chloride content of 10 ppm in the *water phase* of any final carry-over entrained in the scrubbed flash steam – not 10 ppm related to the total mass flow of the slightly wet steam. The chloride content from the entrained water droplets from the flash vessels was about 4550 ppm. Thus the scrubbers had to reduce the chloride concentration by 455 times. The assumed efficiency of the mesh separators was $97\frac{1}{2}\%$ – allowing $2\frac{1}{2}\%$ of the liquid phase to pass through and rejecting $97\frac{1}{2}\%$ downwards. The Wairakei low pressure scrubbers were rated for a steam flow of 63 500 kg/h and a carry-over of 3855 kg/h (about 6% of the steam flow) of water was assumed to be lifted from each impingement plate, most of it being captured and thrown back in the next mesh separator above. It was also assumed that the flash steam entering the scrubber contained 0.5% of entrained water (317.5 kg/h). Under the assumptions made it is estimated that about 590 kg/h of wash-water would be necessary to ensure that the entrained water (estimated at only about 0.15% of the total fluid) would contain chlorides not exceeding 10 ppm in concentration. A larger quantity of wash-water would reduce the chloride content of the final carry-over still further.

In Fig. 50 water quantities are shown in plain numbers (kg/h) while chloride concentrations in ppm are shown in brackets. The calculations are as follows:

Upper impingement plate

Water balance $\qquad x + 97 + 3758 - 3855 = \text{overflow} = x$

Chloride balance $\quad 10(3855 - 3758) + 10x = 97y$

$$970 + 10x = 97y$$
$$x = 9.7y - 97 \qquad\qquad (1)$$

Lower impingement plate

Water balance $\qquad x + 8 + 3758 - 3855 = \text{overflow} = x - 89$

Chloride balance $\quad (8 \times 4550) + 3758y + 10x = 3855y + (x - 89)y$

$$36400 - 8y - xy + 10x = 0 \qquad\qquad (2)$$

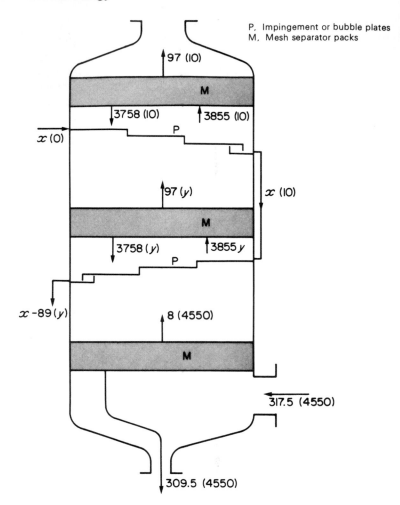

Figure 50   Diagram of low pressure flash steam scrubber as used at Wairakei for the experimental hot water transmission scheme.

Notes: (i) Figures show water flows in kg/h at various points, with the chloride content in ppm in brackets.

(ii) The wash-water x is taken from the station separators and is virtually chloride-free. Its temperature will approximate to that of the entry steam (about $100°$ C), so the steam throughput ($63\frac{1}{2}$ tons/h) suffers no appreciable loss. In fact by draining the wash-water from the intermediate pressure scrubber a slight gain is won from flashing.

(iii) Entrainment of water from bubble plates is approximately proportional to the volumetric steam flow.

Substituting Equation 1 in Equation 2

$$36400 - 8y - 9.7y^2 + 97y + 97y - 970 = 0$$
$$y^2 - 19.175y - 3652.5 = 0$$

Hence $y = 70.778$
and $x = (9.7 \times 70.778) - 97 = 589.55$.

Thus the wash-water flow-rate would be about 590 kg/h and the chloride concentration on the lower impingement plate would be rather less than 71 ppm. Not only has the chloride concentration been reduced to the required level of 10 ppm but the flash steam has also been made drier, by reducing the wetness from 0.5% to about 0.15% by weight.

10.8 Condensers and vacuum

Since no boiler feed is needed with geothermal power plants, condensing turbines need not be provided with costly surface condensers. Relatively cheap jet condensers may be used instead, in which the turbine exhaust directly impinges upon, and mixes with, the cooling water so that the two fluids are discharged together. The cost saving arises mainly from the absence of tubes, while a smaller terminal temperature difference can be achieved by the direct contact. With condensing geothermal turbines the vacuum can be maintained either by direct once-through river or lake cooling or by closed loop cooling through a cooling tower. If there is an abundant supply of natural cooling water available as, for example, is provided by the Waikato River at Wairakei, direct cooling has much to commend it, but it may give rise to pollution troubles (see Chapter 17). Cooling towers have the advantage of requiring no make-up water, other than for starting-up purposes, and can therefore serve a geothermal power plant in an arid location. The reason for this lies in the wet exhaust that results from using steam at low pressures and temperatures. For example, let it be supposed that a geothermal turbine is supplied with dry saturated steam at 7 ata and exhausts to a vacuum of 102 mm Hg (4 in Hg). Assuming a turbine efficiency of 75%, it can be seen from the Mollier diagram that the wetness at exhaust would be about 12%. The total heat of the exhaust, at this wetness, would therefore   be $[(0.88 \times 619.88) + (0.12 \times 51.65)] = (545.49 + 6.20) = 551.69$ kcal/kg. Let it be supposed that the cooling water enters the condenser at $28°$ C and leaves at $48°$ C, at which its latent heat is 570.3 kcal/kg. To evaporate 1 kg of cooling water would require $(48 - 28) + 570.3$, or 590.3 kcal, which could be supplied by 590.3/551.69 or 1.07 kg of turbine exhaust. Thus there would be a mass gain of 7% on the cooling water circulated, and no extra water would be needed – only a purge.

All the major existing geothermal power plants use cooling towers, except Wairakei where there is a large river available for direct cooling. The choice of a riverside site was also originally influenced by the flatness of the terrain and the absence of hot springs beneath the foundations, and also partly by the belief that the centre of the bore-field was only about $1\frac{1}{2}$ to 2 km distant. Subsequent development of the field showed that the more productive bores were rather more distant than had at first been supposed, so that possibly a power plant sited in the middle of the bore field might have been a better choice (despite the poorer vacuum obtainable with cooling towers) since due to the proximity of the bores there would be shorter pipelines and higher available turbine entry pressures (or alternatively bigger steam yields). However, the foundation conditions in the bore field would have been difficult, so the choice of a riverside site may well have been a wise one after all.

There are various possible arrangements of condenser/turbine assemblies that are favoured by different designers, as illustrated in Fig. 51. Arrangement (a) is used at Wairakei: it has the advantage of easy accessibility but the disadvantage of great building height which, in a seismic area, can be very costly in steel. Arrangement (b) seems to be the most popular one: it is adopted at the Geysers (Units 1 to 4), in Japan (Plates 18 and 19), Mexico and the USSR. It has the merits of cheaper building costs, with the turbine at ground level; but the disadvantages of a tall external structure for the condenser and barometer pipe, a large and expensive vapour duct (of corrosion-resistant metal which must be kept drained and may be vulnerable to frost) and a less easily accessible condenser. Arrangement (c) which is used at Larderello, is a hybrid solution that is fairly economical in building costs but which requires a deep pit for the barometer pipe and the use of a warm water extraction pump. Another hybrid arrangement, (d), is favoured at some of the later units at the Geysers: it avoids the deep pit of arrangement (c) but requires a larger condensate extraction pump. Fig. 51 is an over-simplification of the problem, which can be influenced by ground contours and the level at which cooling towers (if used) can conveniently be placed. Considerations such as these will effect the lifting capacity of pumps and the degree to which the vacuum suction can be used to induce the cooling water to flow into the condenser. All arrangements can be used either with direct river cooling or with cooling towers.

Condenser vacuum can be maintained by means of mechanical exhausters or by ejectors activated by steam or high pressure water. Mechanical exhausters are the most economical in energy consumption, but ejectors are usually favoured because of their low maintenance costs and trouble-free operation. At Wairakei, rotary exhausters were at first installed for regular use, with standby steam ejectors. Although the latter consumed about $4\frac{1}{2}$ times as much steam as that required to supply the electricity for the motors of the mechanical exhausters, they soon came to be used in preference to the rotary exhausters, which were subject to a good deal of maintenance trouble. Water ejectors are more efficient than steam ejectors as they involve

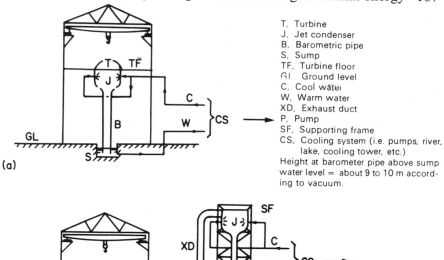

T, Turbine
J, Jet condenser
B, Barometric pipe
S, Sump
TF, Turbine floor
GI   Ground level
C, Cool water
W, Warm water
XD, Exhaust duct
P, Pump
SF, Supporting frame
CS, Cooling system (i.e. pumps, river,
    lake, cooling tower, etc.)
Height at barometer pipe above sump
water level = about 9 to 10 m accord-
ing to vacuum.

(a)

(b)

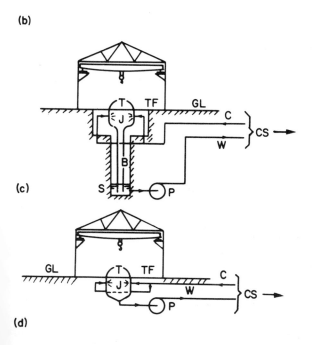

(c)

(d)

Figure 51   Different arrangements of jet condensers and geothermal turbines. (a) High
level turbine (b) Low level turbine: external condenser (c) Condenser pit (d) Use of
extraction pump to save height.

Plate 18 Onuma geothermal power plant, Japan. Note the low level turbine house and high level external jet condenser. See Fig. 51b. (By courtesy of the Japan Geothermal Energy Association.)

Plate 19 Onikobe geothermal power plant (25 MW), Japan. (By courtesy of the Japan Geothermal Energy Association.)

no loss of latent heat, but they are costlier than steam ejectors and require large quantities of water [77]. The use of boiling water ejectors has been suggested [78] but it is not known if this has yet been tried. Two useful studies of mechanical exhausters are given in [71, 79]. The capacity of any exhausting equipment must of course depend not only upon the condenser vacuum but also upon the proportion of non-condensible gases contained in the steam. If this proportion exceeds about 10% or so by weight, condensing machines become uneconomic owing to the high power absorbed by the gas exhaustion process. James [80] makes a case for adopting an economic vacuum of 127 mm Hg (5 in Hg) as an universal optimum for geothermal condensing plants, but his theory has not gained wide-spread acceptance. In general, except where large quantities of river, lake or sea water are available, it is not worth striving for too low a back-pressure, as this can sometimes necessitate very costly cooling towers and high energy-consuming gas extraction equipment.

By comparison with conventional fuel-fired thermal power plants, the amount of heat rejected to waste by condensers is very great. A modern fuel-fired station of, say 33% efficiency rejects about 2 kWh (thermal) for every kWh of electricity generated; whereas a geothermal power plant of, say only about 16% efficiency, would reject $5\frac{1}{4}$ kWh (thermal) for every kWh generated. Hence, the cooling towers installed at geothermal power stations are enormously larger than those at conventional thermal stations of the same capacity.

10.9 Fuelled superheating

The efficiency limitations of geothermal turbines imposed by the relatively low pressures and temperatures normally encountered in hyper-thermal fields could theoretically be improved by using supplementary fuel to impart a high degree of superheat to the steam before entering the turbines. This would have the further advantage of giving drier steam in the low pressures stages with consequent reduction of blade erosion, and could enable higher tip speeds to be adopted (see Section 10.10) and larger outputs per turbine of any given type to be achieved. This was considered at the design stage of Wairakei, but under the conditions then prevailing only a very doubtful economic case could be argued in favour of fuelled superheating – and then only for overload purposes. The idea was therefore not pursued. In 1970 the whole question was re-examined [81] with a view to the possibilities of supplying peak loads geothermally. It was observed that steam at 7.8 ata could yield about 25% more power per kilogram when superheated to 343°C than when saturated – at 170°C. A verdict of *non proven* was returned because the economics of superheating were so dependent upon the cost of fuel, the shape of the load curve, the size of plant and various other factors that no general conclusion could be drawn. It is only fair, however, to say that James [80] took a more optimistic view in 1970. With well reasoned argument he deduced that, under Wairakei conditions at any rate, fuelled superheating would probably have paid at the time when he was writing, though he was not dogmatic in asserting a general case for its adoption. But the spectacular increases in fuel prices since 1973, *vis-à-vis* other forms of inflation, will now almost certainly have ruled out permanently all ideas of fuelled superheating.

10.10 General points of geothermal turbine design

Wood [77] has brought out some factors influencing the design of geothermal turbines. One is the permissible tip speed of the low pressure blades which, for metallurgical reasons (see Chapter 16), must be limited to a moderate value - about 275 m/s – in order to avoid excessive erosion (which is proportional to the square of the tip speed) in the wet environment of the low pressure steam. This in turn fixes a limit to the area of the annualar exhaust ring if the blade length is not to be excessive, and thus to the steam flowrate if the leaving losses are to be kept within reasonable limits. The power capacity of a geothermal turbine will therefore depend upon the pressure drop from inlet to exhaust, the number of exhausts and the rotational speed. Until 1968 the largest geothermal turbines in the world were the 30 MW units at Wairakei, each supplied with mixed flow steam entry at 4.4 and 1.05 ata, with double exhausts to 38 mm Hg back-pressure, running at 1500 rpm; but from that year several 55 MW units were put into service at the Geysers, California, with steam entry at 7.7 ata/179° C, 102 mm Hg back-pressure, double exhausts and a speed of 3600 rpm. With quadruple exhausts still larger capacities become practicable. By 1975 a 110 MW unit had been commissioned at the Geysers, several more units of that rating or slightly larger are under construction or planned, and one unit of 140 MW is due to be commissioned in 1979.

The high wetness factors encountered in geothermal turbines, owing to the low steam entry conditions by comparison with those prevailing in modern conventional thermal plants may necessitate the installation of water separators after partway expansion. At Wairakei, where the turbines are arranged in 'cascade' in three pressure ranges – 13/4.5, 4.5/1.05, 1.05/0.05 ata – separators are incorporated within the station pipework system between each pressure range so that the steam entering the next downstream turbines will be reasonably dry; while internal inter-stage separating facilities are provided within the mixed pressure turbines. In this way, the wetness of the steam entering the condensers is limited to about 10 to $10\frac{1}{2}\%$ under the worst conditions from the mixed pressure sets and about 9% or slightly above from the 11 MW low pressure sets.

In parts of the Tuscan fields – particularly in the Monte Amiata region – newly sunk bores are characterized by very high gas content – sometimes as much as 80% or even more. The only way of economically extracting power from such bores is by means of non-condensing turbines. But in the course of two or three years the gas content falls to perhaps 10% or even less by weight, under which conditions a condensing plant becomes far more attractive economically. Dal Secco [71] has suggested that for gas contents in the 10–50% range the indirect cycle (Fig. 45a) would be the most economical; but for what is likely to be a transitory phase only it would clearly not be worth introducing this cycle specially. The Italians have therefore devised an ingenious two-stage development plant [82]. At first a non-condensing 15 MW unit is installed, which may give perhaps two or three years' useful service. Later, when the gas content has dropped sufficiently, they mount a second low pressure double-flow turbine coaxially with the non-condensing unit and directly coupled to a rotary gas-exhauster. This second unit is provided with a condenser. The non-condensing unit may run continuously while the second unit is under erection, but when the latter is ready for service it is necessary to shut down the first unit for about two or three days only while the two machines are mechanically coupled together, the exhaust from the high pressure unit diverted to the inlet of the low pressure unit, and the lubricating system arranged to serve both units. The result is a 15 MW 2-cylinder condensing machine with greatly reduced steam consumption which may give many years of good service. In this evolutionary way a new field or region may be developed logically and economically in a practical manner.

Blade failures have occurred at Wairakei and Matsukawa due to resonant vibrations. This is a form of trouble by no means confined to geothermal plants: it has also occurred extensively in conventional thermal power stations in many parts of the world where the natural frequencies of blades or blade systems, vibrating in some particular mode, respond to a harmonic of some exciting frequency caused, for example, by the nozzle-passing frequency of the first stage blades. The resulting resonance can cause blade fractures through metal fatigue. Although this is *not* a specific weakness of geothermal turbines, matters may sometime be aggravated in such turbines by the build-up of chemical deposits on the blades, thus altering their natural resonant vibration frequency. This is a hazard to which designers should be very much alive.

The question of turbine blade materials will be dealt with in Chapter 16.

10.11 Geothermal power plant maintenance

It is important that a systematic scheme of preventive maintenance be observed at geothermal power plants, together with rigidly planned periodic overhauls and the stocking of adequate spare parts – particularly turbine blade assemblies, both fixed and rotating. In this way high plant availability may be assured so that a geothermal power plant may operate at very high annual plant factor – often 90% or more [83, 84].

10.12 Extraction of power from geothermal brines

In certain parts of the world – e.g. the Salton Sea area in California – the geothermal fluids consist of boiling brines of high saline concentrations – in some cases exceeding 300 000 ppm – and sometimes of caustic properties. Whilst such brines may offer industrial possibilities for the recovery of salts and other minerals, they can also be used for other purposes including power production. However their chemically aggressive nature, particularly due to the combination of chlorides and H_2S, raise serious utilization problems. Where a power plant can be placed near the centre of a bore-field, appreciable economic gains can be won from short pipework and low pressure drops, but at the cost of losing the valuable scrubbing properties of long pipelines (see Section 9.10). The use of hot brine and flash steam heat-exchangers was therefore tried, but serious scaling troubles were experienced. Fig. 52 shows a plant layout of a binary cycle power installation now under construction in the Salton Sea area for the San Diego Gas & Electric Co. [85]. It is to use isobutane as the binary fluid and scrubbed flash steam (see Section 10.7) as the heating medium for the heat exchangers. Most of the salts will simply pass through the flashing separators, while nearly all of the saline carry-over will be caught by the scrubbers. Rejected heat and brine

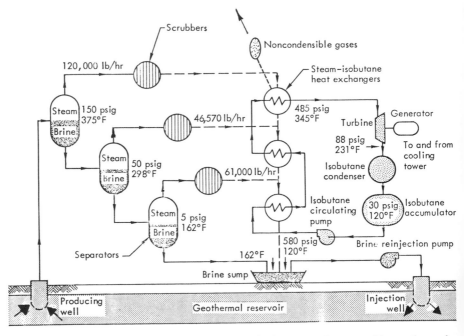

Figure 52 Schematic arrangement of 3-stage flash system for use with geothermal brines in conjuction with an isobutane binary cycle. (After Fig. 36 of [85].) By courtesy of the San Diego Gas & Electric Company.

will be re-injected into the aquifer. The use of downhole pumps (see Section 9.11) and the total flow turbine concept (see Section 10.13.8) is also being considered for the exploitation of hot brines for power production.

10.13 Special geothermal power generators

Although geothermal turbines are usually fairly orthodox in broad conception, a number of special power generation devices have been conceived

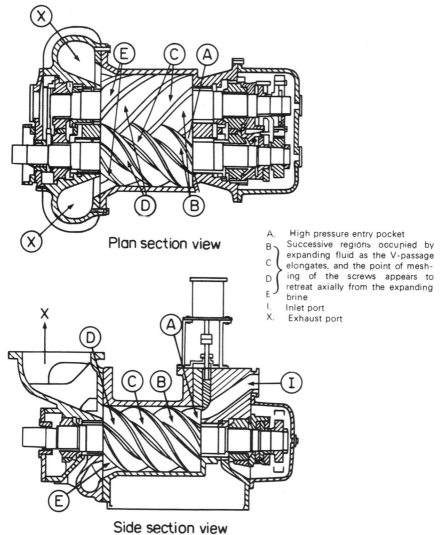

Plan section view

Side section view

A. High pressure entry pocket
B. ⎫ Successive regions occupied by
C. ⎬ expanding fluid as the V-passage elongates, and the point of meshing of the screws appears to retreat axially from the expanding
D. ⎭
E. brine
I. Inlet port
X. Exhaust port

Figure 53 The Sprankle (Hydrothermal Power Company) double helical hot brine expander. (From H.P.C. Brochure.)

for geothermal application. Some of these have only reached the laboratory or experimental prototype stage, while others are mere paper conceptions: but it is of interest to consider them.

10.13.1 *The Sprankle H.P.C. (Hydrothermal Power Co.) prime mover.* A prototype of this machine is actually working in the Salton Sea area, California. It is illustrated in Fig. 53 and consists basically of two intermeshed helical 'gears', [86]. Such a machine, when driven by external mechanical power, can serve as a compressor: when used in reverse, it can act as a positive displacement prime mover. Its principal intended use is for the extraction of power from boiling brines. The hot brine is admitted to the rotors at point A and passes through the helical passages B, C and D, falling in pressure and flashing off vapour as it proceeds, being finally exhausted from E. The pressure exerted by the fluid upon the helical vanes produces a rotary driving force. It is thus a two-phase flow machine. If the brine concentration is at or near saturation, salts will be deposited on the helical and stator surfaces, and the rubbing action is alleged to form a hard, smooth surface of salts which effectively seals the machine against leakage and short-circuiting of the fluids past their intended path through the machine. The makers are offering units of up to 5 MW capacity, and claim that 'the process approximates to an isentropic expansion from a saturated liquid. It is theoretically the simplest conversion process and is thermodynamically optimum'. This is tantamount to a claim of very high efficiency. The author has seen no test results, but it seems surprising to him that the efficiency of such a device could really be very high. The fact that a reasonably good efficiency may be obtained when using the machine as a compressor does not necessarily imply that this would apply when using the machine in reverse. For example, a worm gear drive is quite efficient if the worm drives the worm-wheel, but so inefficient if operated in reverse that it fails to work at all unless the worm thread pitch is very great. It also seems likely that the machine weight per kilowatt of output would be high. More factual test data is needed before a balanced judgement can be passed on the general usefulness of this ingenious conception. It certainly works and for the specific purpose of extracting power from hot brines it could well be useful. It will also be necessary to establish whether the long term effects of internal flashing will be damaging to the machine surfaces in the way that cavitation can damage marine propellers.

10.13.2 *The Robertson engine.* This device was not specifically designed with geothermal application in view: it was intended to operate on a variety of working fluids, but the inventor considers that it should also be suitable for extracting energy from geothermal fluids—water and/or steam. The machine is illustrated in Fig. 54. It has a tapered rotor revolving within a

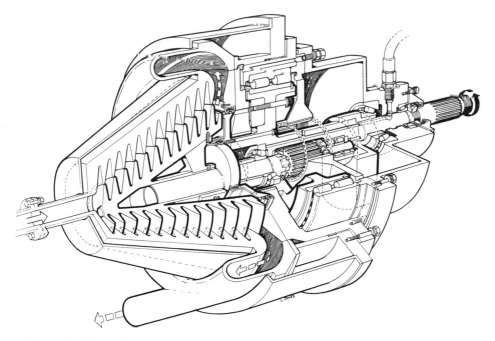

Figure 54   The Robertson Engine. (by courtesy of the inventor)

tapered stator. The rotor has a 'single-start' helix, while the stator has a 'double-start' helix of exactly twice the rotor pitch. The rotor is mounted on a crank, and its rotation on the crank-pin is governed by gears so that each cross-section of the rotor oscillates within the stator as it revolves. This arrangement gives rise to a series of trapped volumes which expand progressively from one end of the machine to the other as the rotor rotates. The number of trapped volumes is half the number of rotor pitches. The taper shape allows the rotor and stator to be lapped by means of grinding paste so as to provide a near-perfect fit, and it allows for progressive adjustment to compensate for wear. As with the Sprankle machine, chemical deposits, precipitated from the water phase as it progressively boils during its passage through the engine, would be wiped so as to give a self-cleaning action. Power loss due to leakage past the helical surfaces would be small, partly because of this self-cleaning action and partly because of the low pressure drops between successive trapped volumes. Gases released from solution would assist the steam in exerting useful work-creating forces on the rotor. One of the main merits of the engine is that it virtually has only one moving part, and as the flow is unidirectional there is no need for valves or timing ports. The volume occupied by the fluid forms a higher proportion of the stator volume than in the Sprankle machine: the Robertson machine should therefore be lighter and more compact per kilowatt of output, though insufficient information has been made available for the relative efficiencies of the two devices to be compared. As with the Sprankle machine, the long term effects of flashing have yet to be proved. The designer, Mr T.J.M Robertson of the Atomic Energy Research Establishment, Harwell, England, has made and tested a small fractional horse-power model which ran on compressed air at 2500 rpm and showed a remarkable sustenance of torque at all speeds down to the stalling point. He plans to make a 40 shaft horse-power prototype to rotate at 3000 rpm with an inlet pressure of 100 psia saturated steam, exhausting to atmosphere – an expansion ratio of 6.8:1. The overall size of the machine is not expected to exceed about 53 cm diameter × 61 cm long, but the designer hopes to improve on this somewhat by optimization.

10.13.3 *Bladeless turbine* [*86*]. An assembly of plain metallic discs is mounted on a shaft, with narrow spaces between the discs. Geothermal fluid – hot water and/or steam – is admitted to the spaces between the discs at the periphery, and is directed to follow an inward spiral path towards a central exhaust port. The driving force is derived from frictional boundary layer drag between the fluid(s) and the disc surfaces. Such a machine would be very cheap to construct and to maintain: the discs could easily be dismantled, cleaned of chemical deposits and reassembled. Nevertheless, its efficiency is unlikely to be high as much energy would be wasted in frictional turbulence, and as centrifugal forces would act in opposition to the centripetal flow induced by the pressure drop between inlet and exhaust.

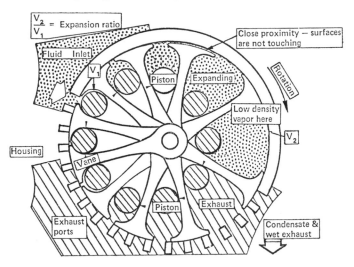

Figure 55 Diagrammatic representation of the KROV (Keller rotor oscillating vane) positive displacement machine. (After Fig. 6 of [*86*].)

10.13.4 *KROV (Keller Rotor Oscillating Vane) machine* [86]. This is a positive displacement machine (as in fact also are the Sprankle and Robertson machines) designed to deal with geothermal fluids of all qualities. Its somewhat fanciful action is self-evident from Fig. 55. Although it should be capable of handling a wide range of fluid conditions over high expansion ratios it is mechanically very complex and its efficiency is unlikely to be high. Nor is the sealing between inlet and exhaust, which would be dependent upon contact between the roller-pistons and vanes, likely to be very sound.

10.13.5 *Armstead – Hero turbine* [87]. The intended purpose of this concept is to provide an extremely cheap turbine capable of handling wet or dry geothermal fluids at the wellheads of newly sunk productive bores, so as to put to immediate use the fluid that would normally be blown to waste. It would thus serve as a pilot plant. It could also be used as an emergency standby unit, for which high efficiency is unimportant, and could even make small contributions to system peak loads. As a pilot plant it could be moved from field to field as early bores in one field are brought into permanent service and as new fields are explored. In principle the machine reverts to the 2000 year old Hero engine, in that it is a reaction turbine of the simplest type. For use with dry steam the arrangement shown in Fig. 56a would be used; steam being admitted through a hollow shaft and discharged through tangential peripheral nozzles provided with a small axial 'rake' to clear the exhaust away from the flat cylindrical rotor. For wet steam, the two phases would be separated within the rotor itself by centrifugal action and the use of a scoop (Fig. 56b) which would enable the water to be discharged at a velocity appropriate to its intrinsic heat energy (which would of course be less than that of the steam discharged at the periphery). The water jet quantity would be self-adjusting owing to the behaviour of boiling water when passing through bell-mouthed orifices (see Section 9.4). Advantage is taken of the moderate sacrifice of efficiency resulting from a large drop in rotational speed to limit the tip speed to a practicable maximum value. Steam friction would probably limit the efficiency to a modest level, but the extreme cheapness of the machine, with its absence of blades and its adaptability to different fluid conditions simply by changing fixed nozzles, should make it economic for limited purposes.

10.13.6 *Gravimetric Loop machine* [88]. This is essentially for use in a binary fluid circuit. Fig. 57 shows the principle. Three parallel fluid columns of considerable height (64 m in a prototype) are arranged to form two loops, the central column being common to both. In the left-hand loop a refrigerant (e.g. freon 11) is pumped counter-clockwise round the loop: in the right-hand loop, water is circulated clockwise by gravimetric forces. The bottom of the water loop is heated by geothermal fluid. The heated water, on contact

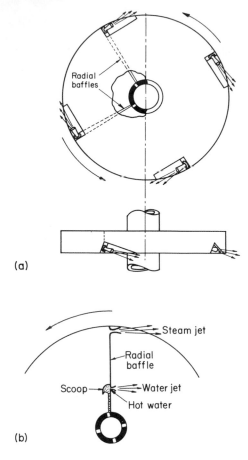

Figure 56 The Armstead–Hero turbine for use at dry or wet geothermal wellheads. (a) As used with dry steam. The recessed nozzles, the number of which is arbitrary are 'raked' so as to discharge the exhaust with an axial component of velocity. (b) As used with water/steam mixtures. The water discharge is controlled by a self-regulating orifice and is placed at a radius suitable to its isentropic discharge velocity. (After Figs. 5 and 7 of [87])

with the freon at the bottom of the central column, causes the latter to boil, and the released freon bubbles give buoyancy to the fluid in this column so that the water and boiling freon rise to the top where the two fluids, being immiscible, are separated. The freon is liquified by means of a condenser and is recirculated round its loop by a pump, while the water crosses over to the right-hand column and descends, passing through a water turbine near the bottom of that column. The different densities of the fluids in the right-hand and central columns provide the driving head for the turbine.

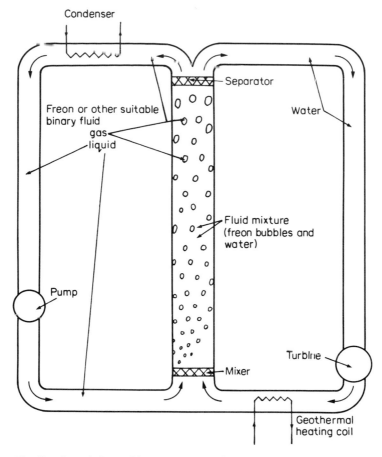

Figure 57   Gravimetric Loop binary system. (After Fig. 1 of [88].)

10.13.7 *E.G.D. (Electro-gas-dynamics)* [89]. This is a concept developed in the USA with a view to the direct conversion of combustion energy into electricity without the intermediary of an engine. The system has certain similarities to M.H.D. (magneto-hydro-dynamics) except that it lacks a magnetic circuit and produces electricity from a blast of ionized particles through an electrostatically charged field. In 1966 the author suggested to the patentees that their principle could perhaps be adapted to the production of power from geothermal fluids, somewhat in the manner shown in Fig. 58a. It depended upon the fact that all known geothermal waters are to some degree saline, and therefore electrically conductive. The water/steam mixture would be expanded, as nearly isentropically as possible, in a convergent-divergent nozzle, upstream of which the fluid would pass through

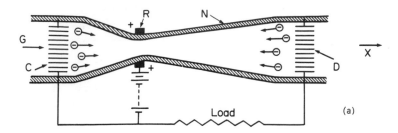

G, high pressure geothermal water/steam mixture with saline water phase.
C, charging grille-cum-atomiser
R, charging ring
N, expansion nozzle of electrically insulating material
D, discharging grille
X, low pressure exhaust
⊖→,charged particle with direction of *pull* (not motion)
    Although d.c. has here been shown to demonstrate the principle, a.c. could be used instead.

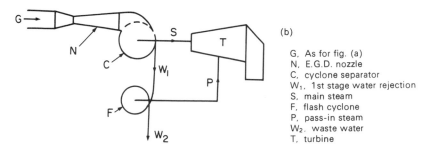

G, As for fig. (a)
N, E.G.D. nozzle
C, cyclone separator
W_1, 1st stage water rejection
S, main steam
F, flash cyclone
P, pass-in steam
W_2, waste water
T, turbine

Figure 58 Electro-gas-dynamics, or E.G.D. (a) Principle of E.G.D. (b) Possible use of E.G.D. in connection with 'bottoming turbine'.

an electrically charged grille, emerging as a stream of charged saline water droplets in an uncharged 'wind' of steam. As expansion proceeds in the nozzle, the droplets would explode into a fine atomized spray and would be carried downstream by their released heat energy against an electrostatic pull from a charged ring at the throat. Upstream of the throat the pull would accelerate the droplets: downstream it would retard them. On reaching the discharge grille, the droplets would give up their charge and (ideally) would have expended most of their kinetic energy and would just 'drop out' at the exhaust, leaving the steam to proceed, but at reduced velocity owing to the backward drag of the water droplets. This device could be arranged in consecutive stages so as to limit heat drops per stage and keep velocities to manageable levels, with a condenser at the last stage. Alternatively the device could be used as a topping unit upstream of a turbine, and the retarded water could be flashed off at a lower pressure to provide pass-in steam (Fig. 58b). There would be formidable difficulties in putting this idea into practice, as E.G.D. is essentially a process involving very high voltages and very low currents (megavolts and milli-amps). Very close spacings – 0.1 mm or less – between the grille plates would be necessary if reasonably high current densities are to be achieved. It is also not absolutely certain that the steam, as well as the water droplets, might not to some extent be ionized; also the mixture might have a low break-down voltage, resulting in flash-backs. Research is needed before a balanced verdict can be pronounced.

10.13.8 *Total flow impulse turbine* [86, 90, 91]. This is simply a proposal for an axial flow impulse turbine to which water/steam mixtures are admitted through expansion nozzles for impingement onto impulse blading. Research is being conducted in the Salton Sea area by the Lawrence Livermore Laboratory. The idea is not new, having already been tried at Pauzhetka in the USSR (90), where it was shown that the power developed from the water/steam mixture from a text bore was virtually the same as that produced from the steam phase alone from the same well, but that the optimal rotational speed was nearly twice as high with steam as with the wet mixture.

10.14 Total flow concept: general

All of the eight devices described in Section 10.13 may be described as 'total flow' devices, in that they seek to handle both the water and the steam from wet geothermal bores. Even the gravimetric loop machine, in common with aother binary systems, can in a sense be regarded as a total-flow device since the whole of the natural geothermal fluid can be used in the heat exchanger to transfer its energy to the binary cycle. However, the remarks that follow apply neither to the harnessing of the total flow by thermal conduction in heat-exchangers, nor to positive displacement machines, but only to those devices in which the energy of the total bore fluid is first converted into kinetic energy before being harnessed – i.e to the bladeless turbine, the Armstead – Hero turbine, E.G.D. and the total-flow impulse turbine. With all but the second of these a fundamental difficulty arises from the fact that the isentropic heat drop, and therefore the ideal jet velocity, of steam will always greatly exceed that of boiling water over the same temperature range. Hence if both fluids are expanded in the same nozzle there will be a drag – the steam accelerating the water and the water retarding the steam. This drag, being frictional, though improving the dryness fraction, *must* reduce the nozzle efficiency. For example, let it be assumed that a 4:1 water/steam mixture is expanded from 149° C to 101° C. The heat drop would tend to produce an ideal steam velocity of 710 m/s and an ideal water velocity of 159 m/s, so that the initial drag would be 710–159, or 551 m/s. If the water were finely atomized, the velocities of the two fluids would tend to become equalized at 269 m/s (the combined momenta divided by the total mass), i.e. $[(4 \times 159) + (1 \times 710)]/5$ m/s.

The total kinetic energy of the two fluids expanding *separately* would be proportional to $(\frac{1}{2}mv^2)$ $[(4/2 \times 159^2) + (1/2 \times 710^2)]/5 = 60\ 522$. But if they expanded together at the equalized velocity of 269 m/s, the total kinetic energy would be proportional to $\frac{1}{2} \times 269^2 = 36\ 180$. Hence, if drag were eliminated by the equalization of the steam and water velocities, the nozzle efficiency could not exceed 36 180/60 522, or about 60% only. Other expansion ranges would produce other efficiencies.

If, on the other hand, each fluid maintained its own isentropic velocity regardless of drag, the trouble would then be transferred to the discs of the bladeless turbine and to the blades of the total flow impulse turbine which would be struck by two fluids at different velocities, so that the discs and blades could not move at an ideal speed to suit *both* fluids. If a reaction turbine were used instead of an impulse machine the trouble would remain; for a high nozzle efficiency would mean that a blade speed suited to the steam velocity would be too high for the water velocity, so that the water would have to be accelerated and would thus act as a brake on the driving force on the blades. Alger [91] quotes laboratory tests, as measured by thrust forces, to show that two-phase fluids can expand supersonically in a single nozzle at coefficients of 0.92 to 0.94, equivalent to efficiencies (proportional to the square of the coefficients) of about 85 to 88%. This seems to suggest that in the throat of the nozzles the two fluids are still moving at very different velocities and that sufficient time has not elapsed for drag to have had much equalizing effect. But between the throat and the turbine blades the velocity differential of the two fluids will have become reduced, with consequent loss of kinetic energy. This, plus the fact that residual drag would cause the two fluids to strike the blading of an impulse turbine at different speeds would suggest a low efficiency turbine even if the nozzles show high efficiency. Similar drag trouble would be encountered with E.G.D. Only in the case of the Armstead – Hero turbine, where each fluid is discharged at appropriately different radii, would this difficulty be overcome.

10.15 Contribution of geothermal power to a composite system
It is elementary knowledge that the cost of producing *anything* is made up of two components:

(a) a *fixed* component incurred in providing and maintaining the means of production, regardless of the quantity of end-product,
(b) a *variable* component that is more or less directly proportional to the quantity of end-product.

With conventional thermal power production the fixed costs are mainly the capital charges on the plant, buildings and equipment required for generation, together with the cost of employing the supervising and operating staff: the variable costs mainly consist of fuel, though they may also include some share of the costs of repairs and maintenance. It is for this reason that electricity is normally the subject of two-part costing with a fixed cost related to the kilowatts and a variable cost per kilowatt-hour. It is also well known that a composite power system can be supplied most economically by a combination of two main types of plant:

(a) *base load plant*, the characteristics of which are high fixed and low variable costs;
(b) *peak load plant*, the characteristics of which are low fixed and high variable cost.

(Admittedly some plants will carry intermediate loads between the base and the peak, but this does not invalidate the broad division of plants into these two principal types).

With a geothermal power plant, for which there are no fuel costs, the usual practice is to regard *all* the production costs as fixed, with zero variable costs. (This view will be challenged on a point of detail in Chapter 15, but is broadly acceptable.) The generally agreed justification for this attitude is that once geothermal steam has been made available by means of capital expended on exploration, drilling and pipework, it may be regarded as 'free' (much as water may be regarded as free in a hydro-power installation once it has been made available at a suitable head by damming). Geothermal power plants would therefore seem to be ideally suited to serve as contributors of *base load* to composite power systems. All the major power plants in existence are, quite rightly, used to supply base load. Moreover, they can achieve annual plant factors of 90% or more – higher than obtainable from any other type of thermal or nuclear plant. Nevertheless, a case can sometimes be made for using geothermal power for non-base load purposes [81, 92] in the following circumstances:

(i)   *As a sole supplier of small systems* [81]. The advantages of 'scale effect' may sometimes offset the disadvantages of operating at reduced plant factor, so that a geothermal power plant of sufficient capacity to meet the system peak may actually produce cheaper kilowatt-hours, despite the lower plant factor, than could a smaller geothermal plant designed to carry the base load only. In the course of time, as the system grows, a point is likely to be reached when it will pay to install peaking plants of some other type: this would permit the base load to grow until it ultimately matches the geothermal plant capacity, when the production costs of that plant will fall to a minimum. Thereafter, the geothermal plant will take its orthodox place in the system as a supplier of base load, while other types of plant will carry the peaks. The objection to this method of operating a geothermal plant – initially at moderate, and later at high plant factor – is that in the early years, when on non-base load duty, it may be necessary to waste steam that cannot be used at off-peak hours. This is because, owing to the bore characteristics, load reduction can only effect a diminution of steam supply if pressures are allowed to rise, and there is a limit to the permissible pressure variation beyond which the bores are liable to become unstable. Hence, the blowing of steam to waste may be unavoidable during off-peak hours. If the field is large by comparison with the heat consumption of a small geothermal plant, this may perhaps be condoned for a few years until the base load has had time to catch up with the plant capacity; but the wastage of even a small part of a finite source of heat is not generally to be commended.

(ii)   *If cheap fuel justifies the superheating of natural steam during peak hours.* This point has already been touched upon in Section 10.9; but the availability of *cheap* heat would be essential.

(iii)   *If adequate thermal storage is available*, [81]. In the unlikely event of a natural storage cavern of suitable capacity and depth being available, it would be possible to store heat during off-peak hours in the form of boiling water, so that a geothermal power plant could operate economically at variable load. A suggestion has been put forward by Charropin *et al.* [93] that natural aquifers could sometimes be used for large scale thermal storage. This raises the possibility of re-injection of bore fluid by pumping (so as to re-compress the steam) into the aquifer during off-peak hours so that fluid need not be wasted, but conserved for peak load duty at the cost of pumping energy. The economics could be doubtful, but the proposal would seem to be worth investigation. If submersible pumps become a practical proposition (see Section 9.11) it would only be necessary to reduce their output at off-peak times and to circulate sufficient fluid through a rejection well to ensure that the bores are kept hot.

(iv)   *For supplying small quantities of secondary peak loads.* (By 'secondary peak loads' are meant small flat-topped strips of load in the duration curve below the extreme peak.) Non-condensing turbines are cheap in capital but extravagent in heat consumption: they therefore have some of the characteristics of peak load plants if heat be regarded as analogous to fuel. The suitability of using such turbines for supplying small quantities of secondary peak loads has been analysed [92] and shown to be sometimes an economic proposition, particularly where fuel is very expensive. The reason why only small quantities of secondary peak loads can normally be supplied lies in the undesirability of blowing to waste steam that cannot be absorbed during off-peak hours, as explained in sub-section (i) above. Of course, if some industrial or other off-peak use for heat can be found, possibly with the help of thermal storage, then there would be no reason why a dual purpose plant should not be devised for supplying peak load power with little, if any, base load component.

(v) *Emergency or standby plant* [92]. Non-condensing geothermal turbines can usefully serve as emergency or standby plants, in much the same way as diesels are sometimes used in conjunction with conventional power plants. Their high steam consumption for the short time required would be immaterial: their cheapness in capital cost would be their great asset. Moreover, in times of temporary shut-down there will be an abundance of steam available, so that no separate bore(s) need be set aside for supplying such power units.

(vi) *Hot rock circulating systems.* If, and when, power can be extracted from artificially fractured rocks by circulating water through the fissures (see Chapter 19), it will be possible to adjust the rate of circulation so as to match the heat supply to the pattern of demand, at any load factor, without introducing problems of bore instability or of fluid wastage at off-peak times. This could prove to be a valuable facility.

10.16 By-product hydro-power

Circumstances can sometimes arise where geothermal waters can be used 'non-geothermally', as it were [92]. In wet geothermal fields large quantities of hot water are discharged with the steam from the bores. If this water contains toxic substances that could endanger downstream drinking water or irrigation supplies it may not be permissible to discharge it into water courses, in which case re-injection (see Sections 10.4 and 17.4) may be necessary. Re-injection might also be advisable to avoid ground subsidence. Very often, however, huge quantities of waste water will be discarded. Where economically feasible, it is clearly advisable first to extract as much heat as possible from the water before throwing it away, either for power generation or for industrial or other purposes. Regardless of whether or not this is done, the discarded water may possess considerable potential energy if the site of the geothermal field is high above sea level; and if the terrain is suitable much of this potential energy may be recoverable by passing the water through hydro-turbines. Indeed, at Wairakei, the bore water is discharged (regrettably, hot) into the Waikato River and finds its way to the sea through a 'staircase' of hydro-power plants arranged in series along the river bed. In this way about $2\frac{1}{2}$ MW of continuous base load is being earned as a by-product of the geothermal installation – a bonus of about 2% on the geothermal energy output, equivalent to an annual saving of about 5000 tons of oil fuel (or its coal equivalent). Elsewhere, according to the configuration of the land, discharged bore waters could perhaps be fed to a separate hydro plant, even though it might not be practicable to use the full drop to sea level. The hydraulic energy thus generated could be used either for base load or for peak load purposes. As with all hydro projects, the potentialities would depend largely upon the terrain. Some geothermal fields are situated in mountainous regions – e.g. El Tatio, in the high Chilean Andes – and substantial hydraulic dividends could perhaps be won in such cases.

There are three ways in which byproduct hydro power could be used:

(i) *for base loads*. The discharge of the bore water is continuous and at a fairly steady rate. Without providing any storage beyond some small forebay pondage it would be possible to generate continuous base load, (Fig. 59a).

(ii) *for peak loads*. By providing storage at the level of the geothermal field for, say 20h flow, and by letting the water down through the turbines for 4h during the peak load time, six times as many kilowatts could be

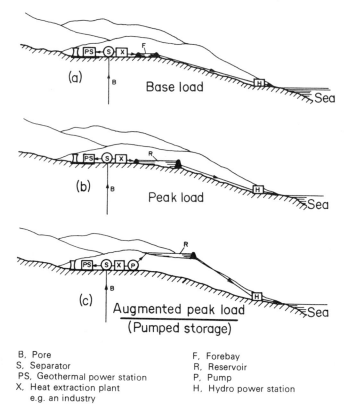

B, Pore
S, Separator
PS, Geothermal power station
X, Heat extraction plant
  e.g. an industry

F, Forebay
R, Reservoir
P, Pump
H, Hydro power station

Figure 59 Byproduct hydro-power plant from wet geothermal field. (After Fig. 4 of [92].)

generated, though without any change in the number of kilowatt-hours, (Fig. 59b).

(iii) *for augmented peak loads*. If the terrain is such that the construction of a high level reservoir is possible, the principle of pumped storage could be used to give still more peak kilowatts at the cost of some base load kilowatt-hours, (Fig. 59c).

10.17 What is a viable geothermal power field?

The question is often asked, especially during the exploration stage, as to what is a commercially viable geothermal field for power generation. Clearly there can be no precise answer, as so much will depend upon the costs of alternative power sources and on other local factors. Nevertheless, as a rule-of-thumb, one well known firm of consulting engineers lays down the following broad conditions as an approximate guide (as in 1976):

(i)   The fluid temperature at the bottom of the bores should be at least 180° C.

Plate 20   The Geysers field, California. Interior view of building housing turbine units Nos 9 and 10 (2 × 55 MW). (By courtesy of the Pacific Gas & Electric Co.)

Plate 21 The Geysers field, California. Power station and cooling towers serving units Nos 5 and 6 (2 × 55 MW). Note the difficult terrain of incompetent rock at the construction site. (By courtesy of the Pacific Gas & Electric Co.)

(ii)   This, or greater temperatures, must be found at depths not exceeding 3 km.

(iii) The yield from a 24½ cm (9 5/8-in) bore should be at least 20 tons/h of *steam*.

10.18 Descriptions of existing geothermal power projects

For details of such projects the reader is referred to bibliography references:

Cerro Prieto, Mexico [164, 189].
Geysers, California [190, 191] and Plates 20–23.
Hatchobaru, Japan [70].
Larderello, Italy [192, 193].
Matsukawa, Japan [194].
Otake, Japan [148].
Paratunka, USSR [72].
Pauzhetka, USSR [90].
Wairakei, New Zealand [68, 195] and Plates 10–17 and 24–26.

Plate 22 The Geysers field, California. Power station and cooling towers serving units Nos 9 and 10 (2 × 55 MW). Photographed in dry weather. Note that the picture was taken before the enforcement of the anti-pollution laws. Note also the ejector discharge stack above the turbine house roof. (By courtesy of the Pacific Gas & Electric Co.)

Plate 23 The same as Plate 22, but photographed in humid weather. Note the greatly increased quantity of vapour by comparison with Plate 22. Note also the 'rigid' expansion loops in the steam pipes. (By courtesy of the Pacific Gas & Electric Co.)

Plate 24 The A nd B geothermal power station buildings, Wairakei, New Zealand. Note the clean lines and absence of the unsightly features associated with fuel-fired power stations, such as chimney stacks, smoke, fuel and ash-handling equipment and fuel storage areas. (By courtesy of the New Zealand Ministry of Works and Development.)

Plate 25 Interior view of the A station turbine house, Wairakei, New Zealand [195]. Note the high-pressure and intermediate pressure turbosets to the left: these six units are all back-pressure sets. In the far background a glimpse can be caught of the low pressure condensing units. In the middle distance are the intermediate pressure steam manifold and inter-stage separators. (By courtesy of the New Zealand Ministry of Works and Development.)

Plate 26 30 MW, 50/0.5 psig pass-in turbo-alternator unit in the B station at Wairakei, New Zealand [195]. (By courtesy of the New Zealand Ministry of Works and Development.)

11 Geothermal space heating, domestic hot water supplies and air conditioning

11.1 Present situation

A brief outline of the development of geothermal space heating and domestic hot water supplies has been given in Chapter 1. The rate of world development in this application cannot be shown quantitatively, as in the case of power generation, owing to the lack of published figures, but some informative data are shown in Table 8 concerning two countries that are at present in the lead in this activity. Although the Hungarian figures are small by comparison with those of Iceland, it should be remembered that Hungary is a fairly recent new-comer in the field of geothermal space heating whereas Iceland has had about half a century of development. Moreover Iceland lies in the seismic belt: Hungary does not.

In 1975 3,500 dwelling apartments, having a total space of some 710,000 m³, were served with geothermal heating in Hungary, and a 10-fold increase is planned for completion by 1985 [95]. The Hungarian Government forces the pace of development simply by withholding oil and gas supplies from their new state-built dwellings in the thermal areas [95].

Well over half the population of Iceland is now served with geothermal domestic heating [7] and it seems likely that within a decade or so nearly all the dwellings in that country will enjoy this service except a few isolated villages and homes.

The oil fuel saved by this particular application already amounts to about 350 000 t annually in Iceland, and is likely to exceed half a million p.a. by 1980. The value of such savings to the national balance of payments in a country almost entirely devoid of indigenous fuel is immense. In Hungary, which possesses some oil and gas reserves that can now be diverted to other purposes, geothermal domestic heating was already saving the nation about 36 000 t of oil p.a. in 1975, and this figure will rise steadily.

Smaller geothermal domestic heating supply systems are to be found in Japan [96, 97], New Zealand [98], the USSR [99], the states of Idaho [9] and Oregon [100] in the USA, and even in the sparsely populated regions

Table 8 Some quantitative data concerning domestic geothermal space heating and hot water supplies.

Country	Category	Heating demand (*MW, thermal*)	Approximate annual heat consumption (*GWh, thermal*)	Annual load factor
Iceland [94]	In service in 1976	557	2581*	
	Under construction or firmly planned before 1980	267	1237*	52.9%
	Tentatively planned beyond 1980	289	1339*	
	Totals, in service, planned and contemplated	1113	5157*	52.9%
Hungary [95]	In service 1975	57.5	262.8 ⎫	52.2%
	Planned for completion by 1985	517.5	2365.2† ⎬	
	Totals, in service, and planned	575	2628†	52.2%

* Estimated on the assumption of a constant annual load factor which, in 1975, has been deduced at 52.9% from the year's energy consumption in Iceland for space heating purposes of 2200 GWh(thermal) and demand of 474.7 MW(thermal) [7]. It is of interest to note that the annual load factor thus deduced is very nearly the same as that attained in Hungary in 1975.
† Estimated on the assumption that the known load factor of 52.2% in 1975 will be maintained in future years.

of the Himalayas, India [101]. An interesting development has recently taken place at Melun in the Paris Basin, France, where heat is extracted from a *non-thermal* area by using abandoned (unsuccessful) oil borings, sunk to depths of 1.5 to 1.8 km, into the hot permeable 'Dogger' aquifer in which waters at 55 to 70° C are to be found [102]. These waters are raised to the surface by displacing them with partially cooled 'used' waters pumped back into the ground at some distance from the production bores, so as to maintain a circulation system. 2000 dwellings were served in this way with heat in 1975, and the system is now being extended to supply 3300 dwellings at Melun. (There could perhaps be potentialities for using abandoned or abortive oil wells in this way in other parts of the world).

All these smaller systems will doubtless expand fairly rapidly in the near future and there will certainly be new space heating developments elsewhere. In fact the great success of the Melun scheme has already led to similar developments in other parts of the Paris Basin (see Section 11.2) while very ambitious plans are afoot for other heating systems of this type with the ultimate aim of serving 400 000–500 000 dwellings in various other parts of France. By comparison with the present world production of geothermal electricity, which probably exceeds 10 000 GWh (thermal) p.a. and uses well over 100 000 GWh (thermal) in the process, the level of development of domestic geothermal heating is now small; but the new emphasis upon non-power applications of earth heat may well cause this gap to narrow rapidly.

11.2 Geothermal fields and non-thermal areas exploited for domestic heating

From Fig. 2 (Chapter 2) it will be seen that 80° C is a recommended convenient temperature for domestic heating supplies, though lower temperatures can and are being used. For example, Einarsson mentions temperatures of only 50 to 55° C and upwards [7]. Much depends upon the climate and the quality of the building insulation, and there is also the possibility of boosting the heat in exceptionally severe spells of cold weather by means of supplementary fuel or electricity, or even by heat pumps. Allowing for a reasonable temperature drop in transmission, wellhead temperatures should be higher than that of the delivered heat; but it is generally true to say that good *semi-thermal* fields can be used for domestic heating, as well as *hyper-thermal* fields. Both types of field are in fact used, while in the Paris Basin even a *non-thermal area* is being exploited by virtue of the depth of the drillings.

In Iceland, thermal waters for domestic heating are mostly drawn from below ground at temperatures ranging from 80 to 120° C from semi-thermal or from slightly hyper-thermal fields such as are conveniently encountered close to, and even within, the capital city of Reykjavik. In one case a field has been operating for over 30 years at an effluent temperature of only 56° C, though this is mainly used for mixing with hotter waters. In general, those fields of low or moderate temperature produce waters of fairly low mineral and gas content, and it is of interest to note that although new thermal fields have been brought into service to cater for the expanding demand growth, the Icelandic fields originally exploited are still providing heat at undiminished output after nearly half a century of service [7]. In some of the newer contemplated Icelandic heating projects hyper-thermal fields are to be tapped, containing some rather aggressive saline fluids at 167° C or thereabouts: in such cases dual purpose power/heating projects are being planned, for which heat exchanger/mixers will have to be used so that reasonably pure water may be circulated for district heating purposes [103].

In Hungary, which lies at some distance from the seismic belt, semi-thermal fileds are being used for domestic heating. It is estimated that such fields underly some 40% of that country's area [95], and in 1970 the recoverable heat from the Hungarian plains alone was reckoned to be about half as much as that contained in all the world's then known reserves of oil fuel [104]. Temperature gradients in Hungary are frequently about 50 to 60° C/km, while the heat flow may range from about 2.0 to 3.4 μ cal/cm² sec [95], both sets of figures being about double the world averages. The semi-thermal fields beneath Hungary are known to extend into Czecho-slovakia, Romania, Yugoslavia and Austria – even into Switzerland, though no serious exploitation of their heat has yet begun in any of those countries except for balneology. Hot water issues from the Hungarian fields at 60 to 100° C [95] from depths of about 1 km, though the aquifer extends downwards to about 3 km in depth. Below these semi-thermal fields some geopressurized fields are known to exist in places, with temperatures up to 200 to 300° C, and although these could perhaps be exploited for various purposes including power generation, they have not yet been developed. The semi-thermal fields of Hungary, for the most part, contain waters of low salinity, [95].

In Japan waters at about 70° C are being transmitted from semi-thermal fields over distances of up to 12 km [105]; and various types of pipe laminated [97] with synthetic materials, are used in preference to steel pipes so as to avoid corrosive troubles.

At Melun, France, as has already been said, heat is being recovered from a non-thermal area having a temperature gradient of only about 30° C/km. Owing to fairly high salinity contained in the subsoil, the hot waters may contain from 8 to 30 g/l of salts and traces of H_2S. Heat-exchangers are therefore used so that clean secondary water may be circulated through the public heating system [102]. The further developments in the Paris Basin referred to in Section 11.1 will, as a first stage, provide heating for about 25,000 dwellings, a quantity of greenhouses and 50,000 m² of floor space in offices and public buildings in four suburban areas around Paris [198]. These schemes have several interesting features. First, like Melun, they are all in non-thermal areas having temperature gradients of about 30 to 35° C/km. Secondly, unlike Melun, they will bear the cost of their own drillings (as no 'free' oil borings are available); and yet they are expected to save from 16 to 20% in annual costs by comparison with fuel heating. Thirdly, in one of the schemes (Creil), *heat pumps* are being used – not to raise the feeding temperature but to augment the hot water supply. This is achieved by re-cycling part of the used water, which leaves the radiators at about 40° C, and pumping into it the heat content of the remainder so as to raise the temperature of the recycled water to 70° C and lower the temperature of the rejected water to 7° C. An arrangement of three heat pumps in 'cascade' enables a performance ratio of 5:1 to be achieved. Fourthly, by placing the production and reinjection bores side by side at Villeneuve-la-Garenne, and by splaying out the drillings in the manner shown in Fig. 18 so as to strike the aquifer at points separated by about 1 km, a reasonably long life of the system is assured and great economy of surface pipework effected: also, since the wells, heat pumps and other ancillary equipment can be accommodated within an area of about the size of a tennis court, land use is minimised and opportunities are offered for aesthetic concealment behind shrubs. These Paris projects are expected to provide at least 70% of the heating requirements; the balance being supplied at peak times by old existing fuel-fired boilers. It is of interest to note that the geology and temperature gradients of the Paris Basin are somewhat similar to those of West Hampshire in England, where hopes have been expressed [138] of winning low grade earth heat.'

At Boise, Idaho, water at 77° C is found at a depth of only about 130 m [9] while at Klamath Falls, Oregon, well temperatures range from 38 to 110° C at depths of 27 to 580 m [100]. Such temperatures suggest the presence of hyper-thermal fields, though it is not known whether the high gradients persist to greater depths. While the Boise field is chemically fairly pure, that of Klamath Falls is too aggressive for direct use, so that downhole 'hairpin' heat-exchangers have to be used.

At Rotorua, New Zealand [98] hyper-thermal gradients are exploited to depths of 15 to 366 m, yielding temperatures of 49 to 177° C – the latter at a pressure of about 13 ata, which suggests artesian conditions, as this pressure exceeds the saturation pressure at that temperature. Here again, heat exchangers are necessary owing to the aggressive nature of the fluids.

Economically exploitable semi-thermal fields are believed to underlie 50 to 60 per cent of the vast area of the USSR [18], although inadequate temperatures and/or high salinities throw some doubt upon the aptitude of the word 'economically'. Domestic geothermal heating has been developed in that country [99, 106, 107]. Heat exchangers often have to be used in the USSR on account of the chemical aggressiveness of the natural thermal fluids [99] while the use of heat pumps, electricity and supplementary fuel has been proposed in certain instances [106] for peaking purposes or for boosting the temperature of the recirculation water in two-pipe systems. The use of heat pumps of course reduces the amount of heat discarded to waste.

As explained in Section 5.13, semi-thermal fields of 100° C or less will not flow spontaneously except where artesian or geopressurized conditions occur, though thermo-syphonic action may sometimes sustain a pump-initiated flow. Hot water for space heating often therefore has to be lifted from the subsoil by means of submersible pumps which, at moderate temperatures, are a practicable proposition. In some instances, however, the water will spout of its own accord under natural forces.

Aquifers in hyper-thermal, semi-thermal and even non-thermal areas are now being used, as shown above, for domestic heating in several countries.

Many semi-thermal fields are known to exist elsewhere, and more hot aquifers will almost certainly be discovered at depth in non-thermal areas. Bearing in mind also that artificial fields may well become available soon (see Chapter 19) the future outlook for this particular application of earth heat would seem to be very promising. It is emphasized that, although in this chapter 'domestic' heating has generally been referred to, the heating of commercial, administrative and industrial buildings is also of course included.

11.3 The technicalities of geothermal district heating and hot water supply

These technicalities have been admirably expounded by Einarsson [7, 105]. Space does not permit a very full presentation of all their aspects in this chapter, but some of their principal features are described below.

11.3.1 *Duration curve*. Fig. 60 is a typical annual duration curve which shows the statistical distribution of daily temperatures in one particular climate. The optimum room temperature in a dwelling is represented by t_r – say 20–22° C – while the design room temperature will be t_d, a figure less than t_r by a small amount Δt known as 'free heat' – i.e. heat lost by the occupants, by electric lighting and appliances, sunshine, etc. The area OKLB will then be a measure of the 'imported' heat necessary to ensure comfort, while the area LFE represents (in degree-days) either the amount of

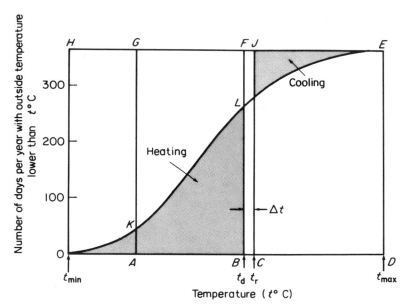

Figure 60   Space heating and cooling annual duration curve. (After Fig. 2 of [105].)

discomfort felt in hot weather or alternatively the amount of heat to be removed by artificial cooling if comfort is to be maintained throughout the year. For the present, heating only is under consideration. To effect the required heating wholly by geothermal means would necessitate the provision of a system of capacity proportional to OB (integrated for the entire number of buildings in the heated district, and allowing for transmission losses) operating at an annual load factor of $OKLB/OHFB$ – or about 30% as drawn. Alternatively a smaller geothermal system, with fewer wells, proportional to AB, and operating at an annual load factor of $AKLB/AGFB$ – or about 42% as drawn – could be installed. The higher load factor would clearly introduce great economies, but the smaller system would leave the problem unsolved of how to deal with the area of extreme cold OKA if people are not to shiver in the depths of winter.

11.3.2 *Peak heating.* One way of dealing with this area OKA is to provide supplementary heating by means of fuel. The actual amount of heating to be added is small – OKA is only about 7% of the total required amount of artificial heating required, namely the area $OKLB$, as drawn, so that even if fuel is expensive per tonne the total fuel bill would not be intolerably high. The important point is that the capital cost of boiler plant in terms of its peak thermal output is much lower than that of the additional wells, pumping equipment, etc., that would be required if the equivalent maximum heat demand were to be met from geothermal heat. Hence a combination of geothermal base load heating to supply the area $AKLB$ with fuel heating to supply the area OKA for peak demands can achieve an optimum economic arrangement (as with base load and peaking power plants). Supplementary electric heating or even heat pumps could sometimes provide alternative means of meeting the peak heating load, but much would depend upon local energy prices. Fuel will usually prove to be the best solution.

The amount of supplementary heating can be reduced by making use of a characteristic of the aquifer, which permits over-pumping for short very cold periods to draw excess heat from the underground reservoir by lowering the level of the water table, and allowing it to recover later when the heat demand is lower. This, however, as well as the use of large storage tanks for containing hot water, is more of a short-term provision for covering *diurnal* variations of heat demand rather than seasonal variations. In Reykjavik, supplementary fuel, storage tanks and overpumping are all used to achieve an economic optimum for the entire heat supply system.

11.3.3 *Domestic central heating.* In Iceland the temperature of the water entering a dwelling is usually maintained at about 80° C by mixing the waters from the hotter and the less hot bores (and with recirculated waters – see below) in suitable proportions. Some of the lower temperature waters flow by gravity from hot springs into storage tanks, while the hotter well waters are raised by means of submersible pumps so as to prevent flashing (with consequent chemical precipitation problems) until after mixing has been achieved and the temperature reduced. In earlier times in Reykjavik a once-through single pipe system was used, in which the cooled water after fulfilling its task in the heated buildings, was discharged into the sewers. More recently a two-pipe system is being used, in which the cooled (but still not cold) water is recirculated and mixed with very hot well water for re-use. The heat requirements of buildings will of course depend upon their purpose, the local climatic conditions, the quality of thermal insulation of buildings and the proportions of window to masonry and roof areas, the thermal insulation of foundations, whether a building is detached, semi-detached, terraced or a flat in an apartment block. Within a dwelling the living room will usually require more heating than bedrooms, passages, kitchen, etc. Einarsson [105] quotes an overall figure for Reykjavik of 19 kcal/h m^3 with an external temperature of $-10°$ C and an average internal temperature of 20° C. This is equivalent to about 22 W/m^3. In a more temperate climate the requirement would be less.

11.3.4 *Domestic hot water.* Tap water for baths and washing is of course supplied by the once-through single pipe system, and the used water is discharged to waste.

11.3.5 *Transmission.* Although this subject should perhaps have been covered in Chapter 9 it is thought to be rather more appropriate to this chapter. All system pipework in Iceland is underground. Pipes are of welded black mild steel laid in concrete channels and insulated with rock wool or aerated concrete if the pipe diameter is 75 mm or more. For smaller mains and house service connections the pipes are insulated with polyurethane foam in the annular space between the steel pipe and an outside protective jacket of high density polyethylene pipe [105].

11.3.6 *Metering.* Domestic heat consumption can be metered either simply by volume of water entering the premises, assuming a constant entry temperature, or alternatively by means of a heat meter which integrates the product of temperature and flow, so as to give a more accurate measurement of heat.

11.3.7 *Sundry points.* The above technical descriptions have of course been over-simplified so that the main principles may be demonstrated. Such secondary considerations as stored heat within the fabric of a building acting as a 'smoother-out' of demand, the load density within the area of supply the automatic achievement of temperature control by varying the mixture proportions of new thermal waters of different origins and of recirculated water according to demand and climatic variations, the variations of 'free heat' from day to day according to the hours and intensity of sunshine, the effects of high winds upon building heat losses, the decreased efficiency of transmission and of radiators with increased load if fed with water at constant temperature; all these factors, which cannot here be discussed for lack of space, must of course be taken into account when designing a public heating system.

11.4 Environmental advantages of geothermal heating systems

One feature common to all geothermal public heating systems is the virtual freedom from pollutive problems. Reykjavik, for example, is a smokeless city enjoying clean air. The aesthetic possibilities of locating production and reinjection wells side by side have already been mentioned in connection with the Paris suburban developments described earlier.

11.5 Air conditioning

As a corollary to space heating, certain hot climates need air conditioning of buildings for the maintenance of comfort. In former times in such countries as India, the human discomfort suffered in the very hot season was slightly mitigated by the use of *punkhas* or electric fans which did no more than stimulate evaporation from the human skin, or of *khaskhas tatties* – curtains of a specially fragrant grass suspended over windows and doorways and sprayed with water from time to time. Such devices could only help to bring the human environment near to the wet-bulb, rather than the dry-bulb: temperature, and thus act as a slight palliative. In very dry climates, where the difference between these two temperatures was small, these devices became more or less useless: in fact the effects of a fan could give rise to greater discomfort by drying the eyeballs than could be offset by the very slight relief it brought to the temperature. From the 1920s mechanical air conditioners have largely replaced these crude amenities in the larger tropical cities and more prosperous homes in outlying places, and have undoubtedly done much to mitigate the trials of life in intensely hot climates. Most of these air conditioners work on the mechanical compressor system, requiring external work usually supplied by electricity; but some work on the absorption principle which requires a *heat* supply. Geothermal heat can provide that requirement.

11.5.1 *Duration curve.* Fig. 60 can apply equally to cooling as to heating, except that in a hot climate the ordinate CJ (or t_r) will be shifted to the left: in fact in some tropical places such as Bombay, Colombo, Singapore, etc., where artificial heating is never, or almost never needed, the ordinate CJ will not be far from the left-hand edge of the diagram, OH, and the cooling work area FLE will be very much greater than in Fig. 60 – the heating work area OBL more or less vanishing altogether. In variable climates of extreme winters and summers, such as New York, the ordinate CJ will occupy a fairly central position in the diagram, so that both heating *and* cooling will be needed for comfort.

11.5.2 *Small units and central systems.* Geothermal airconditioning has hitherto been effected by means of small, or fairly small, units for cooling a single building, apartment or perhaps a small complex of buildings. It has been proposed by Einarsson [7] that whole tropical cities could be served by a public cooling network much in the same way, but 'in reverse' so to speak, as the Reykjavik district heating system. When Managua, Nicaragua, was almost totally destroyed by the devastating earthquake of 1972, Einarsson suggested that such a system might form a feature of the reconstructed city.

11.5.3 *Geothermal cooling in the USSR.* It is reported [18] that lithium bromide absorption machines with a capacity of $2\frac{1}{2}$ million kcal/h of 'cold' – rather less than 3000 kW thermal – using geothermal heat are being mass-produced in the USSR, capable of two-way operation so as to give heating winter and cooling in summer.

11.5.4 *Geothermal air conditioning in New Zealand.* The International Hotel at Rotorua is served by a plant that provides the whole building with all its heating and cooling requirements from a single geothermal bore – hot tap water for baths and washing, central heating in winter and air conditioning in summer [6]. The climatic range to be contended with is from $-4°$ C to $+30°$ C. The maximum heating load is 0.5 Gcal/h (about 585 kW thermal) while the lithium bromide absorption cooling unit requires an input of 0.575 Gcal/h (about 668 kW thermal), requiring 1.47 heat units per single heat unit of cooling. The bore serving the whole of this complex produces water at $150°$ C and about 7 ata. The heat is transferred in a heat-exchanger to fresh water closed circuits which are heated to $120°$ C for the heating system and for the absorption unit.

12 'Other' uses for geothermal heat

12.1 General

It so happens that earth heat has up till now been exploited mainly for electricity generation and for district heating. In terms of energy, these two applications have accounted for the overwhelmingly greater part of the geothermal heat that has, until now, been harnessed in the service of man. Nevertheless, as stressed in Chapter 2, there is a growing awareness of the immense versatility of this form of energy and of its potential economic market, which people have been slow to recognize until fairly recently. It is the present predominance of electricity and district heating that has justified the devotion of a separate chapter of this book to each of these two applications. It would clearly be impracticable to treat all other applications in the same distinctive way, for even now the variety of other possible uses is very great, as hinted at in Chapters 1 and 2. Limited though the present use of earth heat may be for purposes other than electricity generation and district heating, there are already in existence too many types of geothermal installation for more than limited space here to be devoted to them, and the list of new applications is continually growing. For this reason, and since every application involves its own specialized technology, all these 'other' applications must share a single chapter in which little more than a cursory commentary can be bestowed. Broadly, these uses may be classified as follows:

Industry
Farming
Miscellaneous

Each of these broad classifications embraces a wide spectrum of temperature, as can be seen from Fig. 2 (Chapter 2), but whereas most (but not all) industries require moderately high grade heat in the form of steam, many farming and miscellaneous applications can be served by lower grade heat in the form of hot water. The distinction between the three groups is also not always clearly defined: for example, is food processing a branch of farming or of industry? (It has here been treated as 'industry'.)

12.2 Industry

The criterion on which the suitability of earth heat for an industry can be judged is its degree of heat intensiveness, which may be expressed as the number of kilogram-calories absorbed in producing finished product to the value of $1 US, having due regard to the *grade* of heat required. Lindal [8] draws a parallel between the great hydrolectric developments of the early twentieth century which attracted electricity-intensive industries from far and wide to places where very cheap power was available (c.g. the aluminium industry) to the probable appearance within the near future of large geothermal developments which will produce such cheap and abundant heat that it will pay to establish heat-intensive industries close to the available heat more or less regardless of location. This argument which is expanded in Section 15.15.1 may indeed be sufficiently powerful to overcome the one disadvantage from which geothermal energy suffers when compared with hydroelectricity, namely, the fact that it is seldom economic to transport heat over more than 20 or 30 kilometres – at least in the quantities of interest in present times. Electricity, on the other hand, can be economically transported over hundreds of kilometres (in sufficiently large blocks). Thus, whereas the siting of an electricity-intensive industry can compromise to some extent with a source of cheap power, a heat-intensive industry must perforce have to be sited close to a geothermal field, wherever it may be. This fact could perhaps be turned to social advantage in that it could lead to the habilitation of unpopulated waste lands that are now regarded as uninhabitable, and thus help to solve to a small extent the problem of the world's population explosion. In the past, there has been a regrettable tendency to write off remote geothermal fields as worthless owing to the difficulty of heat transportation; but if heat is sufficiently abundant and sufficiently cheap, thermal fields in the most unlikely and remote places could become economic for certain very heat-intensive industries. If this is true, it follows that less heat-intensive industries could be economically justified if the source of heat is close to the required raw materials and labour.

Lindal [8] lists 35 random industrial processes, mostly chemical, which show a ratio of about 45:1 in their heat-intensiveness (which he expresses in pounds of steam required to produce product to the value of $1 US. Although his actual figures will doubtless now have been overtaken by inflation, their relative proportions are probably still valid. Nineteenth in his list, in descending order of heat-intensiveness, was the Kraft process for producing paper pulp; and yet this process is being used in the world's largest existing geothermal industrial installation – the mills of the Tasman Pulp & Paper Co. at Kawerau, New Zealand. This is an illustration of a less heat-intensive industry being economic where the raw materials (forests in this case) and labour are available close to a thermal field. Many chemical industries are several times as heat-intensive as the Kraft pulp process – e.g. the production of caustic soda, ethyl alcohol, acetic acid, lactose and heavy water.

If we should ever succeed in winning very high temperature geothermal steam from great depths, in natural or artificial fields, the scope of application will be further extended despite the inefficiency of power generation as such, because certain industries which need either high grade heat or electricity cannot be served by the relatively low grades of heat now available.

The number of industries that could probably be served with geothermal heat is immense, and is likely to increase with time. To record them all would in itself involve a major work of research. The list which follows therefore makes no claim to being complete – in fact many of those mentioned by Lindal [8] do not even appear in it – but is indicative of the very great scope offered. Many of them are already practised: others have only been considered.

12.2.1 *Chemical industry*. Many chemical industries are heat-intensive without requiring very high grade heat. Some of them are as follows:

(i) *Extraction of sodium and magnesium salts from sea water* [108–114].

(ii) *Extraction of valuable minerals from geothermal fluids* – e.g. lithium [113], bromine and the chlorides of potassium and calcium [8, 111] from 'normal' geothermal waters. A wide variety of other useful substances is believed to be commercially extractable from the highly concentrated hot brines found beneath the Imperial Valley, California [114]. Geothermal gases may also contain useful constituents such as H_2S, CO_2, CH_4, B, NH_3 and other vapours [115, 116].

(iii) *Recovery of elemental sulphur deposited in volcanic craters, fumaroles and hot springs* [110]. See also Section 1.1.3.

Note: The value of substances recovered from sea water, geothermal fluids and geothermal surface manifestations is often of an intermediate nature. Some of them, of little intrinsic value in themselves, can be used for the production of much more valuable materials such as metallic sodium, magnesium and titanium, also sulphuric acid and chlorine (which can be used for a host of other chemical processes [114] when further treated with chemical additives, electricity and additional heat. Multiple-effect evaporation is often used for items (i) and (ii): very hot geothermal fluids are self-evaporative.

(iv) *Heavy water production.* This was once proposed for Wairakei as part of a dual-purpose project, using a counter-flow water distillation process for producing the heavy water. There are other methods of production, of which the hydrogen sulphide-water isotope exchange process is favoured as the most economical by Valfells [117]. The potentialities of this application will largely depend upon 'fashions' in the nuclear power industry.

(v) *Sulphuric acid from elemental sulphur and H_2S occurring in geothermal discharges.*

(vi) *Fermentation of molasses to produce ethyl alcohol, butanol, acetone, citric acid and other products* [8].

(vii) *Production of proteins, vitamins and ammonia* [18]

12.2.2 *Mining and upgrading of minerals*

(i) *The production and refinement of diatomaceous earths.* This is being practised as a highly successful industry at Lake Myvatn, near Námaf-jall, Iceland [8].

(ii) *Production of boric acid by applying steam heat to ores.* This method is being practised at Larderello superseding the older, less economical method of recovering the boric acid from the fumarole vapours [8].

(iii) *The facilitation of mining operations in permafrost areas* – e.g. in Siberia, where the climate is so severe that without the introduction of extraneous heat mining operations would be impossible [18].

(iv) *The drying of peat.*

(v) *The production of alumina from bauxite by the Bayers process* [8].

12.2.3 *Food processing*

(i) *Cane sugar production*, thus releasing the bagasse, formerly used as a fuel, but which is now a valuable raw material used for other manufacturing purposes.

(ii) *Sugar beet manufacture*, requiring 3300 to 5500 kcal/kg of sugar [118].

(iii) *Production of powdered coffee.*

(iv) *Production of powdered milk.*

(v) *Crop drying and dehydration of foods.*

(vi) *Fruit and juice canning.*

(vii) *Production of cattle meal from Bermuda grass.*

(viii) *Rice parboiling.*

(ix) *Fish drying and fish meal production.*

(x) *Freeze drying of foodstuffs* [119].

12.2.4 *Various industries*

(i) *Rayon manufacture.*

(ii) *Other textiles.*

(iii) *Pulp and paper manufacture* [8, 118].

(iv) *Timber seasoning* [8, 120], as practised in New Zealand.
(v) *Veneer manufacture* [120], as practised in New Zealand.
(vi) *Production of cotton seed oil.*
(vii) *Manufacture of aggregate cement building slabs* [8].
(viii) *Brewing*, as practised in Japan [8].
(ix) *Synthetic rubber manufacture* [18].
(x) *Refrigeration and gas liquification* [121].
(xi) *Dye industry.*
(xii) *Lurgi process for total gasification of coal*, if sufficiently high pressure steam is available.
(xiii) *Sinter extraction* (alum) [96].
(xiv) *Plastics industry.*
(xv) *Process steam* for many types of industry.

12.3 Farming

This includes all forms of husbandry in the widest sense – agriculture, horticulture, the raising and breeding of livestock, wool production, etc. (food *processing* has been treated as an 'industry' under Section 12.2. A brief general review of this application of earth heat has been given in Section 1.1.6, where it was pointed out that only low grade heat (up to about 80° C maximum) is required. There could be a tremendous future for geothermal farming, particularly if the ultimate goal of recovering earth heat at any desired place on the globe should ever be realized; for it could render habitable immense areas of the world that have access to fresh water but which are too cold to support farming activities unaided, thus bringing great relief to the ever-growing pressure of the population explosion. One of the principal farming applications is greenhouse heating, and this is really just a form of space heating; but as the heating requirements of humans and plants differ considerably, both as to light and heat, it is more convenient to treat greenhouse heating as a branch of farming. Just as there is an overlap in the classification of greenhouse heating, so does refrigeration fall within the scope of industry and of farming in that it can be used both for industrial purposes and also for food preservation. The principal applications of earth heat to farming are as follows:

12.3.1 *Greenhouse heating* (8, 18, 95, 96, 101, 107, 110, 121, 122, 123 and 124). This is widely practised in Hungary, Iceland, Japan, New Zealand, USA, the USSR and even in the Himalayan recesses of India and probably elsewhere. With the aid of earth heat, fruit, vegetables, flowers, potted plants and even cacti can be raised under glass or plastic, which admits sunlight and at the same time retains much of the solar warmth as well as the artificially introduced geothermal space heating. Double glazing (or double plastic covering) would doubtless improve the efficiency of the process but it has yet to be shown whether this would be economically worthwhile. In Iceland, not only have such produce as tomatoes, cucumbers, lettuces and mushrooms (commonly grown under natural conditions in temperate climates) been raised, but also tropical and subtropical fruits such as bananas (Plate 5), papayas, pineapples, melons and grapes (Plate 4): also exotic flowers like orchids. The annual quantity of heat required will of course depend upon the crop to be raised, the quality of the greenhouse construction and on the local climate. In Iceland, greenhouses generally consume a maximum of 300 to 350 kcal/h/m² of floor area and impose an annual load factor of 30 to 35% [121]. Under British climatic conditions a *mean* figure of about 52 kcal/h/m² has been quoted [138] which is considerably lower than the implied mean for Icelandic conditions (about 90 to 120 kcal/h/m²). This is doubtless due in part to the less rigorous climate, and perhaps also to the proposed cultivation of less ambitious crops. As with space heating, fuel-fired peaking heaters may sometimes give better economic results than catering for the very worst conditions by reliance upon geothermal heating alone. Since it is possible, by means of geothermal heat, to create virtually any climate desired – even sunlight can be simulated by means of electric lamps capable of emitting light of the required wavelength – seasonal limitations that apply in nature can be overcome, so that almost any crop can be raised at any time of year; although naturally it will be more economic to raise crops at times when the local climate can offer the greatest possible encouragement. By 1975 1 700 000 m² of greenhouse area were geothermally heated in Hungary, and 140 000 m² in Iceland. Hungary has a much colder winter and a much hotter summer than Iceland, so the annual load factor of greenhouse heating can be expected to be lower in Hungary. At present, agriculture consumes about $8\frac{1}{2}$ times as much geothermal heat as domestic space heating in Hungary, which emphasizes the importance attached by the Hungarian Government to geothermal farm aids.

Gutman [123] has described how a small hydroponic installation has been established in North-Eastern California at a height of 1250 m above sea level, where the winter temperature sometimes falls to minus 33° C, aided by natural artesian thermal spring yielding hot water just below atmospheric boiling point. Cultivation is effected beneath corrugated plastic giant barrelled cloches placed parallel with one another at 13 ft (about 4m) spacing so that they do not shade one another from the sunlight. Humidifiers are necessary in summer. Some idea of the economic importance of geothermal greenhouse heating may be gathered from the fact that in 1967 Iceland produced about 64% of her vegetables and 98% of her cut flowers and seedlings with the aid of earth heat. In a country where the potato is the only vegetable that can be grown naturally, this application of geothermal heat is of immense importance. The same could doubtless apply to many other places in the world having inhospitable climates.

12.3.2 *Soil heating* [107]. Artificial soil heating can be effected either by means of submerged hot pipes or by the use of warmed irrigation water, or both The example of Agamemnon has been cited in Section 1.1.6., but more closely controlled use of warm thermal waters for irrigation is practised in Japan, Italy, the USA and the USSR. Soil heating can of course be combined with greenhouse heating. It is not known whether soil *cooling* has ever been attempted by means of geothermally activated cooling plants, but the author suggests that this might well prove to be an attractive proposition for growing fruits and vegetables (naturally raised in temperate climates) in hot tropical countries in partial shade – e.g. strawberries.

12.3.3 *Nursery gardens and botanical gardens.* By the joint use of greenhouse heating and soil heating (and/or cooling) the growth of seedlings, saplings, flowers, shrubs, etc. could be encouraged by means of geothermal heat in climates that would naturally to hostile to their requirements. Several botanical gardens have been established in Japan [96] in this way.

12.3.4 *Soil sterilization.* The heating of soil to sufficiently high temperature, by means of geothermal heat, can effectively sterilize it against insect and bacterial pests before the soil is used for cultivating seedlings. This is practised in Japan [96].

12.3.5 *Crop drying* [8, 125]. Grass, seaweed and other crops can be effectively dried by means of geothermal heat. This is practised in Iceland, the USSR and elsewhere.

12.3.6 *Animal husbandry* [8, 96, 110, 126]. Farm buildings, such as cattle stalls, pigsties, stables, dairies and hen coops may be geothermally heated in winter for the greater comfort and health of farm workers and animals. This is particularly effective in poultry raising. Komogata *et al.* [110] claim that the yield of poultry meat can be raised $2\frac{1}{2}$ times by weight in terms of fodder consumed, by means of geothermal heating. Pig food may be cooked from various wastes, milk can be pasteurized, chicks can be incubated, biodegradation of organic wastes may be effected, wool may be washed and dried, – all by means of geothermal heat.

12.3.7 *Fish breeding.* Hatcheries for trout, eels and other fish may be served by means of only slightly heated waters – down to about 20° C. Cooling water effluent from geothermal power plants could be used for this purpose unless they are chemically contaminated.

12.4 Miscellaneous
Several other useful functions can be served by geothermal heat which do not fall within the categories covered by Chapters 10 and 11 or of the

second and third sections of this chapter. Some of these functions are mentioned as follows, but doubtless other applications will be thought of in time.

12.4.1 *Distillation* [127, 128, 129]. With the increasing rates of population growth and of water consumption per head, due to rising living standards and increased industrialization, the supply of adequate quantities of fresh water has long been becoming a matter of growing urgency in many parts of the world. Suitable sites for rainwater catchment reservoirs are becoming more difficult to spare in highly populated countries like Britain. Other densely populated countries like California can at present rely to a limited extent on long distance transportation of water from remote catchments. Newly found prosperity from oil wealth has led to great population growth in almost totally arid areas in the Middle East. Recourse is becoming increasingly necessary to the desalination of sea and brackish waters. Desalination is a heat-intensive process that can be most effectively achieved by means of multiple-effect or flash evaporators. Since a high performance ratio plant costs much in capital but little in heat, and *vice versa*, it follows that where heat is expensive a very high performance plant can be justified, but where heat is cheap a lower performance ratio is acceptable economically. Geothermal heat is generally very cheap, so that distillation plants of low capital cost could be used in conjunction with it. Desalination requires heat of moderate grade – say 120° C – so that hyperthermal fields would be more or less necessary for efficient geothermal desalination. The chances of the coexistence of a hyperthermal field and an arid, but highly populated, area are not great, but such places could well be found in California, Mexico, North Africa, the Canary Islands, Greece and perhaps in parts of the Middle East. Moreover, if and when we ever succeed in creating artificial thermal fields in semi-thermal or even non-thermal areas (see Chapter 19) the opportunities for geothermal desalination will become very great indeed. It is even conceivable that we shall one day succeed in distilling sea or brackish water economically, not merely for drinking purposes but also for *irrigation* – a feat that has not yet been achieved.

12.4.2 *Balneology and crenotherapy*. These applications have been touched upon in Section 1.1.1. Chiostri[130] states that in 1971 in Italy alone 15 million people were treated at more than 200 thermal clinics for various ailments, while in Russia more than 10 millions are medically treated annually with thermal waters. The Japanese [96] have an even more impressive record, claiming that about 100 million visitors annually frequent the hotels located near the 1500 (approx.) hot spring resorts in their country. To what extent crenotherapy, or the healing of certain ailments by the external or internal application of thermal mineral waters, owes its popularity to faith cannot be determined, but in general the medical profession has given its blessing to the practice. At any rate it is an undisputed fact that thermal balneology and crenotherapy are responsible for a gigantic tourist trade of immense commercial value to those countries that are fortunate enough to possess the necessary facilities. The fact that many thermal areas occur in scenically attractive places may account for part of this trade. Geothermally heated swimming pools can also be a good source of revenue to well situated communities and hotels.

Plate 27 Alligator-breeding in the Atagawa tropical gardens, Japan, by means of geothermal heat [196]. (By courtesy of the Japan Geothermal Energy Association.)

12.4.3 *Zoology*. In Japan (Plate 27), the breeding of alligators and tropical fish in geothermally heated waters is proving to be a great attraction to tourists in Peppu, Atagawa and Nagashima. In combination with geothermally heated greenhouses displaying tropical flora, these alligator and fish farms are becoming a great sight-seeing attraction.

12.4.4 *Bottled mineral waters*. Many hot springs in widely scattered parts of the world contain minerals with alleged medicinal properties. A large international trade in these waters, after bottling, has been practised for more than a century and is still growing.

12.4.5 *Cooking*. In parts of Japan, and in the island of Lanzarote (Canaries) cooking is performed on a small scale by means of geothermal steam and vapours.

Cooke [131] describes a modern adaptation of the ancient Maori *hangi* – a steam cooking device for use with natural steam. These are extensively used in New Zealand for cooking food both for humans and for animals.

12.4.6 *Prevention of freezing of fire-fighting water*, and *the melting of snow on roads* in winter. Both these adaptations of geothermal heat are practised in Japan [96, 100].

12.4.7 *Scenic attractions*. Many natural geothermal phenomena possess both scenic beauty and great curiosity value. Tourists are thus attracted from far and wide, bringing revenue to the district – e.g. Yellowstone Park, Wyoming, USA; the geysers of Iceland; Rotorua, New Zealand; the thermal areas of Japan and elsewhere.

12.5 Conclusion

It is hoped that this rather cursory coverage of 'other uses' may serve to underline the great versatility of earth heat that was emphasized in Chapter 2.

There is one important consideration that must never be lost sight of with all forms of geothermal exploitation, namely the somewhat limited transportability of heat by comparison with that of electricity or of piped oil or gas. There can be no precise answer to the question 'How far may heat be economically transported?'. So much depends upon the temperature of the heat to be transported, on whether the heat is conveyed by means of steam or hot water, on the local climatic conditions, on the cost of alternative energy sources, on the chemical ingredients of the transported fluid, and on such factors as the nature of the terrain to be traversed. Hot water can generally be transmitted over far greater distances than steam. For example, although the permissible velocity of steam is about 20 times that of hot water, and although the enthalpy of dry saturated steam at, say 180° C, is 3.642 times that of water at the same temperature, a given pipe could transport 2 37 times as much heat at that temperature in the form of water as in the form of dry saturated steam, because the water would have a density 172.6 times that of the steam. Also the proportional heat loss from piped hot water is far less than from piped steam, for a given pipe and given temperature. Furthermore the loss of pressure, which can be avoided or limited quite cheaply by pumping when transporting water, can result in a serious loss of energy when steam is piped over long distances. Steam for geothermal power installations has been transported at about 14 ata over 3½ km at Wairakei, while hot water for district heating has been transported over 18.3 km in Iceland [132]. These distances are not necessarily economic limits, as will be seen from Section 15.14 and from Fig. 67. In fact Bodvarsson [133] even talks of hot water transmission being conceivable over 50 to 100 km under Icelandic conditions for space heating, though he gives 10 km as a probable maximum economic distance for transporting steam. It seems likely that for our present requirements it would not pay to transport geothermal heat, even by means of hot water, over more than a few tens of kilometres in quantities of practical interest and at the maximum temperatures now attainable. It is possible that in future we may have access to far hotter fluids and that we may need them in much greater quantities than at present, in which case the economic transportable distance will increase; but it could never compete with electricity or piped fuels in this respect. Even now this need not be regarded as a serious obstacle if the heat is sufficiently cheap, and it will become totally unimportant if we ever succeed in winning deep-seated heat in non-thermal areas wherever we will: the real hope for geothermal energy lies in its *cheapness* (see Chapter 15). Scale-forming chemical ingredients in thermal fluids can sometimes impose a fairly stringent limit on the transportable distance of geothermal heat.

13 Dual and multi-purpose projects

13.1 Scope for dual or multi-purpose projects

The potentialities of dual and multi-purpose projects for the efficient exploitation of earth heat were briefly mentioned in Chapter 2. There is not really a great deal to add to what has there been said. It is of course the very wide spectrum of usable temperatures and also the frequent availability of two-phase fluids that provide the great theoretical scope for such projects, as shown up to some extent in Fig. 2 (Chapter 2), and the range of possible combinations of different uses of earth heat could be immense. More multi-purpose plants have hitherto been talked about than have actually been realized, though a few are now in the process of development. It is nevertheless only fair to recognize that despite their obvious ideal advantages in principle there are often considerable difficulties in establishing them in practice. The fact that the two or more end products must be produced in the same locality can raise difficulties of relative marketability, and perhaps of raw material availability in the case of industries. Disparity of demand fluctuations can also introduce problems of storage or of inefficient heat usage.

13.2 Present examples

These, at present, are almost non-existent. At Larderello boric acid is being produced in addition to electric power, but this is virtually a *separate* industry since the same steam is not used for the two purposes. Likewise at Namafjall in Iceland, power is generated and diatomite is mined and processed, but again by separate steam. Neither of these dual industries, therefore, is a true example of dual-purpose geothermal plants. In Japan, greenhouse heating is combined with the raising of alligators. This is a true dual-purpose enterprise, though in terms of heat not very impressive despite its originality. At Kawerau, New Zealand, the exhaust from a 10 MW back-pressure turbine is fed into the pulp and paper processing plant [8]: this too is a true dual-purpose plant.

13.3 Planned and discussed examples

In contrast with the very few working examples of dual-purpose plants, several projects have been planned, or at any rate considered: some of them are understood to be under construction. One of the first, which would have been impressive had it materialized, was the original project for a joint chemical/power project planned for Wairakei in the mid-1950s. High pressure (h.p.) steam was to have been expanded in back-pressure turbines (Fig. 61) through a pressure range of 13.25/4.4 ata to generate 13 MW of electric power. The exhaust from these turbines was to have been used in a twin tower chemical distillation plant to produce heavy water. This chemical plant was to have absorbed a pressure drop from 4.4 to 1.05 ata, yielding

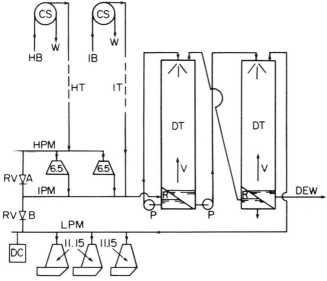

HB, h.p. bore fluid	Only one of many bores of each type shown
IB, i.p. bore fluid	
W, waste hot water	
HT, h.p. steam transmission system	
IT, i.p. steam transmission system	
HPM, h.p. steam manifold (13.25 ata)	
IPM, i.p. steam manifold (4.4 ata)	
LPM, l.p. steam manifold (1.05 ata)	
DT, heavy water counter-flow distillation towers	
R, re-boiler	
P, pump	
V, rising vapour	
DEW, deuterium-enriched water to heavy water concentration plant	
R.V., reducing valves	
DC, dump condenser	

h.p., high pressure
i.p., intermediate pressure
l.p., low pressure

Total power generated		
2 h.p. units × 6.5 MW	=	13 MW
3 l.p. units × 11.15 MW	=	33.45 MW
		46.45 MW
Originally proposed...		
2 i.p. units installed in place of heavy water plant		22.30 MW
		68.75 MW
Further turbines ordered later		123.45 MW
Present capacity		192.2 MW

Figure 61 Simplified diagrammatic arrangement of heavy water/power dual purpose plant originally proposed for Wairakei.

low pressure (l.p.) steam that was to have generated further power in condensing turbines which would have produced $33\frac{1}{2}$ MW of additional electricity. Unfortunately the heavy water plant was cancelled after the power plant had been ordered, owing to changed conditions in the heavy water market, and the middle pressure range was absorbed by installing 22.3 MW of additional back-pressure turbines so that the project became a 'power only' enterprise, which was at the same time expanded from the originally planned capacity of $46\frac{1}{2}$ MW to 192 MW by the addition of about 123 MW of further turbine capacity.

At Paratunka in Kamchatka, USSR, there are plans to use exhaust steam from electricity-generating turbines to heat 80 000 m² of greenhouses that are expected to yield 2000 to 2500 tons of vegetable produce annually [121]. Tikhonov *et al.* [18] discuss the possibility of combining space heating with mineral recovery. Wong [134] and Werner [114] propose to combine desalination with mineral recovery from geothermal brines in the Imperial Valley, California. Behl *et al.* [101] describe a plan to develop the fruit growing industry in the Parbati Valley in North India by installing a geo-thermally activated 100-ton refrigeration plant for preserving the fruit, at the same time using the waste heat for space heating and for a swimming pool in an establishment that will be frequented by tourists and pilgrims. Kunze *et al.* [9] talk of combining space heating in Boise, Idaho, with greenhouse heating and warmed irrigation. The possibility of combining power generation with desalination has been raised on more than one occasion [127, 128] and other dual or multi-purpose projects have been mooted. Thus, although there is yet little to show for it, a great deal of thought has been, and is being, devoted to the multiple use of earth heat. It may confidently be expected that such activities, though they have been slow in starting, will be much in evidence within a few years.

13.4 Flexibility of choice

In theory there is a large degree of flexibility in the sharing of a geothermal resource between two or more modes of exploitation. High pressure steam could be expanded through two, or even three, temperature ranges, each devoted to different end products, while water/steam mixtures offer an even greater variety of manipulation. The steam could be used for one or more purposes: so could the associated water. Pass-out or back-pressure turbines could provide a wide range of adjustment in the relative quantities of steam available at different temperatures and pressures for different purposes. If the demand for hot water is less than its availability in association with the required steam supply some of the surplus could be flashed into steam in one or two stages, and the flash steam could be fed back into the steam system for power generation or other purposes. The low grade heat finally discharged from the major processes could often still be used for fish hatcheries, swimming pools, de-icing purposes, soil-warming, fermentation, or perhaps, if of sufficiently high grade, for domestic or greenhouse heating. The possibilities are endless.

13.5 The problem of demand balancing

If the patterns of demand for the various processes in a dual or multi-purpose project should differ from one another, it will be necessary to provide either storage or bypassing facilities, or both. The latter would be wasteful of power potential and would involve the degrading (but not necessarily the *loss*) of heat. For example, Fig. 61 shows the originally proposed Wairakei power/heavy water project reduced to its simplest terms. The fact that the hot water was to have been rejected to waste at the wellheads need not here be considered except to note that such rejection of a fluid that could – technically if perhaps not commercially in the particular context – be used for power generation or other useful application is a regrettable waste of energy: a waste which alas still persists. At the plant both the turbines and the distillation plant were to have been operated at the highest possible load factor, so that power generation and heavy water production would have been virtually continuous processes. The pressure ranges and plant capacities were so chosen as to balance one another exactly. Under normal operating conditions maximum use would have been made of the steam. The fact that shipments of heavy water would have been intermittent would have been provided for simply by means of *material* storage: no *energy* storage would have been needed. At times, however, a turbine would have had to be shut down for maintenance or due to a breakdown: so too would the distillation plant have had to be taken out of service from time to time. In so far as *planned* 'outages' are concerned, the shutting down of both types of plant simultaneously would have made for the most efficient operation. But in the event of an *unplanned* outage, or if one type of plant required a longer time than the other for maintenance, means had to be provided for keeping as much as possible of the serviceable plant in operation when only one part of the plant was shut down. This requirement was provided for by means of bypass reducing valves between the h.p. and intermediate pressure (i.p.) steam manifolds, and from the i.p. to the l.p. steam manifolds, (valves A and B respectively), and by a dump condenser for bypassing a l.p. turbine. If, for example, an h.p. turbine were shut down, the use of valve A would have enabled full steam supply to be maintained for the continued working of the distillation plant and the condensing turbines. Likewise, the shutting down of the distillation plant need not have affected the power output if sufficient steam were bypassed through valve B. If one of the l.p. turbines had to be taken out of service there would have been insufficient means of 'swallowing' the full quantity of exhaust vapour from the distillation plant. Both the output of the h.p. turbines and of the heavy water plant would have had to be reduced if no means had been provided for disposing of the excess l.p. steam over and above the requirements of the two surviving condensing turbines. This surplus could have been blown to the atmosphere, but this would have caused nuisance close to the plant. A dump condenser was therefore provided capable of absorbing the steam requirements of one condensing turbine, so that the unwanted steam could be unobtrusively rejected as water to the river. This dump condenser is still in service with the 'power only' Wairakei installation. The use of the reducing valves would have involved the degradation of heat, while the use of the dump condenser would have incurred a total loss of heat. In the example chosen, neither the reducing valves nor the dump condenser would be used except on comparatively rare occasions and these causes of inefficiency would have been a small price to pay for the ability to keep a maximum of plant of either type in service at all times.

This chosen example is a particularly simple one as it involved two base load processes. But where a base load process is associated with another process of fluctuating demand, or where two or more processes having fluctuating demands of different time incidence patterns are to work in combination with one another, the inefficiencies of bypassing or of dumping could be appreciable except where it is possible to provide storage facilities to absorb the fluctuating demand differences of the component processes. *Energy* storage, except on a small scale, can be difficult to provide; but *product* storage can be easier (as in the case of heavy water cited above). The end product of one component process of a multi-purpose project might, for example, have a seasonal demand; but if this end product is a material capable of storage it might be possible to provide sufficient storage space to enable the production process to operate at a demand pattern exactly matched to that of its associated process(es), so that differences of output and demand could be absorbed by the product storage capacity. Energy wastage resulting from imperfect matching of process demand patterns, however, is likely to be small by comparison with the wastage likely to result from single-purpose projects, so that the possibilities of establishing dual or multi-purpose projects should always be given serious consideration.

14 The control and safety of geothermal installations

14.1 Control: general

As shown in Chapter 8, every geothermal bore will have its own individual characteristic relating mass flow and pressure: no two bores will behave in exactly the same manner. Where several bores simultaneously supply thermal fluids to a utilization plant, whether for power generation or for other purposes, it is necessary to devise a system of control whereby the bores, despite their different characteristics and allowing for possible changes in these characteristics, will collectively provide the required quantities of fluids at the required pressures and with a minimum of waste, and whereby any unwanted fluids are automatically disposed of. In a dry field there may also be variations in the degree of superheat, which can generally be accepted as a dependent variable of little importance requiring no special means of control. In a wet field, the dryness fraction of the bore (yet another variable) is normally dealt with by wellhead separators from which the two component phases are separately controlled. Thus the control of steam and the control of hot water may here be treated separately. Armstead and Shaw [63] have shown in some detail how both fluids may effectively be controlled so as to enable a utilization plant to function smoothly and correctly. If the use of submersible pumps or the adoption of two-phase fluid transmission (see Sections 9.11 and 9.12) should become regular practice, the problems of control should become greatly simplified. The description which follows therefore deals with the most arduous conditions that need be considered.

14.2 Steam control

With a *conventional* steam power plant the main task of the steam control system is to ensure that the turbine receives whatever quantities of steam it may need for meeting the load with minimum variation of temperature and pressure at the turbine entry. This is achieved by varying the rate of fuel feed and the quantity of combustion air accordingly. In contrast with a boiler, a geothermal steam bore cannot be controlled in this manner. Not only is there no fuel nor combustion air to be regulated but it is impossible to vary the rate of steam flow from a bore without at the same time altering the wellhead pressure and, as already mentioned, each contributing bore will have its own individual pressure/flow characteristics. How then is it possible with a number of bores working in a parallel, each having different (and perhaps changing) characteristics, to ensure that the plant receives the amount of steam it needs at the required pressure? Within this primary problem three subsidiary problems may be discerned, namely:

(i) That of primarily ensuring that the bores deliver the required quantity of steam at adequate pressure.
(ii) That of primarily ensuring that a more or less constant pressure is maintained at certain points.
(iii) That of ensuring an economic balance of steam flows.

14.2.1 *Steam quantity primarily assured.* The upper part of Fig. 62 shows a hypothetical steam transmission system in which four bores A,B,C, and D are connected to a pipeline feeding a plant X, which may be a power plant or some other steam-consuming process. Let it be assumed that for its full performance the plant requires a quantity of steam Q at a pressure P, and that the characteristics of the four bores are such that they can just provide these requirements – no more and no less. Under these conditions the pressures at the plant, at the bores and along the main and branch pipelines would be somewhat as shown by the full line (1) in the lower part of Fig. 62. Now let it be assumed that one of the four bores, say C, is disconnected. This

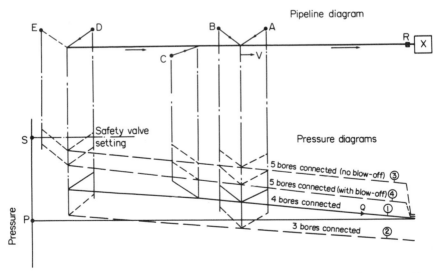

Figure 62 Steam supply control: steam quantity primarily assured. (From Fig. 1 of [63].)

would have the effect of lowering the pressures at all points of the system somewhat as shown by the broken line (2) in Fig. 62. Although the reduced pressures at the bores would cause some increase in the steam yields from each of the remaining three connected bores, the total steam delivered at the plant would necessarily be less than when all four bores had been connected. Thus the supply of steam to the plant would be deficient both in quantity and in pressure.

If C be now restored and a fifth bore, E, connected to the system, the pressures at all points would rise. In order to limit both the admission pressure and the quantity of steam at the plant to the required values it would be necessary to throttle the steam by means of some pressure-reducing device R. As this device would have some inherent 'regulation' some small rise in the admission pressure above P would have to be tolerated. The pressures throughout the system would now be somewhat as shown by the broken line (3) of Fig. 62. It could well happen that the bringing in of the fifth bore would raise the pressures at some of the wells above the setting pressure S of the wellhead safety valves, thus causing these valves to lift. This would be undesirable for the reasons given in Section 9.6, and should be prevented except in the last emergency by blowing off a controlled quantity of steam through a suitably designed vent valve V placed at some convenient point on the pipeline (Plate 13). By thus blowing off a sufficient amount of steam the pressures throughout the system could be reduced to such levels that no wellhead safety valve need lift at all (see the broken line (4) of Fig. 62). The more the blow-off, the lower will be the pressures at all points. Obviously it would be highly improbable in practice for a precise number of bores to give exactly the required amount of steam at the required pressure (as in line (1) of Fig. 62). In order to ensure that the plant may receive what it needs it will always therefore be necessary to connect rather more bores than are theoretically needed to achieve this, particularly if allowance is to be made for the changing of bore characteristics with time and for the possible loss of a bore through blockage, its removal for testing or maintenance or for any other reason. Hence it will nearly always be necessary to throttle the steam at entry to the plant, to blow off some steam to waste, or both. Of these unavoidable procedures, throttling is the less wasteful in energy. The provision of vent valves automatically controlled by the pressure in the steam pipeline close to their point of installation can also prevent excessive pressure rises if the utilization plant steam requirements should suddenly fall – e.g. if a turbine in a power plant should trip off load.

The throttling device R must be designed so as to restrict the steam flow as soon as the downstream pressure rises above a predetermined (and adjustable) level. With a turbine this can quite simply be incorporated with the governing gear: it would have the effect of limiting the steam flow virtually to the normal full load consumption regardless of any excess upstream pressure, and thus preventing the overloading of the alternator. The vent valve should preferably be sited close to the steam main but fairly far from the utilization plant, because if the whole plant should suddenly shut down through some mishap the vent valve would have to blow off the full load plant steam requirements into the atmosphere, and this could be tolerable only at some fairly remote point. Silencers should be provided to muffle the noise from vent valve discharges.

The purpose of the vent valve is solely to prevent the lifting of wellhead safety valves. Its setting should be such that the following requirements are fulfilled:

(i) When the plant is at full performance, either no steam at all, or a bare minimum of steam, should be discharged to waste. Whether it is possible to avoid altogether the discharge of steam will depend upon the number and characteristics of the connected wells and on the pressure drops along the various parts of the transmission system.

(ii) When the plant is shut down, though with the full quota of connected wells as for full performance, the vent valve must be capable of discharging the whole of the bore steam without the pressure at any point in the system rising to a level where a safety valve would lift.

Further details of the setting and performance of vent valves have been given by Armstead and Shaw [63].

14.2.2 *Constancy of steam pressure primarily assured.* Such a requirement could apply to a steam receiver (Fig. 63) *into* which is fed exhaust steam x from some plant unit X, and *from* which is drawn a steam feed y to some other plant unit Y. (An excellent practical example would be the i.p. steam manifold in Fig. 61). Constancy of pressure in the steam receiver at the required level P can only be assured if the bores supply $y - x$ of steam directly to the receiver. Under steady conditions let it be assumed that this constancy of pressure and balance of flows is achieved by feeding steam from four bores as shown in Fig. 63: the pressures along the pipework and at the bores would then be as shown in the lower part of that figure by the full line (1). Now let the balance be disturbed by a sudden cessation of x. The only way of preventing a fall of pressure in the receiver would be to raise the steam flow from the field to the plant to a value of y, thus restoring the balance of flows; and the only way of bringing about this higher flow would be to raise the pressures in the bore field so as to impel more steam along the pipeline. However, higher field pressures would result in *lower* steam yields from the bores, so the additional steam would have to be found from elsewhere. This could be arranged by connecting a vent valve to some convenient point along the pipeline. If, under normal conditions (line (1) of Fig. 63) this vent valve were discharging to waste a quantity of

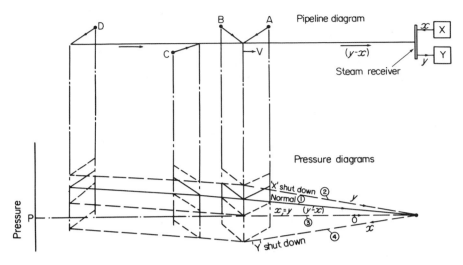

Figure 63 Steam supply control: constancy of steam pressure primarily assured. (Fig. 2 of [63].)

steam somewhat in excess of *x*, and if this valve were to close as soon as the inflow from X is cut off, pressures in the field would rise, as shown by line (2) of Fig. 63, and steam would be diverted from waste to the plant. Similarly, if Y were suddenly shut down the pressure in the receiver would rise unless the whole of the quantity *x* could be diverted elsewhere. This could be done by opening the vent valve sufficiently wide to enable the field pressures to fall, as shown by line (4) of Fig. 63, thus bringing about a reversal of flow along the pipeline from the plant to the field. If at any time *x* and *y* became equal to one another, no balancing flow would be needed to or from the field and the vent valve would have to discharge enough steam to ensure that the pressure adjacent to it was equal to P, as shown by line (3) of Fig. 63.

Although only the extreme conditions of the total shut-down of X and Y have here been considered, a satisfactory control system would have to cater for any variation of either *x* or *y*, however small. Control may be effected by using the pressure at the steam receiver to actuate the position of the vent valve so that a fall or rise in pressure may cause the vent valve to move towards the closed or open position, as may be required. Under transient conditions some momentary rise or fall of receiver pressure would be inevitable, but the characteristics of the control system must be such that the pressure deviation is kept within acceptable limits. If the vent valve were sited at the plant (and the only objection to doing this would be the nuisance of large volumes of escaping steam), simple proportional control with a very sensitive setting would be satisfactory, but if it were sited remotely from the plant this type of control would be unstable owing to the time lag inherent in a long stretch of pipeline. A 'zero error' two-term type of control with a small proportional and a large integral term can give satisfactory control. The addition of a derivative term could give rise to hunting. The temporary pressure deviations at the receiver will depend upon the magnitude of the disturbance to the balance of flows.

The hypothetical example here chosen would involve an unacceptable waste of steam under normal conditions (something more than x), but it would seldom be necessary to cater for such a drastic condition as the total shutdown of X, or at least it might be possible under such a condition to curtail the value of y. It would then be possible to arrange for satisfactory control with only a moderate 'normal' discharge through the vent valve. If so, the control would give a greater range of operation in one direction than in the other, for if the valve is normally only slightly open it cannot move very far in the closed direction. If a disturbance exceeded the capability of the vent valve to deal with it, automatic action would have to be supplemented by manual adjustments. For further details of this system, see [63].

Vent valves may consist of single or multiple units arranged to operate in series or in parallel with one another. This applies both to those required for the assurance of steam quantity and of constant steam pressure. At Wairakei, where both systems are adopted, each is provided with two valves operating in parallel. This allows for limited control if one valve should be taken out of service for maintenance, after suitably adjusting the setting of the other valve remaining in service.

14.2.3 *Economic flow balance assured.* This question has already been considered to some extent in Section 13.5, and has been described at greater length by Armstead and Shaw [63]. 'Economic' flow balance implies that as far as possible the removal or curtailment of any one plant unit in a multiple assembly of plant units shall not affect the performance of any other plant unit. The required balance of flows can be obtained by manual manipulation of reducing valves and dump condensers, but more speedy operation can sometimes be achieved to some extent by automation. For example, the opening of reducing valves could be effected by proportional control from the downstream pressure of a parallel plant unit. Back-pressure governing could also prevent dangerous pressure rises occurring in low pressure pipework, at the same time helping to avoid the lifting of safety valves at the plant. There will generally be a limit to what can be achieved automatically: the bringing in of a dump condenser for example, would require too many operations for automatic actuation. A combination of automatic and manual operational control will usually be needed.

14.3 Hot water control [63, 67]

The problems of hot water control are of two kinds. First there is the automatic disposal of unwanted hot waters collected at the wellheads in wet geothermal fields, and this has been covered by Section 9.4. The more difficult problem of controlling piped hot water at near-boiling temperatures will now be discussed.

In Section 9.9 the dangerously explosive nature of superheated water was emphasized. This alone would be sufficient to justify very careful control when transmitting such fluids, but further controls are necessary to ensure that supply and demand of the hot water are continuously reconciled. As mentioned in Section 14.1, the advent of a successful submersible pump which would enable superheated water to be transmitted at pressures far exceeding the saturation pressure corresponding with the temperature would greatly simplify the problem of control: so too would the adoption of two-phase fluid transmission (Section 9.12). The only known case where slightly pressurized water at 200° C or more has been transmitted successfully over $1\frac{1}{2}$ km or so was the experimental pilot hot water scheme described in Section 9.9, where the needs for pumping, attemperation and controlled valve movement speed were emphasized. The method of controlling that transmission system is thought to be of sufficient interest for it to be briefly described here, as it is always possible that a similar scheme may be introduced elsewhere in the future, should hopes of submersible pumps and of two-phase transmission not be entirely fulfilled.

The Wairakei experimental scheme is illustrated in Fig. 64 which shows one (of several) h.p. and one (of several) i.p. water collection drums, with the steam spaces connected to their appropriate steam transmission systems. A low lift pump, X_1 raises the h.p. water into the 'floating' head tank HT, (mentioned in Section 9.9) which serves not only as a means of pressurizing the water transmission pipe to the power plant and of taking up short term fluctuations of hot water yields, but also of providing means of actuating the control of the flow rate of hot water into the first-stage flash vessel at the power plant. Hot water from an i.p. collection tank is injected into the hot water transmission line by a pump, X_2 having higher delivery head than X_1 (equal to the sum of the static pressure from the head tank (Plate 17) plus the pressure difference between the two steam systems). Both pumps are of the self-regulating cavitating type, like the extraction pumps normally provided in conventional power plants for removing the condensate from the condensers, which will deliver into the hot water pipe whatever quantities of water are supplied to the collection tank, up to the maximum pump rating. Should this capacity be exceeded, the water level in the collection tanks would rise and spill the excess water to waste by means of a float-actuated relief valve HLR. At the plant end of the transmission main is an

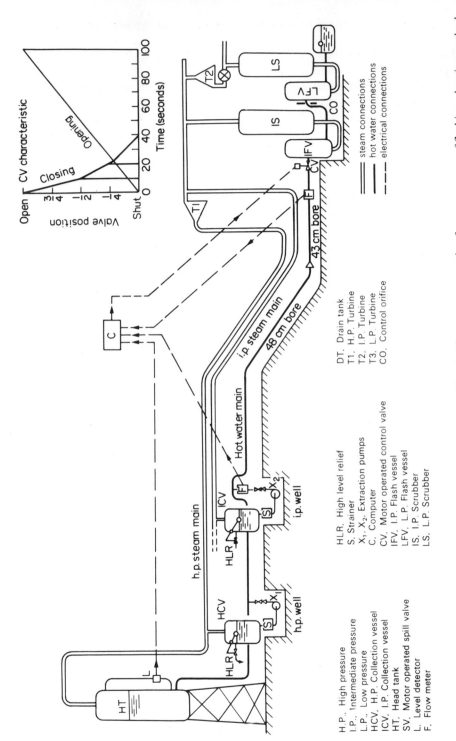

Figure 64 Experimental hot water transmission scheme tried out at Wairakei for power generation from two stages of flashing, showing method of control. (From Fig. 4 of [67].)

H.P., High pressure
I.P., Intermediate pressure
L.P., Low pressure
HCV, H.P. Collection vessel
ICV, I.P. Collection vessel
HT, Head tank
SV, Motor operated spill valve
L, Level detector
F, Flow meter

HLR, High level relief
S, Strainer
X₁, X₂ Extraction pumps
C, Computer
CV, Motor operated control valve
IFV, I.P. Flash vessel
LFV, L.P. Flash vessel
IS, I.P. Scrubber
LS, L.P. Scrubber

DT, Drain tank
T1, H.P. Turbine
T2, I.P. Turbine
T3, L.P. Turbine
CO, Control orifice

automatic control valve, CV, with a flowmeter, F, just upstream of it. Under steady conditions, this control valve will take up a position that allows the passage of exactly the total quantity of hot water, h.p. and i.p., being pumped into the system: the level in the head tank will then remain steady. As soon as a difference arises between the hot water *supply* from the pumps and the hot water *demand* as determined by the position of the control valve, the water level in the head tank will rise or fall accordingly. This change of level is detected by a level-sensitive device L, which sends a signal to a computer C, which automatically adjusts the control valve position in such a manner as to restore the balance between inflow and outflow. This is effected by calibrating the head tank water level so that under steady conditions its height is proportional to the flow through the control valve. A rise or fall of head tank water level will therefore immediately cause the computer to detect a discrepancy between the *apparent* flow, as registered by the head tank level, and the *true* flow as registered by the flow meter. Any such discrepancy causes impulses to be sent to the control valve causing it to change its position until the discrepancy disappears and a new steady state is achieved with a different head tank level. The relationship between the detected discrepancy and the corrective control valve movement is a two-term one with a proportional and an integral element. (As with the constant pressure steam control system described in Section 14.2.2 above, the addition of a derivative term could give rise to 'hunting'.) A special mechanical drive ensures a characteristic valve movement as shown in the inset to Fig. 64, so as to limit surge heads induced by the valve movements to acceptable values. The rate of opening must be much slower than the rate of closure because the negative surge heads induced by water acceleration tend to cause boiling in the pipeline. The rate of closure may generally be more rapid, but this too must be restricted as the valve approaches the closed position because the reflected surge heads which follow the moment of closure will depend upon the rate of valve movement just before that moment.

As the transmitting capacity of the pipeline will depend upon the quantity of attemperating water injected into the system, it is necessary to curtail the volume of transmitted water if the amount of attemperation should fall below a certain critical figure. Hence a further signal must be sent to the computer from a flow meter that summates the total flow of i.p. water, and when this signal indicates an insufficiency of attemperating water an appropriate sequence of tripping off h.p. well pumps is initiated so as to ensure adequate curtailment of the total hot water flow according to the degree of attemperation. A hot water transmission system which does not rely on attemperation, but only on pressurizing, could of course dispense with this refinement.

A final measure of control is needed to ensure that the utilization plant can make use of all the hot water delivered to it. On arrival at the plant the hot water is flashed into steam (in the Wairakei experimental scheme) and if and plant is unable to use all of the flash steam it would be necessary to spill some of the hot water to waste. This can be done by means of a remotely controlled spill valve connected to the head tank. The operator can adjust the amount of spill by remote control from the power plant, or alternatively the spilling may be automatic by the actuation of the spill valve from the pressure of the flash steam, which would tend to rise if it were being generated more rapidly than it could be used. The hot water spill valve has a similar function to that of the vent valves in the steam control systems – that of trimming the supply to match the demand.

14.4 Safety precautions: general

In common with all installations that make use of fluids at high temperatures and pressures a geothermal installation must be provided with the necessary means for protecting personnel against injury and plant against damage, both under normal and under emergency conditions. In Section 9.3 the means have been described whereby dust, grit and rock particles ejected with the hot fluids from the bores may be intercepted at the wellheads, while in Section 9.6 the provision of safety devices to prevent the occurrence of excessive pressures at wellheads and in the fluid transmission system has been described, as well as the precautions to be taken in wet fields against the entry of dangerous quantities of hot water into the steam transmission system. These steps, however, do not exhaust all the precautionary measures necessary in a geothermal installation.

14.4.1 *Pressure and vacuum limitation.* Although the bursting discs and safety valves at the wellheads will protect the entire steam collection and transmission system against excess pressures, they can do nothing to safeguard downstream plant and equipment designed to operate at lower pressures. For example, in Fig. 61 it would be necessary to protect the intermediate and low pressure steam manifolds against excess pressures by means of safety valves. Moreover the low pressure manifold must also be protected against *vacuum* conditions; for if the supply of exhaust steam from the heavy water plant (or from the i.p. turbines that later took its place) were to fail, the low pressure turbines would continue to 'suck' steam and would induce sub-atmospheric pressure in the i.p. manifold – even down to condenser vacuum level in the event of total cessation of inflowing steam. Low pressure pipework is of necessity of large diameter owing to the high specific volume of steam, and would normally have thin walls: it would thus be liable to collapse under external atmospheric pressure. The provision of stiffening rings and internal struts could give protection against a moderate fall of internal pressure, but it may also sometimes be necessary to provide vacuum-breaking relief valves to admit air if the pressure drop should exceed a safe limit. The operation of such relief valves should be avoided except in the last extremity, as it could lead to corrosion troubles.

14.4.2 *Water entrainment.* Despite the ball check valves (Fig. 38; Section 9.6) it is always possible that one of them may be defective and that large gulps of water may enter the steam pipework system. At all costs this water must not be allowed to reach the exploitation plant (turbines being particularly vulnerable). If the entry of water is only moderate, the normal pipeline drain traps (Section 9.10) should be adequate to remove it during its passage from the wells towards the exploitation plant; but if excessive, these drain traps may lack the necessary removal capacity. A back-up defence system of strategically placed water-detectors should therefore be provided at intervals along the main steam pipelines. The presence of water at the detector most remote from the utilization plant would simply send an alarm signal to the plant operators to alert them of a potential hazard. If the drain traps can deal with the water present, the trouble may pass and normal conditions restored; but if water should reach the next downstream detector the resulting signal should be used to disconnect *part* of the exploitation plant (e.g. trip off one or two turbines in a power station). This would slow down the velocity of steam in the mains and would give the drain traps a better chance of removing the remaining water in the mains. Should the water still reach subsequent downstream detectors, further plant components would have to be disconnected consecutively. Finally, if water should reach a point fairly close to the exploitation plant entry, *all* plant should be tripped off so as to halt the steam flow altogether. A plant shutdown, though very inconvenient, would be preferable to a disaster. A system of this sort takes progressive and discriminate action, avoiding 'panic' action until the last possible moment. The presence of water detectors at the junctions of branch steam lines with the steam mains would facilitate the location of the source of trouble and thus shorten the time taken to deal with the offending well. Water carry-over into the mains could be due to too small a waste discharge orifice coupled with a sticking ball check valve, or by the failure of the high level spill device in the head tank of a hot water transmission system.

14.4.3 *Gas detection.* Nearly all geothermal steam contains hydrogen sulphide – a lethal gas when present in quite modest proportions. This gas is rendered more dangerous by the fact that it paralyses the olfactory nerves when fairly strongly concentrated. Thus its characteristic and rather unpleasant smell, which is noticeable when it is present in harmlessly low concentrations, may fail to give warning just when such warning is most needed. As the density of H_2S is 18% greater than that of air at the same temperature, it is apt to accumulate in low-lying pockets such as building basements. Methods of limiting the escape of H_2S into the atmosphere will be dealt with in Chapter 17, but as a precaution against possible imperfections of those methods it is essential that all possible 'gas traps' be adequately ventilated. It may also be advisable periodically to take gas samples at all potential danger spots, and perhaps to install gas detectors coupled to alarms; and also to limit the length of time permitted to personnel to stay in danger areas.

14.4.4 *Overspeeding.* The steam turbines of a geothermal plant will normally be designed to run safely at an overspeed of about 10% above synchronous speed (as in conventional steam stations). With a mixed system, however, they may have to operate in parallel with hydro turbines designed to run safely at speeds of up to 40% above synchronous speed. In the event of a system disturbance in which a large proportion of the load is lost, there may be a danger of a synchronized combination of hydro and geothermal plants running away to speeds which, though safe for the water turbines would be dangerous to the steam units. To protect the latter from such a contingency it may be necessary to provide 'overspeed protection' which will trip off the geothermal plant if the hydro-plant tries to run away with it. The setting at which this tripping occurs must be below the safe overspeed limit of 10% but high enough to prevent unwanted tripping under minor system disturbances. About $7\frac{1}{2}$% overspeed is of the right order. In systems containing mostly steam plant, with only a small element of hydro, this precaution may not be necessary; but in New Zealand, where hydro plant greatly predominates – or *did* predominate when Wairakei was built – this overspeed precaution was very necessary.

14.4.5 *Standby corrosion.* Protection against this hazard will be discussed in Section 16.8.

14.5 Monitoring
At some central control point it is necessary to monitor certain essential measurements so that overall operating control may be exercised over a geothermal installation, or group of installations, by means of local and/or remote control. Typical of the basic indications required for this purpose are:

(i) Steam pressures in various pipe manifolds (e.g. HPM, IPM and LPM, Fig. 61) and at key points in the steam transmission pipework, such as at vent-valve locations.
(ii) Steam flows into the plant, summated where several pipes operate in parallel.

(iii) Total kilowatt output, if the utilization plant is a power station.
(iv) Position indicators for reducing valves and certain other important steam valves.
(v) Head tank level and control valve position, and perhaps water temperatures at certain key points, for a hot water transmission system controlled in the manner described in Section 14.3.

In addition to indications, facilities will be required to enable the operators at the plant to take action by remote control, where necessary, for such operations as the following:

(i) Varying the lifting pressure and operating range of the vent-valves for a steam control system as described in Section 14.2.1.
(ii) Varying the zero point and the proportional and integral settings of the vent-valves for a steam control system as described in Section 14.2.2.
(iii) Varying the opening pressure and degree of sensitivity of automatic reducing valves.
(iv) Varying the proportional and integral settings of a hot water control system such as that described in Section 14.3.

Alarms related to safety precautions must also be brought to the control centre to indicate, for example, entrained water in the steam pipes, high pressure in steam manifolds, high water level in the head tank of a hot water transmission system, etc.

As to the bores themselves, the cost of monitoring conditions such as valve positions, steam and water flows, operation of water interceptors, the tripping of hot water pumps, etc., can seldom be justified. Experience of geothermal bores suggests that so long as steady conditions are maintained at the utilization plant the behaviour of individual bores generally remains remarkably constant and the bores can be left unattended for long periods. Operators soon acquire a 'feel' of a geothermal installation and can infer from pressures and steam flows whether, for example, the bursting disc of a bore has ruptured, in which case the field maintenance staff can be instructed to locate and remedy the fault. Routine inspection in the field, with regular logging of pressures and flows, is cheaper than remote instrumentation and usually just as effective. However, good communications by telephone or radio between the control point and the field are imperative, so that any abnormality sensed at the control point can be located and rectified by the field attendants, and so that any abnormalities detected by the field staff can be reported immediately to the control centre and appropriate instructions may be issued. A mimic diagram at the control centre can be useful, so that operators can see at a glance which bores are in service or under maintenance, which valves are open or closed, and what plant units are operating. Such a diagram need not, however, be automatic: hand setting will usually suffice.

14.6 Maintenance

Although briefly mentioned in Section 10.11, it is important here to repeat that with a geothermal installation – especially with an integrated *system* of installations as at Larderello or in the Geysers area – it is most advisable to have a carefully planned scheme of preventive maintenance. It also pays to keep an adequate stock, at convenient places, of spare parts – particularly complete turbine blade assemblies, both fixed and rotating – so that plant outages may be kept to a minimum. Ricci and Viviani [83] and Matthew [84] have described the arrangements adopted at the Larderello and Geysers fields. Limited workshop facilities are available on site at all major geothermal power plants in the world, while for major overhauls it may sometimes be necessary to send faulty plant to the nearest city having the necessary facilities, though such steps seldom have to be taken. At the Geysers a new well-equipped workshop is just about to be established in the field area, capable of dealing with all (even the heaviest) maintenance likely to be needed. Corrosion hazards, a serious cause of troubles in the early days of geothermal development, have become almost negligible by the adoption of suitable resistant materials for the duties required (see Chapter 16). At the Geysers, turbine units are overhauled normally at 2 to 3 year intervals (by comparison with 5 to 8 years for conventional steam plants), but the outage time averages only 2–3 weeks (by comparison with 8–12 weeks for conventional steam plants). Thus the planned availability of geothermal turbines would seem to be substantially better than for conventional turbines.

For the maintenance of bores it is constantly necessary to log the fluid outputs so that the pattern of declining yields may closely be observed. Timely reaming of bores liable to show calcitic blockage can sometimes restore a bore yield to much higher outputs, and a study should be made to determine the economic period for such reaming. A maintenance rig should always be held available to cover the contingency of an unforeseen blockage or casing failure, while the sinking of replacement bores should be carried out at an adequate rate to cover the ultimate decline of older bores to uneconomic yields. Sometimes the life of a bore may be extended by perforating the casing at well chosen depths.

14.7 Man-power, automation and decentralization

In the early days of geothermal power generation the very novelty of the development imposed a cautious manning level at the power plants, as confidence had not yet been won through extended experience. Even at Wairakei, after half a century of experience at Larderello as a guide, the manning strength when the station was first operating at its full potential output was approximately as follows, for 192 MW of plant made up of 13 generators:

	Power station	Field	Total
Supervisory and shift staff	37	13	50
Maintenance staff	14	18	32
Totals	51	31	82
Men per MW installed	0.27	0.16	0.43

These figures included allowance for absences on leave and sickness, but excluded watchmen and messengers and those responsible for the upkeep of the residential staff village. There was also a certain amount of flexible give-and-take, between the station and field staffs, which were not rigidly separated into 'water-tight' groups. Now at the Geysers, with 522 MW installed in six separate power station buildings with a total of eleven generators, the whole plant is controlled and maintained by 45 men only [84] including clerical staff – i.e. about 0.09 men/MW, or one-third of the Wairakei figure despite the spread of the power plant in six separate buildings. (The field staff at the Geysers is not known, as the steam supply system is owned by different authorities, and not by the Pacific Gas & Electric Co. which owns the power plant; so comparisons of field staff with Wairakei cannot be made).

This great reduction in man-power has been achieved partly by the growth of confidence born of experience and partly by a high degree of automation which is a necessary consequence of the decentralization of the power plant over several scattered buildings. The primary purpose of decentralizing the Geysers plant was to economize in field pipework, so that the average distance between bores and power plant could be minimized. However if each power plant were to be fully manned after the manner of Wairakei (where all the 192 MW is concentrated at one site – albeit in two adjacent turbine houses), this economy could well have been nullified. A high degree of automation and remote control thus became an economic necessity. In the Tuscan fields 421 MW of plant are spread over 17 sites, so that in Italy too, decentralization and automation have become the natural consequence of expansion. (At Wairakei there has of course been *no* expansion since the original installation: otherwise the same trend would probably have been followed there too). Now that plenty of satisfactory working experience has been gained, it is fairly certain that the larger of the thermal fields yet to be developed in future will follow the same pattern of decentralization and automation.

In the Geysers field all plants have been designed for unattended operation with 'shut safe' devices that warn an operator at a control centre some

40 miles away when they act. A single, 2 man, 8 hour shift of operating/ maintenance personnel is provided at each plant for 5 days per week only, so that for 76% of the time the plant is unattended. After the installation of the sixth power unit it was thought advisable to provide in addition a 24 hour roving operating service as a 'back-up'. Soon, a computer-based information and control centre will be established and manned continuously. The system despatcher will take appropriate action in coordination with local staffs for starting up, shutting down and effecting load and voltage adjustments, based on transmitted signals and alarms. If a unit trips off load, the connected wells are allowed to blow to waste through silencers until it has been ascertained that the unit cannot soon be restarted.

In fairness to Wairakei it should be pointed out that there were two good reasons for the concentration of the whole power plant at a single point. In the first place there was an abundant supply of cool river water and a convenient flat riverside site available. Secondly the original intention (see Section 13.3) was to have a dual purpose plant which would have necessitated the placing of the power plant close to the distillation plant.

14.8 Annual plant factor

The reliability of controls, the keeping of an adequate stock of spares, the enforcement of a strict preventive maintenance programme, together with the inherent reliability of thermal fluid flows have collectively ensured that geothermal power plants can attain higher annual plant factors than any other form of generating plant. A similarly high degree of high availability for non-power applications of earth heat can also be assumed. Matthew [84] mentions that during 56 turbine-years at the Geysers there have only been 12 enforced outages averaging 3 weeks each – i.e. 1.24% of the time – though there have admittedly been unintended load *curtailments* that have raised the total effective time of enforced outages to 6.2% which is greater than the corresponding figure of 2.5% for conventional reheat turbo units. On the other hand the availability of geothermal steam supplies is considerably greater than that of boilers, so that the overall availability of complete geothermal installations is generally better than that of conventional steam power plants.

Annual plant factors for geothermal power plants exceeding 90% are not uncommon. In fact No. 1 unit at the Geysers, which achieved 82% even in its first year of service, has since attained 95% in a single recent year.

At Wairakei where 41 MW of the 192 MW installed are provided as spare capacity – unmatched by a correspondingly adequate steam supply system, thus leaving only 151 MW available for normal service – an annual plant factor of 98% was achieved in 1975/76 if reckoned on the 151 MW figure. This, however, may be rather misleading as it is probable that some of the 41 MW of spare plant was sometimes used to consume some surplus steam capacity over and above the amount required for the 151 MW which it is normally intended to serve. The annual plant factor based on the total installed capacity of 192 MW was 77% for the same year, but this too is an unfairly low figure, as the steam supply system would have been incapable of producing 192 MW.

In expanding systems such as those in Tuscany and in California, annual plant factors of only 76 or 77% have regularly been recorded, but these figures considerably underrate the potential utilization of geothermal power plants because they are distorted by new plants coming into service and operating only spasmodically for part of their first year of operation. 'Steady state' plant factors have not yet become available for such expanding systems. In general, it would be safe to assume that a completed geothermal power plant could regularly attain annual plant factors of not less than 85% and often at least 90%.

Some economic considerations

15.1 General

There is of course more to economics than mere costing, although costs must always be important. Such factors as national self-sufficiency, environmental impact and energy conservation may now often overshadow considerations of cost. Nevertheless, costs can never be ignored. It is also important to recognize at the outset that the days of very cheap energy have probably gone for ever. Those days ended in October 1973, when politically motivated swingeing increases in oil prices rocked the economies of many nations. The alarming rates of inflation that subsequently afflicted so many countries, frequent movements in the currency exchange markets, and great instability in the rates of interest on borrowed money have all contrived to make it increasingly difficult for rational comparisons to be made between geothermal energy costs in different countries (or even between the costs of different forms of energy in the same country) when the foreign currency component differs greatly between one form and another. Moreover when making energy cost comparisons between one country and another it is important to bear in mind the national average income *per capita* as a yardstick of acceptability.

Difficult though the task of advance costing of geothermal energy may be, it is one that cannot be shirked; for decisions of whether or not to proceed with geothermal development will always have to be made at some stage or other against a background of many uncertainties. Unfortunately the author forgets who it was that once aptly likened the task of electricity cost accountancy to that of a merchant who buys by the cubic metre and sells by the ton a commodity whose specific gravity is both difficult to determine and constantly changing. The task of geothermal cost estimation in advance of actual development is just about as forbidding.

15.2 Unpredictability of geothermal costs

Even in stable times, so many uncertain factors can arise between the moment of intent to develop an energy source and the moment when the energy starts to flow, that the precise forecasting of costs for *any* form of energy has always been a precarious business. However, with geothermal energy, the task of predicting costs has always been rendered far more difficult by a number of other uncertainties peculiar to this form of energy. The instabilities of modern times have aggravated the difficulties. Exploration costs can seldom be foreseen with any degree of accuracy. A sum of money may be devoted to exploration, based on a rational plan of action, but until quite a lot of data have been collected and analysed it is virtually impossible to predict how much money should be spent on each of the various exploration techniques available. The costs of exploratory drilling are particularly hard to foresee, as the success ratio can vary so greatly. Also it is hard to foretell the depths to which the bores must be sunk to reach productive horizons (and, as has been shown in Section 7.15 and Fig. 21, drilling costs are extremely sensitive to depth); nor can the quality of rock to be penetrated by the drill be known with any certainty beforehand. Again, we can never know in advance either the bore characteristics, or the practical bore density per square kilometre, which will greatly influence the fluid transmission costs. There are other uncertainties such as the foundation conditions for a plant, the siting of which cannot be chosen until the pattern of productive bores is at least approximately known; or the ability or inability to discharge waste fluids into water courses without risk of pollution. All these uncertainties, however, should not be used as an argument against geothermal development, for somewhat paradoxically the very monetary troubles from which so much of the world is now suffering are tending to *improve* the economic prospects of geothermal energy and to make the taking of financial risks more worthwhile. Geothermal exploration always involves some measure of risk, but experience has shown that where the risks have been taken on the basis of reasonably promising evidence, it has nearly always paid handsome dividends. Moreover, exploration techniques are continually improving and the risks are steadily being reduced. A useful analysis of the problems of geothermal costing was made by Leardini in 1970 [135].

15.3 Geothermal cost experience until 1973
When the rate of cost inflation was lower, it was possible to make a reasonable attempt to rationalize geothermal costs, both in terms of heat and of electrical energy. Such an attempt was made in about 1970 [136]: the results were reasonably valid for the 1960s, although they have since become outdated by inflation. Nevertheless, it was the *comparative* costs between geothermal and alternative energy sources that were of greater significance than the absolute figures. In those days of cheap fossil fuels it would have been approximately true to say that geothermal heat from exploited fields

could be expected to be of the order of 12 US cents/Gcal at the wellhead. After allowing for the cost of collecting and transmitting the thermal fluids from the bore field, and of separating and (usually) rejecting the hot water at the wellheads, the cost of geothermal heat from a wet field piped to a single delivery point for use in some exploitation plant might have been approximately quadrupled to, say, 50 US/cents Gcal. In those same years, the cost of fuel heat ranged from about 100 to 250 US cents/Gcal. If this heat was required in the form of steam or hot water, the price (after allowing for the capital charges, operation, repairs and maintenance incurred by the necessary boilers or heat exchangers) was raised to anything from about 150 to 400 US cents/Gcal according to the location and scale of the enterprise. A fair average figure would have been about 275 US/cents Gcal – *about* $5\frac{1}{2}$ *times* the cost of geothermal heat. This, of course, is a very broad statement, and individual cases would have differed widely, but it may be taken as a statement of a general trend. In the case of small industries or micro-power plants that could be served by a single bore without costly collection and transmission pipework, the cost advantage of geothermal heat could have been very much greater. These approximate cost comparisons were based upon oil fuel as the alternative source of heat. Heat from coal or lignite could sometimes have been considerably cheaper, but not nearly cheap enough to cover this greatly advantageous factor of about $5\frac{1}{2}$. In Reykjavik, Iceland, even after adding the extensive city heat reticulation system, the cost in 1961 of geothermal heat delivered to domestic consumers was only about 58% of the cost then prevailing of heat supplied from oil burned in small domestic boilers [137], and according to Einarsson [7] this percentage had fallen to 25–30% by 1974.

In the case of electric *power*, the advantage in the 1960s in favour of geothermal energy, though still positive, was less striking. This was because conventional power plants could operate at very much better thermal efficiency and could make use of power plant costing much less per kilowatt than geothermal power plants owing to the higher temperatures and pressures and larger capacities that could be used. Nevertheless, it was probably broadly true to say that geothermal power in those days cost only from 50–60% of conventionally generated thermal power at the same annual plant factor.

15.4 Effects of inflation

From about 1970 world price inflation started to accelerate, and in late 1973 oil prices rocketed, with other costs rising sharply, though less markedly than oil. By 1976 the price of oil fuel had at least quadrupled since the end of the 1960s and other fuel prices had risen in sympathy, though not so steeply. However, the fact that fuel prices have risen more rapidly than material and labour costs has rendered geothermal energy still more attractive *relatively* than conventional thermal energy, despite the higher interest rates on investment. This can be demonstrated by means of the following arbitrary, though not unrealistic, example.

Let it be assumed that ten years ago a geothermal power plant could generate electricity at 55% of the cost of that generated by a conventional thermal power plant of the same capacity and serving the same duty, and that the proportional composition of the production costs were as follows:

Conventional		Geothermal	
Fuel	61%		
Operation, repairs and maintenance	12%	Labour and materials	20%
		Capital charges	80%
Capital charges	27%		
	100%		100%

Let it further be assumed that fuel prices have quadrupled, that labour and material costs have doubled, and that capital charges have risen from 9% to 14%. The relative production costs from new plants would now be:

Conventional

Fuel	$61 \times 4 = 244$
Operation, repairs & maintenance	$12 \times 2 = 24$
Capital charges	$27 \times \dfrac{14}{9} \times 2 = 84$
	352%

Geothermal

Labour and materials	$20 \times 2 = 40$
Capital charges	$80 \times \dfrac{14}{9} \times 2 = 249$
	289% of 55%
	$= 159\%$

Relative production costs	$\dfrac{\text{geothermal}}{\text{conventional}} = \dfrac{159}{352} = 45\%.$

Thus, whereas geothermal energy was costing about 55% of conventional thermal energy about ten years ago, it would now be costing only about 45% – equivalent to a relative improvement of (55–45)/55, or about 18%. This is probably a cautious deduction, for exploration costs have become relatively more moderate with improved methods, though they too have of course inflated.

This comparison, while making no claim to accuracy, clearly illustrates the relatively improved economic position of geothermal energy resulting from the fact that fuel costs have inflated more rapidly than other cost components. In so far as industrial heat is concerned, the improved position of geothermal energy should be considerably greater owing to the higher fuel content of the cost of conventional heat.

15.5 Nature of geothermal costs

It has earlier been stated that geothermal costs are virtually fixed. Traditionally, it has been the custom to regard them *all* as fixed, but in the author's opinion it would be logical to assign a small *variable* cost component arising from the exploration costs. This is because the fruits of exploration, if successful, are a finite quantity of energy, which may be expressed in MW-years (see Fig. 14; Section 5.15) or in teracalories. If the resources of a geothermal field were infinite, there could be no justification in regarding the exploration costs as anything but fixed, but it is generally believed that each field has a finite extractable heat content, though that content may be difficult to assess with any degree of accuracy. This finite quantity of heat may be squandered quickly in a large plant, or it may be eked out slowly for a longer time in a smaller installation. It may be used wastefully in an inefficient plant, or economically in an efficient plant. It may be used 'now' or, like fossil fuels, it may remain stored in the ground for future use. The general belief is that the infeed of magmatic and/or radioactive heat is small by comparison with the rate of draw-off in all exploited fields, so that the commercial development of a thermal field may be likened to the mining of·a finite seam of fossil fuel. When we exploit a geothermal field, we are tapping a capital store of energy that has been gradually built up over thousands of years. The energy, except for a small fraction, is not strictly 'renewable' unlike hydro-power which is seasonally, or at least periodically, replenished.

If a geothermal installation is operated at a fairly constant annual plant factor, the assessment of a variable cost to cover exploration expenditure would be of little significance, once we have decided upon the method of exploitation. However, when we come to consider different ways of exploiting a field, the existence or non-existence of a small variable cost component assumes some importance. For example, it becomes significant when we consider the relative merits of using condensing or non-condensing plants and whether the latter could be used for carrying peak loads.

The actual evaluation of the variable cost component defies precision because of the inevitable uncertainty as to the thermal capacity of a field, and also because 'exploration' is a process that may well be continued spasmodically for some time after a field is first exploited. Still in theory the variable cost could be assessed by dividing the total costs incurred over the years of financing the exploration work (including interest and amortization of borrowed money) by the estimated total exploitable energy capacity of the field. Both numerator and denominator – especially the latter – are difficult to assess. In terms of generated electrical energy, where appropriate, the variable cost assessed in this manner is likely to be small – probably not more than, say US 0.5 mill/kWh* for condensing plant and US 0.85 mill/kWh for non-condensing plants (which would consume about 70% more steam per kilowatt-hour than condensing plants after allowing for differences of admission pressure and condensation losses arising from the fact that condensing plants would be sited at a central station at some distance from the bore field while non-condensing plants would only be used if sited close to the bores (see later). The corresponding variable cost component of *heat* at the wellhead would be of the order of 3.5 US cents/Gcal.

When it comes to considering the use of non-condensing plants for peaking purposes, about 10% should be added to the variable cost component to allow for scavenging steam to exclude air from the turbines when not in service.

These figures are intended to be no more than very roughly illustrative, and probably err on the high side (at 1977 levels). By comparison with the variable cost component for conventional thermal power plants, the geo-thermal variable costs would be extremely small.

15.6 Relative economics of using condensing and non-condensing turbines

The normal method of exploiting a thermal field for power generation is to pipe the steam from a number of bores to a central power station, where electricity is generated by means of condensing turbines. This practice is followed because of the relatively greater efficiency of condensing plants compared with non-condensing plants. An alternative method of exploiting the field would be to instal a group of relatively small non-condensing plants close to the wellheads and to interlink them electrically. Such plants would require minimal pipework, the simplest of weather protection and foundations, no condensers and no cooling water system. Moreover, by siting the turbines close to the bores, the heat losses inherent in long steam mains can be reduced and higher turbine admission pressures can be adopted or, more well steam could be extracted by using the same turbine admission pressures. It is true that their higher steam consumption would mean sinking

*1 US mill $= 10^{-3}$ US\$ or $\frac{1}{10}$th of a US cent

about 70% more bores in order to achieve the same output, but this can be more than offset by the economies effected. It would not in fact be unreasonable to expect that the capital cost per kilowatt of the non-condensing plant arrangement might perhaps be from 12 to 14% lower than for the central condensing plant alternative, and that the production costs per kilowatt-hour (despite the higher variable cost, as suggested in Section 15.5) could be about 10% lower.

This would seem to pose a paradox: if power can be generated more cheaply by means of small non-condensing plants scattered around the bore field than by means of a central piped-up condensing plant, how is it that all major existing geothermal power plants make use of condensing turbines? The answer is that production costs per kilowatt-hour, though important, are not the only criteria on which the economic choice of the type of a geothermal power plant should be based. Another criterion of even greater importance is *energy conservation*. This illustrates the truth of the opening sentence of Section 15.1.

Let it be assumed that the assessed thermal capacity of a field, based on the use of condensing turbines, be estimated at 3400 MW-years. Since non-condensing plants would consume about 70% more steam, the capacity of the field based on the use of non-condensing plants would be only about 2000 MW-years. In other words, by using condensing plants the life of the field would be about 70% longer for a given installed capacity. Would it be right to squander about 1400 MW-years for the sake of saving perhaps 10% in production cost per kilowatt-hour? There is a difference between the *cost* and the *value* of a kilowatt-hour. One way of looking at the problem is to compare the geothermal production costs with those of the most nearly competitive *alternative* source of base load energy. To take a hypothetical example, let it be assumed that the production costs of the alternative source are US 17 mill/kWh while those of condensing and non-condensing geothermal plants are US 10 mill and 9 mill/kWh respectively. Over the life of the field, the savings effected by using geothermal energy rather than the alternative source would be:

For a central condensing plant
$$3400 \times 1000 \times 8760 \times \frac{17-10}{1000}$$
$$= \text{US \$208.49m.}$$

For non-condensing plants
$$2000 \times 1000 \times 8760 \times \frac{17-9}{1000}$$
$$= \text{US \$140.16m.}$$

Although the latter would save US 8 mill/kWh, as against US 7 mill for the former, by comparison with the alternative source of base load, the larger saving could be applied to a much smaller number of kilowatt-hours, so that the *long term profitability* of the costlier condensing plant would be *US $68.33m* greater than if the cheaper non-condensing plants were used. This, of course, is an oversimplification, for it assumes a constant money value over the life of the field. In an inflationary market the savings in favour of the condensing plant would probably be far greater if the alternative energy source were fuel-consuming. The example serves to show, however, that the normal considerations of 'merit order' generally applied to thermal plants in terms of their variable cost component per kilowatt-hour would usually have to be over-ruled if any non-condensing geothermal plants were connected to the system; otherwise such plants would tend to be offered base load. There would, however, be two exceptions to this over-ruling – pilot plants and non-condensing plants using geothermal steam of high gas content.

15.7 The economics of geothermal pilot plants

In the early stages of developing a geothermal field it is customary for the first few successful bores to be blown to waste at high output for a considerable time in order to study the bore behaviour and the field characteristics. A pilot plant could put this steam to good use. Since the steam thus used would otherwise be wasted, it could be of greatest value if consumed by a plant in almost continuous operation, so that it could contribute its mite towards the system base load. Pilot plants are of an experimental nature, installed to foster confidence and to gain experience. They should be cheap, as their function is comparatively ephemeral, so non-condensing units would be suitable for the purpose. They could be quickly installed, and as they would make use of a single bore, or very few bores, they would be of fairly small capacity – say 2 to 5 MW. Since they would only use steam that would otherwise be wasted, they should not be expected to bear any share of the costs of the bores or wellheads, these being primarily provided to serve a later and larger, more permanent plant. Nor, for the same reason, need they be burdened with any variable (exploration) costs (see Section 15.5). However, if a pilot plant is to enjoy these exemptions, it would be unfair to compare its production costs (as computed on the basis of these exemptions) with the *full* production costs of other plants. It would be more logical to compare them only with the direct incremental savings earned by the generated output of the pilot plants. For practical purposes, these savings may be taken as the value of fuel saved in any conventional thermal plants that may be contributing energy to the integrated electricity system. As pointed out in the Introduction, *every kilowatt of pilot plant capacity, if supplying base load, can save about 2 t of oil fuel per annum* (or equivalent for other fuels) that would otherwise have to be burned elsewhere in the system. Even under 1977 conditions of inflated costs it should be possible to operate a non-condensing geothermal pilot plant at an annual cost of perhaps US $75 kW or thereabouts. Such a plant would not only pay for itself but would also contribute some modest excess revenue if the cost of fuel oil exceeded US 37\frac{1}{2}$ t — a condition likely to be found in non-oil producing countries.

The fact that a pilot plant may have served its original purpose within 2 or 3 years of its installation need not mean that its life is finished after that period. It could be shifted to a new field for further pioneering work, retained *in situ* as an emergency generator, or even used for peak load purposes (see Section 15.8).

15.8 The economics of geothermal power for contributing to peak loads

It is well known to power system planners that cheap kilowatts can some-times be as valuable as cheap kilowatt-hours, since it is necessary to cater for peak loads as well as for base and intermediate loads. Non-condensing geothermal power plants could sometimes provide relatively cheap kilo-watts, but since considerations of bore stability make it difficult to switch geothermal turbines on and off as theoretically convenient, the potential usefulness of such plants for peak load purposes is subject to constraints. Ideally, if a use can be found in a dual purpose plant for off-peak steam for industrial or other profitable application, possibly with the help of thermal storage, it could conceivably pay to use the power production element of such a dual purpose installation to contribute to the system peak and inter-mediate loads more profitably than to the base load. However, the problem is not a simple one. It has been analysed at some length elsewhere [92], where it has been shown that any possible usefulness of geothermal power for contributing to system peaks should preferably be confined to 'secondary peaks' rather than to the extreme peaks. By 'secondary peaks' are implied flat-topped slices of load in the duration curve just below the extreme peaks.

15.9 Scale effect

With conventional thermal power plants the capital cost per kilowatt installed is sensitive to what is generally known as the 'scale effect'; that is to say, a very large plant will tend to cost less per kilowatt than a small plant of similar type. An approximate formula for this effect, which broadly applies to conventional thermal plants, is:

$$\text{Total capital cost} \propto \text{kilowatt capacity}^{0.815}.$$

Thus, if plant A has 10 times the capacity of plant B, its total cost would be about 6$\frac{1}{2}$ times that of plant B and the cost per kilowatt of plant A would be about 65% of that of plant B. This advantage in favour of the larger plant is partly due to the spread of overheads over a greater number of kilowatts, partly to other general economies of large scale manufacture, and partly to the fact that the larger plants can use higher pressures and temperatures which are conducive to higher efficiencies, to more compact blading and the accommodation of a larger number of stages on a single shaft.

With geothermal plants the scale effect is much less pronounced (with the reservation that follows below) for the following reasons:

(i) Drilling costs are more or less directly proportional to the installed plant capacity (except that since an integral number of bores must always be used, the proportion of a fractionally unused bore will tend to be higher for very small installations).

(ii) Turbo-generator costs are less subject to scale effect, partly because there is a more limited choice of the maximum plant unit size and partly because admission pressures and temperatures are usually modest, being influenced by factors other than mere plant size.

The net result is that for geothermal power installations the scale effect formula would be more like:

$$\text{Total capital cost} \propto \text{kilowatts}^{0.875}$$

Thus if plant A has 10 times the capacity of plant B its total cost would probably be about $7\frac{1}{2}$ times as great, and its cost per kilowatt would be about 75% that of plant B. The advantage of size is therefore less marked with geothermal than with conventional thermal plants. However, this is true only if the costing theory outlined in Section 15.5 – that of assigning a variable cost component per kilowatt-hour to cover the exploration costs – is accepted. If the exploration costs were treated simply as capital expenditure, the scale effect of a geothermal plant would be *greater* than for a conventional thermal plant, and the formula would become something like:

$$\text{Total capital cost} \propto \text{kilowatts}^{0.7}.$$

That is to say, if plant A were 10 times the size of plant B, its total cost (including exploration) would be about 5 times as great, and its capital cost per kilowatt would be about 50% of that of plant B. This formula, however, should be treated as indicative only, since exploration costs can vary so widely from project to project.

It is clearly more logical to treat exploration costs as a variable cost per kilowatt-hour so that the scale effect is less marked. The alternative of treating exploration as capital expenditure means that if only a small installation were first adopted, the exploration costs (unless partly allocated to a suspense account to be charged against expected future development – which cannot always be assured with certainty) would have to be fully recovered on that first installation and that a large quantity of residual

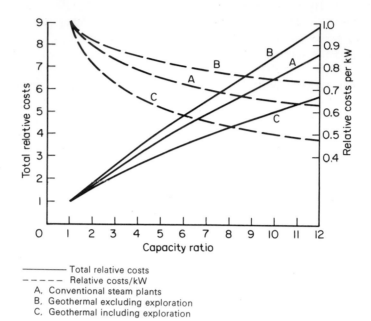

Figure 65 Approximate cost scale factor.

free heat' would remain in the ground for use in future developments which, if realized, would quite illogically be exempted from any share of the exploration costs.

The three approximate formulae quoted above are illustrated in Fig. 65.

15.10 Drilling costs
Quite apart from inflation and variations in international exchange rates, it is virtually impossible to be at all specific about drilling costs for the following reasons:

(i) A drilling contractor must incur substantial 'mobilization costs' for acquiring his equipment, transporting it to where it is needed, preparing his drilling sites, providing access thereto, hiring a team of skilled men and providing them with temporary housing and other social facilities in out-of-the-way locations. He may also have to provide water supplies and construct wellhead cellars, which may, or may not, be included under his contract obligations. Hence, very substantial sums of money are necessarily spent before even a single metre of borehole has been sunk. The essence of reducing the total costs of perforating a field with bores is *time*, so that maximum total penetration may be effected within a given period, with a minimum of 'idle' or non-productive time. Delays due to breakdowns of equipment, moving from site to site, the need to plug zones of lost circulation, held up supplies of casings, mud, bits, drill-stem tubes, etc., can all affect the overall costs of drilling enormously

(ii) Exploration holes may be two or three times as expensive as production holes of the same depth, because of the need to obtain continuous cores and to take frequent downhole measurements and samples.

(iii) As has been explained in Section 7.15 and illustrated in Fig. 21, drilling costs are very sensitive to depth. It is often loosely said that drilling costs vary exponentially with depth. Fig. 21 shows that this is more or less true for depths exceeding about 2500 m, but for shallower depths this is an understatement. Fig. 21 was based on cost statistics covering about 25 millions of metres oil wells in one particular year, and there is no reason why the *proportions* should differ very much in more recent years despite the effects of inflation on absolute costs. Fig. 21 suggests that mobilization costs *on average* are about three times the cost of drilling the first 100 m. Although this may be true as a worldwide average it is clear that on a site where very few holes are drilled the mobilization costs *per hole* would be far higher.

(iv) Drilling costs depend greatly on the nature of rock to be penetrated. For example it has been stated [138] that a hole 3 km deep, if drilled in granite, would cost about three times as much as if drilled in sediments. Since Fig. 21 is based on such a broad selection of sites (though mostly in sedimentary formations, as it relates to oil wells) this dependency has largely been smoothed out, but in practice a hole of a given depth could vary from site to site by a large factor according to the type of rock encountered.

At Wairakei, in the 1950s, the cost of drilling 8 inch nominal bores ranging in depth from about 250 to about 1000 m (average about 600 m) was on average about £35 000, or US $100 000 each at the then prevailing exchange rate. At Broadlands, New Zealand, in the 1960s, the average cost had risen by a factor of about 3.7, but the average depth there was about double. According to Fig. 21 the doubling of depth would account for a cost factor of about 2.5, so the excess was partly due to inflation and higher interest rates and probably partly due to different local conditions. New Zealand costs were generally regarded as being high by comparison with other countries, mainly because of the high quality of amenities offered to their drilling staff. For example the average cost of the Tuscan bores, averaging about 1000 m in depth, was only about US $50 000; though these were admittedly sunk a few years earlier.

More recently (1976) the following estimates have been quoted [138]:

Country	Depth in sediment- ary rocks	in granite	Remarks
UK	⎧ 3 km £200 000 ⎨ 4 km £300 000 ⎩ 8 km £1000 000	£600 000 £1000 000 £3000 000	
USA (Los Alamos)	3.66 km	£290 000 ($500 000)	0.73 km is in sedimentary rocks and the remainder in granite. US costs are generally lower than in the UK.

It will be noted that these costs do not conform to the pattern of Fig. 21, which emphasizes the difficulty in quoting consistent figures.

There is not very much published data about geothermal drilling costs, which are known to be somewhat higher than for oil wells, mainly due to the harder rocks generally encountered. Such data as are available often seem to be contradictory and are not always supported by particulars of depths and diameters. Costs in oil-producing countries, where drilling rigs and personnel abound, tend to be considerably lower than in countries where no drilling has hitherto been practised and where equipment and personnel must be imported. Hence the hopelessness of trying to produce standardized information on drilling costs.

Fig. 66, which is a derivation of Fig. 21, is not without interest. It shows the world trend in average drilling costs per metre with varying depth. Obviously at zero depth the cost per metre must be infinite. The average cost falls to a minimum at about 500 m depth and rises rapidly at great depths.

The cost of *directional* drilling, as used in the Paris suburban district heating schemes (Section 11.2) has been estimated at 28% greater than that of vertical bores reaching the same depth [198]. This should not be regarded as a universal figure because it is largely influenced by the degree of divergence and hence the additional length of casing. As the object in Paris was to gain the greatest possible separation between the bottoms of two adjacent bores, this percentage is probably an approximate maximum. The extra cost is small by comparison with the savings arising from having a single drilling site and pad and from reduced surface pipework.

15.11 Scope for cost improvements

It is important that we ask ourselves in what way can we seek to *reduce* the costs of geothermal exploitation. There is possible scope in several directions.

15.11.1 *Exploration costs.* In the early days of geothermal exploration, when our knowledge of the nature and performance of geothermal fields

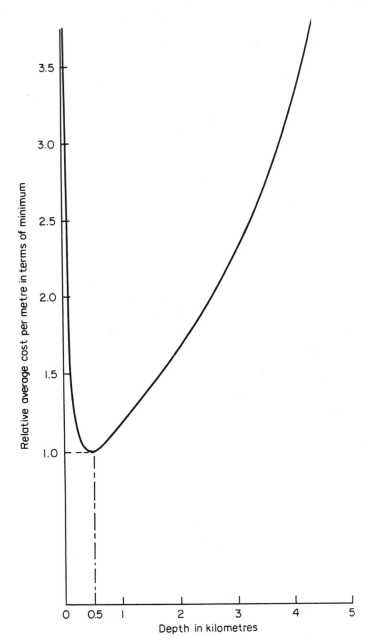

Figure 66 Typical proportional average drilling cost per metre. (Derived from Fig. 21.)

Note: As this curve is derived from petroleum well cost data, it should be used with reserve for geothermal drilling: it is the *shape* of the curve that is significant.

was rudimentary, the search for exploitable earth heat was largely a 'hit-or-miss' operation. There was much 'wild-cat' drilling; bores were sunk in the vicinity of surface thermal manifestations without any proper understanding of underground processes. Drilling for steam or hot water was largely a gamble that sometimes did, and sometimes did not, pay off. During the last quarter century, however, exploration techniques have become greatly refined and improved, and we have learned much about subterranean processes. The skills of the earth scientists and the drilling engineer have become so coordinated as to enable exploration to become much more systematic than in former years. Methods still continue to improve, and the risk element, though not yet entirely removed, is steadily being reduced. The chances of finding productive drilling sites have been much improved, so that the costly sinking of abortive bores now forms a far smaller part of the total exploration costs than formerly.

15.11.2 *Two-phase fluid transmission and submersible pumps.* The technical feasibility of using submersible pumps and two-phase fluid transmission has been touched upon in Sections 9.11 and 9.12. If either should become standard practice for wet fields the economic savings could be very great, not only in wellhead equipment and fluid transmission costs, but also from the fact that by bringing the hot water right up to the plant the exploitation of its flash potential would be so greatly simplified. Moreover, the scale effect of supplying and maintaining few very large separators at the plant, by comparison with having to provide very many small separators scattered all over the bore field, would also be considerable.

15.11.3 *Binary fluids.* As already shown in Section 10.2.6, the use of binary fluids has both advantages and disadvantages. It is too early to draw positive conclusions concerning its economic aspects. It is possible, however, that in certain circumstances appreciable economic gains might be achieved. It is important to understand that the use of binary fluids does not imply any improvement in cycle efficiency as such: it simply enables use to be made of heat that would otherwise never enter the turbine cycle at all, being rejected to waste.

15.11.4 *Dual and multi-purpose plants.* Such plants have been discussed in Chapter 13. Their potentially great economic advantages are self-evident.

15.11.5 *By-product hydro-power from wet fields.* This question has been discussed in Section 10.16. Clearly by-product hydro-power could, where practicable, bring economic advantages, though these would be too individual to particular projects for any generalizations to be made. The question would arise of whether a value (variable cost) should be assigned to the water.

This could be quite a logical allocation, as indeed it would be if the heat content of the bore water were first usefully exploited. In any case, any *variable* cost which may be assigned to the steam in accordance with the arguments presented in Section 15.5 should be appropriately reduced, since the charges on the exploration costs should be spread over *both* fluids – logically in proportion to the utilized energy content of each.

15.11.6 *Novel drilling methods.* These will be mentioned in Chapter 19. It may simply be stated here that these are other possible sources of economy and extension of scope in geothermal development.

15.11.7 *Heat pumps.* The use of heat pumps is by no means exclusively relevant to geothermal development. Nevertheless, where low grade waste heat must unavoidably be discarded from a geothermal exploitation plant it can *sometimes* be economic to upgrade the heat at a fairly high performance ratio so that the natural fluid is rejected at a lower temperature and more heat extracted therefrom, while at the same time providing a reasonably large quantity of higher grade heat that could perhaps be used for some other practical purpose. In the USSR [99] there is a project for combining low grade heat with a heat pump for space heating, while at Creil in the Paris Basin heat-pumps are already being used for boosting the input and improving the efficiency of a district heating scheme (see Section 11.2). Too much optimism should not, however, be placed upon the potentialities of the heat pump, which has a seductive appeal arising from the often misunderstood conception of 'performance ratio'. To the unwary, the heat pump is apt to suggest that something can be had for nothing. The heat pump is, of course, only a device for *transferring* heat from one place to another, and the amount of energy required to effect the transfer may be considerably less than the amount of transferred energy.

If electrically driven heat pumps are used there is no net gain in useful heat unless the performance ratio exceeds the reciprocal of the *delivered* efficiency of the electricity (i.e. of the product of the power station, transmission and distribution efficiencies). For example, if the power station efficiency is 35% and the combined transmission and distribution efficiency is 80%, the performance ratio would have to be greater than $1/(0.35 \times 0.8)$, or 3.57. Hence the 5:1 performance ratio achieved at Creil in the Paris Basin heating scheme (Section 11.2) is energy-profitable.

The heat pump certainly has its uses, but generally it is serviceable if the vapour pressure of the refrigerant is not too high; i.e. it is practicable only at warm and tepid temperatures. It is also important not to fall into the common error of proposing the use of heat pumps to upgrade the heat from the warm condensing water from a steam-field power plant, for it can never pay to degrade heat to a low level and then pump it 'uphill' again. In such cases it would be better either to use back-pressure or pass-out turbines, the exhaust temperature of which is sufficiently high for direct application to space heating or some other purpose. The only conceivable exception to this dictum would perhaps arise where two ownerships are involved. If a power generating enterprise has failed to use a back-pressure or past-out turbine in this manner, it might conceivably pay *some other party* to apply a heat pump to the warm water discharged from the condensers of the power plant, even though the overall efficiency of the double process would be less than that of the more correct use of the heat.

15.11.8 *Excess geothermal power capacity*. In Section 10.15(i) mention was made of the possibility of taking advantage of 'scale effect' to operate a geothermal power plant at a reduced plant factor in order to gain in capital costs per kilowatt installed, with consequent improvement in the production costs per kilowatt-hour. This could sometimes bring economic benefit to small systems, where the base load demand is (at present) smaller than the capacity of a reasonably sized geothermal plant, although with very large composite systems in which available geothermal capacity is likely to form only a fraction of the system base demand, this would not apply, [81].

15.12 Actual geothermal power costs

In view of what has been said in the early part of this chapter it will be understood that the author quotes actual recorded geothermal cost figures with reluctance. Apart from the inflationary trend that has inevitably increased both the investment and production costs of recently constructed geothermal installations by comparison with similar ones of older vintage, it is also necessary to recognize the wide range of individuality of developed fields. Differences of geology, aquifer depth, underground permeability, fluid temperatures, fluid corrosiveness, capacity of installation, site accessibility, local wage levels, national statutory constraints and other local factors can result in very wide differences in costs between two or more developments constructed at the same time in different parts of the world. Moreover, costs are often published without specifying the rates of capital charges borne by the enterprise, and these have differed, and continue to differ, widely from one concern to another. Also, some quoted costs cover the entire works and activities involved, including exploration, drilling, fluid delivery, exploitation plant and equipment and perhaps (in the case of power projects) of power transmission lines for delivering the energy to the national or state grid system. This is usually so where a single owner is involved throughout – e.g. the Government of New Zealand or the Municipality of Reykjavik. Elsewhere, as in California, more than one enterprise has been involved in exploration, drilling and supplying the steam, while another enterprise (The Pacific Gas & Electric Co. in the case of the Geysers field) buys the steam in bulk from these suppliers and uses it to generate and transmit electricity into the Californian network. In such cases of multiple ownership it is impossible to arrive at the true overall costs of exploitation, since the price paid for steam by the Pacific Gas & Electric Co. includes the profits of the steam suppliers. Hence the production costs quoted by the company are a mixture of true costs and of an enterpreneur's profit.

For all these reasons it would be unwise to seek uniformity or pattern in the figures quoted below, while comparisons between one set of figures and another could be invidious. Nevertheless, the quoted costs are not without interest.

15.12.1 *The steam purchase agreements of California.* When a steam supplier provides steam to one of the power plants belonging to the Pacific Gas & Electric Co., the price is not fixed but is variable according to an agreed formula relating to the costs of alternative energy sources such as fuels and hydro-power. This linking of steam price to the costs of alternative energy sources is not without its critics, for it is argued that the OPEC countries have only to raise their oil prices and the steam supply concessionnaires of California will immediately collect more revenue for supplying exactly the same quantity of steam without their costs having in any way been affected by the action of OPEC. The argument of the steam supply concessionnaires, however, is that the arrangement offers them a fair reward for the risks they take, that it provides them with funds for further geothermal exploration and that it offers encouragement to other concessionnaires. The purchaser (The Pacific Gas & Electric Co.) is also satisfied with the arrangement because of the incentive it provides to the developers of thermal fields. This somewhat surprising acceptance has its explanation in the historical development of the electrical industry in the States. In the early days of natural gas development, fixed cheap prices were agreed which made natural gas economically attractive by comparison with other fuels, so the whole nation more or less shifted to a gas economy. Although the developers did fairly well out of it for a time, there was no extra reward to encourage further exploration. They therefore lost the incentive to develop more gas and have it rigidly controlled in price, so they invested in other more rewarding ventures or did nothing at all. Suddenly the country found itself geared to a gas economy and running out of gas. Despite clear warnings, the country had failed to develop its coal industry and other energy sources because of the artificially low gas prices. The Pacific Gas & Electric Co. do not want history to repeat itself: they prefer to observe the rules of the market place and to make sure that they will get geothermal steam, for as long as it is available at a competitive price rather than no steam at all. The price formula is such that geothermally produced power is always substantially lower than power from other sources: there is no risk that geothermal steam will price itself out of the market. At one time, when the steam price was low, the developers of the Geysers field showed signs of losing interest in continuing exploration because they found better and more rewarding ways of investing their capital. Furthermore, since some of the steam concessionnaires were also involved in oil prospecting, they diverted all of the then scarce tubular supplies to that more rewarding effort. Another factor that has justified increased revenue for the steam suppliers is the increased costs of drilling. Some of the earlier wells cost only about $50 000 each. More recently, partly because of rising costs and partly because the nature of the field is necessitating much deeper drilling in more difficult formations, the cost of some of the wells is approaching $1 m each. Yet another factor is increasing bureaucracy which has delayed the necessary sanctions for power station construction. In one case it took 3 years to get a permit to build a unit identical to one already in service, and this delay occurred after the developer had completed his exploration for the necessary steam and had given assurance as to its availability. In this particular case the developer will have to wait about 7 years before he receives any return on his very substantial investment. Hence the steam price formula is not a matter of mere profiteering, as might at first be thought by those who are not familiar with the historical facts.

Another feature of the steam purchase agreements of California is that the steam is not metered. It is considered sufficiently accurate, and is certainly far simpler, to charge for the steam at so-much per kilowatt-hour generated. Thus if a power plant were shut down, no steam would be consumed, no power would be generated and no charge would be made by the steam suppliers despite the fact that their costs are more or less fixed and continuous, regardless of the power station output. This arrangement of indirect metering of steam supplies by means of the kilowatt-hour meter is accepted by the steam suppliers in view of its simplicity and in the light of experience as to the probable annual production of electricity from each power plant – i.e. on the assumption of an average expected plant factor.

15.12.2 *Californian power costs.* As explained in the two preceding subsections, it is possible only to quote the costs of the power producer, the Pacific Gas & Electric Co. (exclusive of field exploration, steam winning and supply costs, all of which are borne by other enterprises). Table 9 shows the *actual* capital investment figures of the Pacific Gas & Electric Co. in geothermal power plants in the Geysers field up to and including Unit No. 8, and their *estimated* figures for newer and future units Nos 9 to 15 (inclusive), as prepared in 1976. It may be noted that at the UN Geothermal Symposium held in San Francisco in May, 1975, participants were given a similar schedule showing substantially lower figures for the *estimated* costs. Although it was claimed that these lower figures included allowances for future inflation, market trends during the following year proved that these allowances had been inadequate, thus necessitating an upward revision (which of course did not affect the costs of the first eight units). This clearly illustrates the difficulties, even for a sophisticated and well managed organization like the Pacific Gas & Electric Co., of keeping abreast of inflation when making future estimates. It is by no means improbable that the figures will be revised yet again, but at the time of going to press, Table 9 shows the most up-to-date figures available. These costs include all power plant, buildings and civil engineering works from the point of delivery of the steam supplies at the various power stations, and include step-up transformers, substations and electrical transmission to the nearest convenient point in the state grid.

The figures are of limited significance, partly because of the exclusion of the steam supply system (part of which would be accounted for by exploration and drilling) and partly because of the wide variety of unit capacities and 'repeat' orders. The sudden drop in the cost of units Nos 7 and 8 by

Table 9 Actual and estimated capital costs (as in 1976) of geothermal power plants in the Geysers field, California.

Unit No.	Year of actual or expected Comissioning	Net rating in MW	Investment in US $	Incremental cost per net kilowatt in US $		
1	1960	11	4 010 000	167.1		Actual
2	1963	13				
3	1967	27	7 610 000	140.9		
4	1968	27				
5	1971	53				
6	1971	53	12 756 000	120.3	112.0 average	
7	1972	53				
8	1972	53	10 982 000	103.6		
9	1973	53				Estimated
10	1973	53	14 280 000	134.7		
11	1974	106	16 560 000	156.2		
12	1976	106	21 710 000	204.8		
14	1976	110	28 560 000	259.6		
15	1977	55	17 550 000	319.1		
13	1977	135	29 380 000	217.6		
Totals		908	163 400 000	180 (average)		

Note: The estimated costs of future machines include an allowance for inflation. These future estimates are revised periodically.

comparison with units Nos 5 and 6 is presumably explained by the gains from a repeat order; so it would be wiser to treat these four units together at an average cost per kilowatt of $112.0. It would be difficult to attempt to allow for scale factor because some of the quoted costs are for single units and some are for pairs. It is of interest to note that the estimated cost of a single 55 MW unit in 1977 is $319.1/kW whereas the cost of 53 MW units in 1971/72 was only $112/kW. It would be unwise, however, to conclude that this implies a 2.85 inflationary factor, because the former figure was for a single machine while the latter was for four units of similar size. Nevertheless, the effects of inflation in those 5 or 6 years must have been very great.

Unfortunately the absence of steam supply capital cost figures makes it impossible to estimate the total geothermal investment in the Geysers enterprises: but it is of interest to note that at Wairakei the heat supply system accounted for approximately half of the total costs excluding the electrical transmission. However, the Wairakei heat supply system costs included the experimental hot water transmission system, and the distance from the bore field to the power plant was much greater than at the Geysers. The *gross* capital costs at the Geysers, therefore, including exploration, drilling and steam supply, would probably be considerably less than double the figures quoted above. No idea of the Geysers steam supply investment can be gleaned from the price charged for steam to the Pacific Gas & Electric Co. owing to the flexibility of this price, as explained in Section 15.12.1.

Total *production costs* of electricity delivered to the Pacific Gas & Electric Co.'s system from the Geysers field have been estimated as follows (in $ US mills*/kWh delivered to the grid system:

	Annual capacity factor		
1973 estimate based on delivery from Unit No. 14 (110 MW net), with steam supply at 4.8 mills/kWh plus re-injection costs (see Chapter 19) at 0.5 mills/kWh [139]	70%	80%	90%
	9.7	9.2	8.8
1978 estimate based on delivery from unit No. 15 (55 MW), with steam supply at 12.5 mills /kWh plus re-injection costs at 0.5 mills /kWh [140]		23.4	

It should first be noted that the first set of figures are already outdated by a 75% increase in the estimated investment cost for that unit: this will greatly raise these figures. Nevertheless, at an 80% annual load factor, an increase of 154% is expected in the production costs compared with an estimate made only 5 years earlier, though part of that increase will be explained by an adverse scale factor. This increase represents an annual

*1 US mill = 10^{-3} US$ or 1/10th of a US cent

growth of $20\frac{1}{2}\%$ p.a. in production cost. It is of interest to note that in 1960, when the first Geysers plant was commissioned, the steam price was only US 2 mills/kWh. The expected increase in steam costs over 18 years is therefore $6\frac{1}{2}$ times, equivalent to an average annual steam price increase of 10.7% p.a. disregarding re-injection costs. The average growth in steam costs from 1973 to 1978, as revealed by the above figures has been about 21% p.a.

Despite these alarming growth rates in costs, geothermal power continues to be the cheapest form of energy in the Pacific Gas & Electric Co.'s system. The 14 year *average* cost of geothermal energy in California (1961–1974) has been US 5.612 mills/kWh as compared with US 8.256 mills/kWh for fuel-fired plants [84]. When adjusted to the same annual plant factor the geothermal costs would still only be 89% of   conventional thermal power costs despite the small unit capacities of the earlier Geysers plants. This 89% is admittedly much greater than the 50–60% figures quoted in Section 15.3, but it should be remembered that the steam suppliers are being heavily subsidized. As already stated, the costs to the Pacific Gas & Electric Co. do not reflect the true overall costs.

15.12.3 *New Zealand power costs.* The *Wairakei* capital costs (192.2 MW installed, commissioned 1958–1963) were as in Table 10.

These costs were mainly incurred in the second half of the 1950s. The following factors relate only to Wairakei, and would tend to have raised the costs by comparison with those of other geothermal power projects constructed at the same time:

(i) Owing to a change of design 'in mid-stream' from a multi-purpose (power generation and heavy water production) to a 'power only' project, a complicated arrangement of 13 relatively small units [68] became necessary and undoubtedly cost far more than if a purpose

Table 10 Capital costs at Wairakei.

Item	Cost			
	£m	£/kW	US$m	US$/kW
Steam winning	4.392	22.85	12.298	63.98
Heat transmission	4.975	25.88	13.930	72.46
Power house and plant	7.441	38.72	20.835	108.42
Cooling water system	1.262	6.57	3.534	18.40
Substation	0.669	3.48	1.873	9.74
Total	18.739	97.50	52.470	273.00

The dollar conversions have been made at the then exchange rate of £1 = US$2.80.

built 'power only' plant had been designed from the start, with a different choice of working pressures and turbine arrangement.

(ii) The adoption of river cooling, and the subsequent discovery that the centre of gravity of the field was further away from the river than had at first been thought, led to very long pipelines.

(iii) The heat transmission costs included the cost of the abortive hot water transmission scheme described in Section 9.9, which would not have been undertaken had it been foreseen that the selected wells would 'dry out' so soon.

(iv) No costs have been charged to the electrical transmission line to interconnect the station with the North Island grid.

(v) The New Zealand Government charged only part of their exploration costs to the project, holding the balance in a 'suspense account' which would have been booked to a subsequent extension then planned but in fact not yet carried out. It is not known how they have since dealt with the unbooked balance of this account.

The *production costs* at Wairakei were estimated in 1961 at 4.9 US mills/ kWh [68], but the interest rate was then only 4% p.a. A life of 20 years was assumed for all the assets except the bores for which a 10 year life was assumed. (It is customary with most geothermal projects to depreciate the bores more rapidly than the other assets, as their output tends to decline with time. Thus there will always be a proportion of relatively new high-yield bores together with some older low-yield bores in simultaneous service. In assessing the production costs allowance should be made for the re-use of wellhead equipment and part of the branch pipelines moved from old to new bore sites, while charges are made for the re-adaption and partial wastage of such equipment.) By the late 1960s the Wairakei production costs had risen to over 6 US mills/kWh due to labour and materials inflation.

The *Broadlands* project (due for service in 1981) has been the subject of cost estimates *at 1976 levels*, at NZ $550 /kW capital investment and NZ 0.93 cents/kWh production costs. At the prevailing exchange rate at that time these figures were equivalent to about US $514 /kW and 8.69 US mills/ kWh. Although these costs are 88% and 77% higher than the 1961 Wairakei estimates respectively, the mean annual growth rate of costs amounted only to 4.3% and 3.9% respectively. The Broadlands plant is for a smaller capacity than Wairakei (165 MW versus 192 MW), so scale factor would tend to count against Broadlands, but the new installation will be rationally designed *de novo* without the unfortunate complications that beset the Wairakei plant. It is for this reason that the Broadlands cost estimates exceed the Wairakei costs by much less than could be accounted for by inflation.

15.12.4 *Mexico power costs.* For the 75 MW geothermal power plant at Cerro Prieto, first commissioned in 1973, the total investment was

US $14 400 000, or $192/kW(gross) at the generator terminals, which would probably work out at slightly over $200/kW(net) after allowing for auxiliary consumption. These figures compare favourably with the Geysers figures as set out in Table 9, considering that they include exploration, drilling, steam supply as well as power plant and buildings, etc., and that only two units of 37.5 MW each were installed. The only production costs for Cerro Prieto were those *estimated* in 1974 at 12.56 US mills/kWh generated, based on an assumed annual plant factor of 80%, a 30 year plant life and a 10 year bore life.

15.12.5 *Italian power costs.* In 1964 the *estimated* production costs in the Larderello complex ranged from 2.6 to 2.74 US mills/kWh for non-condensing units and from 2.38 to 2.96 US mills/kWh for condensing plants, based on an assumed interest rate of 6% p.a. [141]. These costs were of course considerably lower than for geothermal power plants in other countries as the Italian plants were constructed earlier and (by comparison with Wairakei) to rather more austere standards. In 1976 the production costs of geothermal power were updated to the figures shown in Table 11 by the Italian authorities themselves [142].

15.12.6 *Iceland power costs.* The only Icelandic geothermal power plant in service at the time of going to press is a small $2\frac{1}{2}$ MW back-pressure unit at Námafjall [143]. The production costs for this plant were rather vaguely estimated in 1970 at 'from $4\frac{1}{2}$ to $5\frac{1}{2}$ US mills/kWh' at 8% p.a. interest rate and assumed asset lives of 7 years for the bores and 20 years for other assets.

15.12.7 *USSR power costs.* The only published costs discovered by the author relate to the small 5 MW plant at Pauzhetka, in Kamchatka, for

Table 11   Production costs in Italy (1976).

Item	Cost			
	Condensing plants		Back pressure plants	
	Lire/kWh	*US mills/kWh*	*Lire/kWh*	*US mills/kWh*
Capital amort-ization and interest	3.7	4.77	1.9	2.45
Research and drilling	2.2	2.84	4.4	5.67
Exploitation and maintenance	2.1	2.70	1.2	1.55
Total	8.0	10.31	7.5	9.67

No Italian capital investment costs are available.

which an energy cost of 7.2 US mills/kWh was quoted [18] in 1970. Considering the small size of the unit and its very remote situation, this can be regarded
as a low figure even for 1970. It was claimed that this production cost was
30% lower than could have been achieved from any alternative energy
source in the area. No capital investment costs for Russian geothermal
power installations are known to the author.

15.12.8 *Japanese power costs*. The most recent Japanese power cost
estimates, quoted in March 1976, [144], are for the Kakkonda 50 MW
installation due to be commissioned this year (1977). They are as shown
in Table 12.

The *production* costs, based on a 20 year asset life and 10% interest,
and on an annual plant factor of 91.3% (8000 h p.a.) work out at $6\frac{1}{4}$ to
7 Yen/kWh, or 20.7 to 23.2 US mills/kWh.

These figures seem to be high by comparison with the geothermal costs
of power in other countries. Nevertheless, the Japanese claim that by
comparison with oil-fired thermal plants the geothermal costs are competitive. With oil fuel at 25 Yen/l the cost of oil-produced energy works out at
7.8 to 8.9 Yen/kWh, or about 25.8 to 29.5 US mills/kWh. Nevertheless,
it is well to remember that at the time when these estimated figures were
quoted, the international exchange markets were very unsettled, and the
conversion rates could have varied almost from day to day. In 1970 an
estimate of the production costs at Matsukawa [145] was given at Yen 2.63/
kWh, the figure at that time being more or less on a level with

Table 12   Production cost in Japan (1976).

Item	Capital costs	
	Yen (millions)	US$ (thousands)
1½ year's basic investigations	250	828
Roads and temporary housing	300	993
Exploration and test boring (1½ years)	1000	3311
Production wells (1½ years)	2500	8278
Re-injection well	500	1656
Power house, steam system, cooling towers etc.	7000	23179
Environmental expenses	500	1656
Totals	12050	39901

i.e. Yen/kW = 241 000   US$/kW = 798

conventional thermal production. This implies an approximate cost increase of $2\frac{1}{2}$ times in 7 years, or an average annual growth rate of about 14% in geothermal production costs.

15.12.9 *Comparative electricity costs.* In 1971 Bowen and Groh [146] produced the following comparative estimates of electricity costs supplied by different types of plant – all at an assumed annual plant factor of 90%, with 14% p.a. fixed charges, and valid for 1970:

	Geothermal	Nuclear	Hydro	Coal
Capital costs in $/kW	110.00	225.00	250.00	150
Production costs in US mills/kWh	4.86	6.5	4.55	5.92

These costs would of course now be very much outdated by inflation in general and the fuel price explosion in particular. It is interesting to note that although nuclear and coal produced electricity are estimated at 34% and 22% more than geothermal, hydro is shown slightly cheaper by about 6%. However, comparisons of this sort should be treated with reserve, as hydro costs depend so greatly on local factors and on the size of installation. The figures, however, are of some historic interest. It is emphasized that the geothermal capital costs are for the power station only and exclude exploration, drilling and steam supply. The *gross* geothermal capital costs per kilowatt would be considerably higher than for a coal-fired plant and more so for an oil-fired plant: it is the extreme cheapness of the *heat* that makes geothermal energy economically attractive.

15.13 Actual geothermal heat costs

Very little reliable, comparative, recent data are available for the costs of geothermal *heat*, except for a useful analysis by Einarsson [7] who quotes typical Icelandic costs of domestic heat of geothermal origin, as in 1974, in terms of the temperature at source, the transmission distance from the source to the market, and the capacity of the system (or size of the market) as represented by the diameter of the transmission pipe. He shows the effect of these three parameters in graphical form (Fig. 67), alongside which he shows the prices at which fuels must be available for them to be competitive if used for domestic heating. The diagram clearly shows that with rising fuel prices it pays to transmit geothermal heat over longer and longer distances before its delivered cost is too great to compete with fuel as the alternative. Einarsson expresses his heat in thermal kilowatt-hours; if

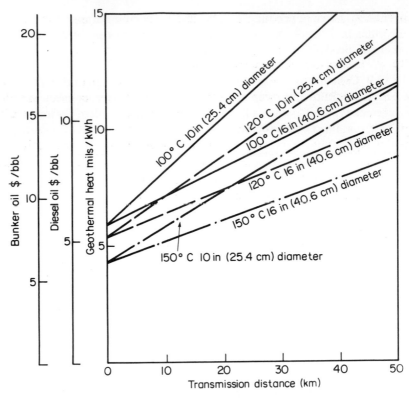

Figure 67　Geothermal heat costs for Iceland, valid in 1974, in terms of transmission distance, pipe diameter and temperature. (After Fig. 7 of [7].)

Note: The two left-hand scales show fuel prices above which geothermal heating becomes competitive. This shows that heat transmission over quite long distances was already economic by 1974.

kgcals are preferred a simple conversion can be made at 1 kWh = 860.5 kgcal. He wisely gives warning that, although conservative, his figures make no claim to great accuracy in view of shifting markets and different local site conditions. The distribution costs are based on those applicable to Reykjavik city, where the space-heating density is about 20 MW/km², and it is assumed that distribution is at 80° C regardless of the temperature at source.

Other recent heat cost information includes the prediction that the new district heating scheme at Creil outside Paris will cost from 16 to 20% less than fuel heating [198], a claim that geothermal space heating in Hungary in 1970 was costing only 27% of coal heating [104] – and this was before the worldwide fuel price rises that followed the 1973 oil crisis – and another claim by the Geothermal Data Bank that greenhouse heating at Galzignano, near Padua, Italy, is effecting a 30% cost saving by comparison with oil heating.

15.14 Comparative heat transmission costs

As is well known, the transmission of steam is very expensive: hot water can be transmitted far more cheaply in terms of heat content, because the greater density of water more than compensates for its lower enthalpy and lower practical velocity. Some idea of the *relative* heat transmission costs through a given pipe may be obtained from the following figures:

Saturated steam at 5 atü average pressure	7.88
Saturated steam at 10 atü average pressure	5.86
Saturated steam at 15 atü average pressure	4.87
Hot water at 100° C average temperature	2.46
Hot water at 150° C average temperature	1.46
Hot water at 200° C average temperature	1.00

These figures are only ratios, and are proportional to the reciprocals of the heat transmissible through an arbitrary length of pipe – 1000 ft (about 300 m) × 20-in (about 50 cm) diameter. A pressure drop of 5% of the absolute average pressure (or about 10% of the pressure at the entry end) is assumed in the case of steam, and a fluid velocity of 7 ft/sec (2.133 m/sec) is assumed for the hot water.

When we consider the relative costs of transmitting hot fluids for power generation purposes, the pattern is entirely different, because of the low power potential of hot water as disclosed in Fig. 47 and because of the relatively high cost of the terminal equipment (pumps, head tank, flash vessels and control valves and equipment) which greatly reduce the relative advantages of hot water as a heat transmission medium. Moreover, the position is complicated further by the fact that flashed hot water makes use of low pressure turbine stages which are intrinsically costlier than the high pressure stages.

The economics of two-phase fluid flow transmission in a single pipe (see Section 9.12) cannot yet be properly evaluated owing to lack of sufficient working experience. It seems probable that if no serious technical difficulties are encountered, it could prove very favourable.

15.15 Further economic points

15.15.1 *Importance of location.* Looking to the future when it may perhaps become possible to win earth heat at any place we choose, we shall of course extract the heat as closely as possible to the markets. At present, however, we are constrained to exploit this form of energy only where it can be found; and sometimes it occurs in most inconvenient places. It is sometimes argued that if a good field should be found in some remote situation far from populated areas, it is necessarily worthless. This is not inevitably true; for even those who quote this theory are obliged to define the word 'remote', which is a relative term. If the field is hyperthermal and of good quality and capacity, the generation of electricity could perhaps economically justify the transmission of power over one or two hundred kilometres or even more. However, if the field is semi-thermal, capable of yielding only low grade heat, it is clear that the word 'remote' would, at best, have to be limited to some tens of kilometres (see Fig. 67) if a 'local' market is being sought. Nevertheless, as implied in Section 12.2, if heat (even of moderately low grade) is sufficiently abundant and cheap and of an adequate grade for some large industry, it could sometimes pay to establish such an industry at the field, to transport the necessary raw materials and labour to the site – even across national frontiers – to set up a community centre there, self-sufficient in food (Section 12.2) and with adequate amenities and high standards of living comforts, and to ship the end products of the industry to the markets. Hence the known or suspected presence of large thermal fields in very remote locations should not be dismissed out of hand as valueless. If the possible scale of their exploitation is sufficiently large, fields in the most 'off-the-map' places could be of great commercial value.

15.15.2 *Value versus cost.* The cost of geothermal heat is not the only criterion of its value. The costs of local alternative energy sources of equal adaptability, and the political and environmental advantages of geothermal exploitation, are factors that must also be taken into account. Judgements on these criteria will not necessarily be valid for all time. In one year it could be concluded that geothermal development is uneconomic, whereas later – perhaps as a result of higher fuel prices or of strained relationships with another country from whom energy has hitherto been imported, or even of the discovery of cheaper methods of geothermal exploitation – the situation could be reversed.

15.15.3 *Diminishing returns.* Until, or unless, universal geothermal energy becomes available, it is only to be expected that the most promising hyperthermal fields will first be exploited and that the economically less attractive fields will follow later, due account being taken of location and the size of the market. This tendency has been historically true of hydro-power development, in which countries have tended first to 'skim the cream' off their hydro-resources and later to develop less attractive projects under changing economic pressures. However, this will apply only for as long as exploitation is confined to present methods. When, as will almost certainly happen one day, large scale deep earth-heat mining becomes economically practicable (Chapter 19), a new and far greater lease of life should follow.

16 Chemical and metallurgical problems

16.1 Chemical composition of geothermal fluids

Hitherto, apart from various passing references to non-condensible gases, geothermal fluids have been treated in this book simply as H_2O – hot water and/or steam. However, these fluids are invariably accompanied by gases and, in the case of wet fields, by water-soluble substances, some of which can be potentially dangerous to the construction components of a geothermal installation and to the environment.

The proportions of *non-condensible gases* vary widely from field to field, ranging from only about 0.1% by weight (in relation to the steam phase of the geothermal fluid) at Hveragerdi in Iceland to figures of 10%, 25% or even more in the Monte Amiata field in Italy. Typical intermediate figures (by weight) are:

Wairakei	from 0.35% to 0.5% (to steam)
The Geysers	from 0.6% to 1.0%
Cerro Prieto	about $1\frac{1}{4}$%
Larderello	from $4\frac{1}{2}$% to 5%
Matsukawa	about 1.1%

Of the non-condensible gases CO_2 always forms by far the largest component – from 80 to 97% by weight – with H_2S as the next largest ingredient – from 1% to 19%. The remaining gases consist of small quantities of CH_4, N_2, H_2 and sometimes traces of NH_3, H_3BO_3 and the rare gases.

The *water-soluble components* in geothermal waters usually consist of chlorides, silica, perhaps metaboric acid, sulphates and traces of fluorides. Waters are sometimes alkaline, as at Wairakei (pH = 8.6), and sometimes acid, as at Matsukawa (pH about 5). Rare cases have been known – e.g. the Tatun field in Taiwan – of pH values as low as 1.5, where the acidity has destroyed bore casings in a few hours and the field has been unusable. Sometimes traces of valuable minerals are to be found in thermal waters.

In wet fields, almost all the non-condensible gases remain with the steam phase while almost all the water solubles are associated with the water phase, though each phase may carry small traces of the chemicals normally to be found in the other.

16.2 Dangers of chemical action on plant components and on the environment

With the presence of all these chemical ingredients in geothermal fluids, it is only to be expected that corrosion or deposition is liable to occur at various places in a geothermal installation unless special precautions are taken. Likewise, since some or all of the chemicals are likely to escape at some stage into the environment, pollution can be caused unless suitable steps are taken to prevent, or at least to mitigate it. This will form the subject of Chapter 17. For the present only the vulnerability of exploitation equipment and plant will be considered. Strict attention must be paid to the metallurgy of all component parts, taking into consideration the chemical and physical conditions of the fluids where they come into contact with them. Wet fields tend to be rather more troublesome than dry fields owing to the greater variety of impurities to be found in the fluids and also to the catalytic action of the presence of water or vapour. In the early days of geothermal development it was common practice, before manufacturing the plant, to submit, for long periods stressed and unstressed test pieces of all materials, proposed to be used in construction to environments of bore steam, air/steam mixtures and bore water, so that careful measurements could be made of corrosion rates and samples could be subjected to microphotographic examination for sub-surface changes of metallic structure. In this way, vulnerable materials could be avoided and resistant materials used for the construction of the plant in appropriate places. Such tests are still carried out to some extent, though so much is now known of the effects of bore fluids upon construction materials that much less empirical testing is now necessary.

16.3 Chemical deposition

This is apt to occur at the following points:

16.3.1 *Calcite deposits in wet bores.* Certain waters rich in calcium salts are apt to give rise to the build-up of hard deposits in wet fields at the points where the uprising hot water starts to flash into steam. Such deposits at first reduce the fluid output owing to the restriction they form, and ultimately they may almost choke the bore altogether. The only practical solution to this trouble is to ream the bore periodically to clear the deposits. If this is necessary at rare intervals, perhaps about once annually, it is a trouble with which one can learn to live; but if calciting builds up too rapidly, it may be necessary to abandon the bore. The frequency of reaming depends largely on the fluid yield when the bore is clear of deposits, as well as the rate of build-up. With bores of very large output quite a large degree of calcite deposition can be tolerated for a bore to become economically unserviceable. The occurrence of calcite blockage of bores is related not only to the calcium but also to the CO_2 content of the bore fluid. Experience in New Zealand fortunately suggests that the rate of deposition tends to decline with time.

Calcite deposition may sometimes also occur in the rock formation if the permeability is not very good, as flashing may take place before the fluid reaches the bore walls. In such cases the fairly early abandonment of the bore may become unavoidable.

16.3.2 *Silica deposits from wet bores.* At Wairakei and other wet fields silica may be present in the bore water to saturation. As the temperature falls from the original value at depth, there will be a tendency for the silica to come out of solution. There are various allotropic forms of silica, of which (fortunately) the commonest is capable of remaining in solution in a state of super-saturation for long enough to prevent precipitation until the thermal water is clear of the plant. Thus at Wairakei, no silica deposition occurs in the bores, wellhead separators or flash-tanks; but in open discharge channels carrying waste hot water which is rapidly cooling, very thick deposits are built up which have to be periodically and laboriously chipped away. In the course of time, the underground waters feeding the base of the bores through rock fissures may slowly deposit silica if flashing occurs in the formation where fluid velocities are low. This can gradually reduce the fluid yield of a bore and, perhaps after a few years, lead to its abandonment. Thórhallsson *et al.* [147] have described the troubles experienced with silica deposition in Icelandic heating systems, and the remedy of dilution with fresh water that has shown some degree of success.

16.3.3 *Turbine blade deposits.* If separation of the steam from the water, in wet fields, at the wellheads is imperfect, and particularly when the bores are close to the generating plant (and the scrubbing effect of pipe lines – see Section 9.10 – is not therefore very effective), traces of solubles may reach the turbines of a geothermal power plant, and may be deposited from the small residual water fraction onto the blades as the temperature falls with expansion. This results in the gradual build-up of hard chemical deposits which can reduce the output of the turbine and perhaps also upset its dynamic balance. This trouble has been experienced at Cerro Prieto in Mexico, where the power station is fairly close to the bores, and has effectively reduced the station's annual plant factor to a figure of only about 70% (as against more than 90% sometimes achieved in other geothermal power plants) during the first two or three years of operation. Blade deposits have there caused a loss of power of 4 or 5 MW (out of 75 MW nominal) and turbines have had to be taken out of service from time to time for the deposits to be removed by sandblasting. Steam-washing with oil has been found to be effective in Italy [83] in reducing the quantity of blade deposits and in facilitating their removal. The risk of blade deposits is a factor that favours the adoption of fairly low working pressures, which are conducive to larger steam passages (see also Section 8.8).

16.3.4 *Iron sulphide deposits.* The action of H_2S on mild steel is to form a protective skin of iron sulphide which fortunately inhibits further attack. Thus, although a black deposit will usually be found on the inside walls of all steam pipes and pressure vessels, this is quite harmless. This has enabled the use of cheap mild steel for all these plant components, which form a high proportion of the whole equipment.

16.4 Corrosion and stress – corrosion
Despite the wide variation in the composition of the chemical impurities from field to field, the broad characteristics of geothermal fluids are similar in quality though they may differ in degree. Table 13 may be taken as broadly indicative of the resistance or vulnerability of the various materials likely to be used in component parts of a geothermal exploitation plant.

16.5 Turbine materials
As shown in Table 13, 13% chrome iron of the type normally used for turbine blades is, in the hardened or martensitic state, susceptible to stress – corrosion cracking in the presence of geothermal steam, but enjoys immunity if in the soft or ferritic state, provided that chlorides are not present except in minute quantities (not exceeding 10 ppm in the residual water droplets which may be present even in very small quantities in the steam). Turbine blades of softened and tempered stainless iron are therefore commonly used. However, softened metal is liable to suffer *erosion* in the presence of water droplets and, owing to the relatively high wetness of the steam at the lower pressure stages of a geothermal turbine (fed with dry saturated steam at inlet), this could be a hazard. It is inadvisable to protect the blades by brazing on erosion shields on account of the risks of local hardening and the vulnerability of spelter. It is therefore customary to limit turbine blade tip speeds to a moderate figure (say 900 ft/sec or 275 m/sec) so as to reduce the erosion risks, which are proportional to the *square* of the tip speed.

16.6 Condenser and gas extraction equipment
The most arduous conditions from the corrosion aspect occur in the condenser and gas extraction equipment of a geothermal turbine. Here, bore gas,

Table 13   Resistance or vulnerability of materials to chemical attack in the presence of geothermal fluids.

Material	Remarks
Category I. Resistant to corrosion	
Good quality close-grained cast iron	
Ordinary mild steel or low alloy steel (in the unhardened, fully pearlitic condition)	H_2S forms a protective coat of FeS, but below 30° C (86° F) this is not so, and the material becomes vulnerable. Also, the presence of oxygen makes the material vulnerable
13% chrome iron in the soft, or ferritic state (but not in the matensitic state). Also the same material, if hardened and tempered at 750° C	This material is suitable for turbine blades. It is essential, however, that chlorides are either absent or present to a minute degree only
Ordinary 60/40 brass	Resistance is poor in the presence of oxygen
Admiralty brass (70% Cu, 29% Zn, 1% Sn)	
Aluminium brass	
Pure aluminium	Vulnerable only at very high temperatures
Hard chromium plate	
Stellites Nos. 1 and 6	
18/8/3 (austenitic) Cr/Ni/Mo steel	
18/8 (austenitic) Cr/Ni steel, columbium or titanium stabilized; (non-precipitation hardening)	
Category II. Vulnerable to corrosion	
Copper and copper-based alloys other than the brasses shown in Category I	
Nickel and nickel-based alloys	
Brazing spelter	Potentially dangerous
Zinc	Unreliable at higher temperatures
Aluminium alloys	*Pure* aluminium is reliable (see Category I)
Category III. Vulnerable to stress–corrosion	
Aluminium bronze	
13% chrome iron in the hardened, or martensitic state	
A.T.V. (an austenitic steel)	11% Cr, 35% Ni, 1.3% Mn, 0.5% Si, 0.3% C
1.5/0.3 Mn/Mo steel	
Hardened carbon and low alloy steels (incorporating martensitic or quasi-martensitic structure)	
Most, if not all, stainless steels if chlorides are present (aggravated if oxygen is also present)	External stress–corrosion has occurred on stainless steel bellows where bore water has leaked onto them
13% chrome iron, even in soft state, if chlorides present except minimally	Hence the need for flash steam scrubbers to protect turbine blades

water vapour and oxygen (released from solution out of the cooling water in jet condensers) are all present, forming a proper 'witch's brew'. The internal surfaces of jet condensers and ejector condensers should therefore be protected with various proprietory resins or with aluminium spraying. Austenitic stainless steels should be used for gas exhauster rotors where the gases are wet, and inter-cooler tubes are usually made of 99.5% pure aluminium, which is resistant to geothermal steam except at high temperatures. Care should be taken not to over-cool the gases in the exhauster system, as this is conducive to attack on exhauster rotors.

16.7 Expansion bellows pieces

The 'concertina' pieces of expansion bellows are made of stainless steel. This material is immune to corrosion from geothermal fluids except when simultaneously exposed to oxygen and chlorides. On the inside surfaces of the bellows, oxygen will normally be absent as the bore fluid is always oxygen-free and as the maintenance of internal pressure will exclude the possibility of air entering from outside. There may, however, be a certain amount of chlorides present in the pockets of condensed steam that will collect in the underside corrugations – especially near the bores before the pipeline scrubbing effect has had time to act to any great extent. Only when a pipeline has been shut down for inspection and maintenance, and air admitted, should there be any danger from these pockets of slightly chloridic water. Before opening up a steam pipeline, it is therefore advisable to flush the pipe with clean water. Bellows pieces at the tops of expansion arches should be altogether removed and quickly flushed out when air has to be admitted into a line. On the *outside* of a bellows piece, the only danger likely to arise is if a leaky flange allows chloridic water to drip onto the corrugations exposed to the air. This has actually led to the failure of bellows pieces at Wairakei on one or two occasions. Great care should therefore be taken to ensure that expansion piece flanges are absolutely tight against leaks.

16.8 Standby corrosion

The internal parts of a geothermal turbine, provided the blading is made of the correct materials, are normally immune against attack from geothermal steam. However, when a machine is shut down, air may enter through drains or glands as the trapped steam condenses as a result of radiation heat losses. This air, in the presence of the non-condensible gases remaining in the turbine, could cause serious corrosion of both stator and rotor. Hence it is advisable to provide an electrically heated hot air blower that will quickly dry out and purge the remaining wetness, vapours and gases, so that all chemically hostile substances may be purged from inside the machine before the top half of the stator is removed if inspection or repairs are necessary. Suitable connecting and venting points must be provided on the turbines. The stored heat of the hot metal of the turbine will speed this drying out and purging effect.

16.9 Chemical attack on equipment from polluted atmosphere
In the neighbourhood of geothermal fields, even before exploitation, there is usually a rather strong smell of H_2S. Although natural selection may have ensured that local vegetation is unharmed by this gas, the same immunity will not extend to all man-made objects after exploitation. In Chapter 17 it will be explained how the source of this hazard – the very presence of H_2S – will be greatly reduced if not entirely eliminated in due course; but for the present it cannot be ignored. Unless special precautions are taken, exploitation will tend to increase the atmospheric concentration of H_2S. Aluminium conductors in substations and on transmission lines will usually take on a protective coating of black sulphide (as in the case of the inside surfaces of mild steel vessels exposed to geothermal fluids) which inhibits further attack. However instruments and relay contacts will almost certainly suffer if they feature exposed copper, as sealing is seldom perfect. Contacts and bare connectors of silver are advisable and, as an additional precaution, control rooms may sometimes be sealed with air-locks and ventilated with purified air. Exciter commutators of copper can be very troublesome, not only because the copper itself is attacked by H_2S but also because the sulphide film causes sparking at the brushes which wear away at an alarming rate. Self-excited generators are now therefore generally used.

16.10 Sulphate attack on cooling towers
Carry-over of bore gases into the condensate and cooling water can cause sulphate attack on the concrete sructures of cooling towers. At Otake, Japan, alkaline river water has been introduced into the cooling system as a means of countering this hazard [148], while at the Geysers, California, a coal tar epoxy protection is provided on concrete surfaces [149].

16.11 Conclusion
The subject of chemical and metallurgical problems in geothermal installations is too vast to be adequately covered by a single chapter in a general book such as this. For further and more detailed information the interested reader is referred to [150–154] in addition to the references that have appeared earlier in this chapter.

17 Environmental problems

17.1 Pollutive masochism

A century ago, except for a few murky industrial areas and slums, the world was on the whole a beautiful place. It is true that a few large cities suffered from choking fogs due to the excessive local burning of coal, but the country-side – whether dominated by untouched primeval features or by a happy blend of nature and human husbandry – was quiet and restful: an infinite variety of flora and fauna flourished as it had for aeons past. The steamship and the railway scarcely affected the Eden-like quality of most of the earth's open spaces. Then came the motor-car, the aeroplane, the industrial 'explosion', the centripetal migration of people from the country into the cities, the oil age, the proliferation and use of weapons capable of ever-increasing devastation; above all, the alarming increase of population (running now at an excess of new-born babies over deaths at a rate of some 160 human beings per minute). These teaming millions consume ever-increasing quantities of food and energy, and produce more and more indestructible wastes. The growing millions of motor-cars require more and larger roads to be carved out of our farming lands, and pour vast quantities of poisonous fumes into the atmosphere. In densely populated countries, ugly housing estates rob us of ever more agricultural land, while 'ribbon development' is tending to turn the greater parts of these countries into a uniform and soulless 'subtopia'. The human thirst for electricity is rendering the country-side hideous with forests of huge 'pylons'. Chemical wastes pollute the rivers killing fish, and occasionally produce clouds of lethal vapours that have taken toll of human and animal life. Leviathan tankers are sometimes wrecked at sea, while many other ships callously and deliberately discharge oil waste into the oceans, spewing hundreds of thousands of tons of oil upon the surface of the waters, endangering bird life and fish and fouling the beaches. Sewage is being poured untreated into the sea, with horrible consequences, including the slow destruction of one of the world's most beautiful cities – Venice. Callous people throw their litter – much of which, like plastic and glass, is indestructible – all over the countryside, while the dismal sight of car cemeteries is becoming commoner every year. The annual combustion of thousands of millions of tons of coal, or the equivalent in other fuels, is pouring noxious fumes into the air and raising the CO_2 content of the atmosphere so rapidly as to alarm the climatologists as to the possible adverse effects on the world climate. In recent years the CO_2 content of the earth has been rising at a compound rate of 0.7% p.a., which implies doubling in a century. This gas has a 'greenhouse effect' in that it admits solar heat but inhibits its escape. The result of rising CO_2 content could be to make the earth hotter and, in the extreme, to approach the intolerable conditions that prevail on the planet Venus.

Predatory men are killing many animal species, such as the whale, the elephant and other wild game, at a rate that makes their early extinction (following the disappearance of the dodo and the moa) more than a possibility. All this has been accompanied by more and more noise, a horrific decline in the quality of the Arts and a growth of philistinism. In short, Man is very busily fouling his nest at an alarming rate.

It is true that there are a few trends in a happier direction. The virtual death of the domestic coal grate has made the 'pea-soup' fogs of London a feature of the past. The heating of Reykjavik by means of earth heat has created a city of almost pure clean air. Many societies have been formed for the preservation of wild life and for the protection of places and buildings of outstanding beauty and interest. The quality of the water in the Thames has recently been so greatly improved that fish life, not many years ago on the verge of extinction, has almost returned to normal. However, these commendable trends, welcome though they be, amount only to a gentle zephyr in competition with the hurricane of self-inflicted pollution that threatens to engulf us.

Until about the middle of this century, despite the warnings of a tiny minority of thoughtful people, the world in general remained quite indifferent to what we were doing to ourselves. Those who thought about the problem at all were mostly apt to dismiss it with a cynical 'après nous le déluge'. During the last two or three decades, however, there have been welcome signs of a belated stirring of the public conscience concerning the growing threats to the quality of our environment.

17.2 How can geothermal development help solve the problem of pollution?

Geothermal enthusiasts have been rather too ready to claim that the exploitation of earth heat is an entirely 'clean' process. Certainly it is far less guilty of pollution than fuel combustion, but it must frankly be recognized that it is not entirely blameless, unless certain precautions are taken: even these precautions cannot *totally* eliminate pollution, though they can reduce it to acceptable levels. The enthuiasts do not serve the cause of geothermal development if they fail to face the facts. Fortunately, the necessary precautions can mostly be, and in some cases are already being, taken with remarkable efficiency and at not excessive cost. The 'laissez faire' attitude that prevailed in the early days of geothermal development has now become very unwise in view of the greatly increased scale of development. It is far better to examine the environmental problems one by one and to consider how they may best be overcome. Experience in California has already shown that pollution from geothermal power development can be reduced almost to zero. Hence, the large scale future exploitation of earth heat that may confidently be expected should go a very long way towards reducing the pollution of our environment.

17.3 American anti-pollution laws, and the cost of their too rapid enforcement

In the USA very stringent anti-pollution laws have been enacted and are now being rigidly enforced at the power plants and steam fields in the Geysers area, not only at all the newer installations but also retrospectively at the older ones. Although these laws are on principle to be welcomed, it is arguable that they have been too suddenly and too rigidly applied. Full compliance with the new laws has had the effect of delaying construction by a year or more, and delay costs money. It has already been mentioned that 1 kW of geothermal base load can save about 2 tons of oil fuel annually. With oil fuel at US \$80/ton the cost of a year's delay would be about \$160/ kW; and a glance at Table 9 will show that a delay of about $1\frac{1}{2}$ years would have the effect of approximately *doubling* the cost of a Geysers power station. It would seem that a more gradual introduction of the anti-pollution laws, with some permissible relaxation over a few years, would have been more in the national interest, especially as the marginal effects of geothermal development in California upon pollution would undoubtedly have been far less than those caused by the combustion of the fuel that could have been saved by avoiding the delays consequent upon the rigid and immediate enforcement of the new laws.

17.4 Re-injection

Before considering, one by one, the various pollutive effects of geothermal development, it is well first to examine an operation that could well provide the antidote to several of these effects, namely re-injection of thermal fluids into the ground after use. The possibility of re-injection was considered many years ago, but was regarded with some timidity because there were many doubts as to the practicability of the process. One of the uncertainties was how the re-injected water could be persuaded to re-enter a pressurized aquifer. This invoked the time-honoured problem of the veterinary surgeon who is trying to administer a pill to a sick horse through a blow-pipe. Would the horse blow first? In more technical terms, would excessive pumping power be absorbed in forcing the unwanted fluid into the ground? Another question was whether the introduction of cooler waters into the permeable substrata would interfere with the useful output of heat from the productive bores. Would the permeability of the aquifer be reduced or even totally eliminated by chemical deposition? Would re-injected waters outcrop elsewhere, thus simply transferring the pollution problem from one place to another? Would re-injection trigger off seismic shocks by inducing cooling stresses in the rock?

At the UN Geothermal Symposium held in Pisa in 1970 some discussion was devoted to re-injection, but there were many sceptics then. By the time the next UN Geothermal Symposium was held, in San Francisco in 1975, much useful empirical data had been gained and, although it is too early to be dogmatic about the pros and cons of re-injection, practical experience now offers promising evidence that it could often (but not necessarily always) be not only practicable but perhaps sometimes even thermally advantageous. For by re-injecting at some strategic point in a field, well removed from the producing bores, it could be possible to impose a warm barrier that delays the ultimate ingress of cold water from beyond the confines of a field. Re-injection could thus become a useful tool, in certain cases, in field management, by re-cycling both water and heat, and could provide the equivalent of 'boilder feed' as an alternative to the wasteful rejection of heat at the surface. Even if the heat of the hot water can be usefully exploited before re-injection, the return of the cooled liquid to the substrata could avoid toxic pollution and could perhaps provide a medium for picking up the heat of the hot rock left behind after the extraction of the original hot water contained in an aquifer. More experience is needed before a balanced judgement can be pronounced, but it is now known that re-injection can be achieved in several fields by gravity alone, without the aid of pumps; and it is already being practised in El Salvador [155], California [156], France [157], and Japan [158]. There could undoubtedly be occasions where local conditions, such as excessive silica or low subterranean permeability, would preclude the adoption of re-injection – at least without prior chemical treatment and/or settlement facilities – but it seems probable that the practice will become increasingly adopted and will thus greatly asist in the prevention of pollution.

Experiments in El Salvador [155] have been performed by entraining tritium tracer with the re-injected water. After two days a very small quantity of this tracer appeared in the discharge of another well about 400 m away, while after a few weeks much smaller traces were detected in two other wells at distances of about 400 m and 1050 m from the injection well. Less than 1% of the tritium re-appeared. As in this case a production bore was used for re-injection experiments, whereas if re-injection were to be practised continuously a far more remote site would probably be chosen, this would seem to show that direct interference – at least in the Ahuachapan field – would not be a significant problem.

The great value of re-injection is that it offers at least a possible solution to all the forms of pollution described in Sections 5, 8, 11, 12, 13 and 17 of this chapter.

Experience at the Geysers shows that re-injection is not very expensive. The 1976 estimates of the charges to be made in 1978 by the steam suppliers to the Pacific Gas & Electric Co. are 13 US mills/kWh for steam supply *including* only 0.5 US mills/kWh for re-injection – i.e. re-injection is expected to account only for 3.85% of the total steam charges, and a far smaller proportion – only 2.14% – of the total production costs.

17.5 Hydrogen sulphide

H_2S is almost invariably present in geothermal fields. Its properties and dangers have already been mentioned in Chapter 14.4.3. Rare fatalities have occurred in the vicinities of fumaroles, but fortunately no incidents have yet been reported from this hazard in geothermal exploitation plants. This can probably be attributed to the provision of adequate ventilation in cellars and basements – a most necessary precaution. H_2S also attacks electrical equipment (see Section 16.9), and it may have adverse effects on crops and on river life.

At power plants H_2S occurs in high concentration at the gas ejector discharge points. Until recently it has been deemed sufficient to place these points at high level and to rely on the high temperature of the discharged gases to produce the necessary buoyancy and thus ensure their wide dispersal. Apart from this precaution, and the provision of good ventilation in buildings, the tendency in the past has been more or less to ignore H_2S on the argument that in any case it occurs naturally in thermal fields. However, this overlooks the fact that exploitation concentrates and increases the release of this gas into the air. Moreover, the scale of development has now grown so rapidly in certain fields that greater precautions are now necessary. It has been estimated that if suitable precautions had not been taken, no less than 28 tons/day of H_2S would have been emitted in the Geysers field [159] in 1975. By the time that 15 units are in service, this figure will have risen to about 50 tons/day. At Cerro Prieto about 55 tons/day of H_2S is being discharged into the air with the present 75 MW installation [160]. By 1983, if the present extension plans are realized, this figure will have risen to something approaching 300 tons/day. Such quantities cannot be ignored indefinitely. American legislation now insists on the reduction of the H_2S content of the air to within the threshold of odour, so that if it can be smelled, the law is being broken. This has led to a curious *reductio ad absurdum* in that Nature herself is guilty of breaking the law in several places such as the Yellowstone National Park! At the Geysers, all regular escape paths of H_2S are being trapped and the gas is being burned to form SO_2, which is then scrubbed with cooling tower water. This causes the precipitation of elemental sulphur which can be filtered out, while the contaminated water is re-injected into the ground [161]. The only 'untamed' escape paths at the Geyesers would be from rogue bores which, thanks to good drilling practice, are unlikely to occur, and from escaping steam from wells under test or from piped steam at times when the plant cannot absorb all of it so that some has to be vented to the air in order to maintain proper pressure control at the plant (see Chapter 14): fortunately such occasions occur infrequently, so that the pollution resulting therefrom is negligible. Where river cooling is adopted, as at Wairakei, some H_2S escapes in solution with the condenser cooling water. With rivers of large flow, as in the case of the Waikato River at Wairakei, this is probably unimportant (though this is arguable), but with small streams the only solution would be conversion to cooling towers and the adoption of the process used at the Geysers, which is known as the Claus process. Where bore water from wet fields is discharged into rivers, the dissolved H_2S could affect fisheries and weed growth: it has been suggested that this may in fact be happening at Wairakei despite the large rate of river flow. If trouble is experienced from this, re-injection seems to be the only practical answer.

17.6 Carbon dioxide

By far the greater part of the incondensible gases from a thermal field consist of CO_2. This can escape into the air or into local water courses. The fact that fuel combustion usually produces far greater quantities of this gas in terms of the heat consumed than does geothermal exploitation has generally been regarded as an excuse for inaction, particularly as the gas is not toxic. However, in certain exceptional fields, such as those in the Monte Amiata region of Italy, the CO_2 discharged to the air may be far greater than from fuel-fired plants of comparable size and duty. As stated in Section 17.1, concern about the rising CO_2 content in the atmosphere is already worrying the climatologists, and it is known that high CO_2 content in river waters can aggravate weed growth. It is most undesirable that geothermal exploitation, especially if this is to bear a major share of the world's energy burden in future, should contribute towards these ill effects even though it may be far less guilty, on average, than fuel combustion. Ideas for the commercial extraction of CO_2 from effluents are being closely studied – e.g. the manufacture of dry ice, carbonic acid for beverages, the production of methyl alcohol – but have not yet proved economic. (Moreover, in each of these cases the CO_2 would be returned to the atmosphere after use!) Meanwhile the emission of CO_2 to the atmosphere seems to be inevitable until some permanent method of chemical fixation can be found. The problem has not yet become seriously urgent, but if geothermal development grows dramatically, as well it soon may, it will have to be tackled seriously before long.

17.7 Land erosion

At the Geysers field, where heavy rains and steep slopes of incompetent rock often cause natural landslides and high erosion rates, the artificial levelling of the ground for the accommodation of field works, roads and power plants has sometimes aggravated these conditions by creating very steep local gradients (Plate 21) and by the removal of local vegetation which normally protects the land from erosion by the binding action of its roots. Closer control, replanting of shrubs and trees, more careful site selection and improved construction methods are helping to solve this problem [159]. The close spacing of several wells within a single levelled area (see Fig. 18) could sometimes be helpful.

17.8 Water-borne poisons

The water phase in wet fields sometimes contains toxic ingredients such as boron, arsenic, ammonia and mercury which, if discharged into water courses, could contaminate downstream waters used for farming, fisheries or human water supplies [162, 163]. Although not strictly 'poisonous', very saline bore waters too can be harmful. Possible solutions to the problem of water-borne poisons include re-injection, disposal into the sea (if not too remote) through ducts or channels, the use of evaporator ponds as at Cerro Prieto [164, 165], storage during the dry season (in 'monsoon' countries) with subsequent release into rivers when in spate during the wet season when dilution would render the offending substances more or less innoccuous, or even chemical treatment [162]. Another possibility is to *desalinate* the bore water, using bore heat for the purpose, where there is a demand for fresh water; and either to re-inject the concentrate of poisonous ingredients or to market it if the recovery of the ingredients should have commercial possibilities.

17.9 Air-borne poisons

From ejector exhausts, from the upward vaporous effluents from cooling towers, from silencers, drains and traps, from discharging bores under test, from 'wild' bores and also from control vent-valves, various harmful substances (beside H_2S) may sometimes escape into the air at geothermal sites. These may include mercury and arsenic compounds and radioactive elements. Certain noxious, though not poisonous, emissions such as rock dust and silica-laden spray may also be air-borne. At Wairakei a certain amount of forest trees were damaged by silica deposition, while in

El Salvador the contamination of coffee plantations was a problem. At Cerro Prieto salt deposition on buildings and agricultural lands caused some trouble [165] while at Wairakei the build-up of hard silica on car windscreens, from the spray emitted from discharging bores and silencers, has caused nuisance. Horizontal or inclined well discharge in a controlled direction, where there are no vulnerable 'targets' can sometimes offer a solution to this problem, but this is clearly no more than a makeshift solution that would seldom be acceptable. Fortunately, air-borne toxicants are seldom present in harmful concentrations (other than silica, which is not truly a 'toxicant' but rather a nuisance); but systematic monitoring is advisable to enable a careful watch to be kept on possible future dangers.

17.10 Noise

The nuisance caused by noise can cause a serious health hazard unless workers on new well sites and at test sites wear ear-plugs or muffs, without which their hearing can be permanently damaged. Though the noise from regular escape paths of fluids can be reduced by means of silencers, there are occasions when noisy discharges cannot be avoided; e.g. bores newly 'blown' or those undergoing test or maintenance. The erection of temporary sound barriers can sometimes be helpful in such cases. Re-injection can eliminate the noise arising from flashing steam where hot water is now being discharged to waste (as at Wairakei). Drilling operations can be somewhat noisy, but no worse than road repairs, and they do not persist for very long. Noise also occurs in and near power plant buildings, from the whirr of machinery, but this is no worse than in conventional thermal power plants and is very difficult to control in machinery halls or near outdoor auxiliaries such as ventilating fans. Control rooms and offices can be sound-proofed, so that most of the attendant personnel are protected. Legislation against noise has been enacted in the USA and other countries, and although its strict enforcement may sometimes be difficult its very existence acts as a powerful incentive to designers to overcome the nuisance.

17.11 Heat pollution

This can sometimes be very troublesome. Apart from being pollutive, it is very wasteful and so every effort should be made to avoid its occurrence. The necessary adoption of relatively low temperatures for geothermal power production must result in low efficiencies and the emission of huge quantities of waste heat in far greater proportions than in conventional thermal power plants of the same capacities. Where cooling towers are used, this waste heat escapes into the air and by way of the surplus condensate which, in the latter case, may be re-injected. With direct river cooling the waste heat simply raises the temperature of the river. This of course is a wastage that differs only in degree from that which occurs in conventional thermal power stations, with the added factor that higher condenser back-pressures are usually adopted in geothermal power plants, so that the water is a good deal warmer. What is far more serious is when the hot water phase in a wet field is discarded to waste, for even if power is generated from the flash steam from this water, the temperature of the water at discharge is likely to be at or near to the atmospheric boiling point. The discharge of huge amounts of such hot water into rivers and streams (as is being done at Wairakei) can damage fisheries and encourage the growth of unwanted water weeds. If, as at Cerro Prieto, the hot water is simply discharged into lagoons, from which it can evaporate, the heat escapes as vapour into the air. The escape of heat and vapour from cooling towers and 'dump' lagoons may affect the local climate by the formation of fog and/or ice [166], although in certain climates the increased humidity could have a beneficial effect. The most damaging form of heat pollution is undoubtedly the discharge of very hot waters into rivers or streams. Possible remedies include re-injection of unwanted hot bore water, and also of surplus cooling tower water; the generation of additional power by means of binary fluid cycles; or, best of all where possible, the establishment of dual or multi-purpose projects which usefully extract low grade heat from turbine exhausts and/ or from rejected hot bore waters. Where rivers of high flow rates are used to receive very hot discharged waste waters, as in the case of the Waikato River at Wairakei, a zone of heated water is apt to be formed along one bank, extending downstream for perhaps half a kilometre until mixing has become complete. Apart from a few casualties, most of the fish learn to avoid the very hot zone. At Wairakei, during normal river flow, the average temperature rise of the river, allowing both for the condensers and the hot bore discharge, is about $3°F$ after mixing, and about $5°F$ at times of low flow. (At a time of drought in 1974, a temperature rise approaching $11°F$ was actually observed). It has been claimed that heat pollution has had an adverse effect on the fish in the Waikato River [167]. With smaller rivers or larger power installations, the effects could of course be worse.

The question is sometimes asked whether heat pollution from our growing energy consumption could ultimately affect the world climate. In Chapter 4 it was mentioned that the *natural* outward flow of earth heat amounts only to about 1/40th of 1% of the heat received by the earth from the sun. At present the annual world energy consumption is about 250 million TJ, equivalent to a continuous steady power consumption of about $8 × 10^9$ kW, compared with a natural outward flow of earth heat of about $3 × 10^{10}$ kW (thermal) – see Section 3.5. Thus the present world energy consumption is only about 8/30ths of 1/40th of 1% of the solar energy falling continuously upon the earth – i.e. about 0.0067%. Practically all of our energy consumption re-appears as heat surrendered to the atmosphere and seas, but this is at present such a microscopic, proportion of the solar energy we receive that we could clearly afford to multiply our energy consumption by quite a large factor without imposing any *thermal* threat to our climate. The question of CO_2 affecting the upper atmosphere (see Section 17.6) is another matter.

17.12 Silica
Silica has already been mentioned under Section 17.9 as a relatively minor form of air-borne pollution, but silica can be extremely troublesome with district heating systems [147]. Re-injection of silica-laden waters could foul up the permeability of the substrata in the aquifer, thus necessitating the frequent changing of re-injection bore sites. Settlement ponds can mitigate this trouble, but various physical and chemical methods of treatment may offer more promising solutions to this problem [160, 162].

17.13 Subsidence
The withdrawal of huge quantities of subterranean water from a wet field can cause substantial ground subsidence which could result in the tilting and stressing of pipelines and surface structures, perhaps with serious or even disastrous consequences. Vertical movements have been observed at certain points in the Wairakei field of up to $4\frac{1}{2}$ m in 10 years [168], together with smaller, but appreciable horizontal movements also. Re-injection is really the best solution to this problem, for although it would not replace the whole of the withdrawn fluid (the steam phase having been used up), natural recharge could well make up the difference.

17.14 Seismicity
Fears have sometimes been expressed that prolonged geothermal exploitation of a field could trigger off earthquakes, especially if re-injection is practised in zones of high sheer stress where fairly large temperature differentials could occur. These fears arise because all existing exploitations of hyperthermal fields are in naturally seismic areas where ground instability is self-evident, and any interference with nature might precipitate earthquake shocks [166]. At the present state of geothermal development these risks do not seem to be great but, if very large scale geothermal exploitation should be undertaken in future, the problem could perhaps become a serious one. Fortunately it is already the subject of study, so that reliable information and means of forestalling the danger should be ready in good time.

17.15 Escaping steam
The large quantities of waste heat escaping from the tops of cooling towers at geothermal power plants give rise to immense volumes of water vapour: the same may be said of dump lagoons, as used at Cerro Prieto. In the dry

climates of California and North-west Mexico this vapour is usually absorbed quickly into the atmosphere (Plate 22) but, in colder and more humid climates, immense clouds of billowing condensing vapour can and do persist long enough to create serious local fog hazards (Plate 23), and in some cases cause ice precipitation. Worse than cooling towers can be the clouds of condensing vapour issuing from the tops of silencers in wet fields such as Wairakei (Plate 11), where vast quantities of superheated flashing hot water are being discharged to waste. Bore steam escaping from control vent-valves can also be troublesome, but if properly adjusted this can be kept to small proportions except on the rare occasions when an unexpected change of plant conditions (such as a tripped-off turbine) gives rise to a large escape of steam through the vent valves. The drifting of large clouds of condensing steam across roadways can give rise to traffic hazards. Warning signs and diversionary routes can of course mitigate the trouble, but the best remedies for dealing with the nuisance arising from discharged hot water are either to use the inherent heat productively for some industry or other application, or to re-inject. These steps would not of course cure the trouble from cooling tower vapour or from vent-valve discharges, or from bores newly blown or under test. In practice, no really serious inconvenience has arisen from this purely local and relatively minor nuisance, with which one must learn to live, after taking whatever precautions as may be practicable.

17.16 Scenery spoliation
It is virtually impossible to be entirely objective in matters of aesthetics, but it cannot be denied that thermal areas often occur in areas of outstanding beauty, highly prized by the local population and frequented by tourists. Conservationists will often oppose geothermal development on the ground that scenic amenities will be destroyed. Though it cannot be denied that man-made engineering works can seldom compete aesthetically with natural scenic beauty, it must be conceded that the power installations in the Geysers field, California, have been most tastefully camouflaged. The pipelines have been painted to blend with the background; scarcely a puff of steam is visible except on the rare occasions when vent-valve operation is unavoidable or when a well is newly blown or under test; the power plants themselves are relatively inconspicuous; and the dry climate quickly absorbs the plumes of water vapour arising from the cooling towers. At Wairakei, New Zealand, where the scarred ground surface has been rehabilitated by careful 'landscaping' and damaged trees have been removed so that only the healthy forest can be seen, visitors flock to see the geothermal development in greater numbers than those who frequented the area before exploitation. It could even be claimed that the Wairakei scene has a certain majesty of its own, where the billowing steam from the wellhead silencers–

itself a form of 'pollution'–contributes a dramatic aesthetic quality (Plate 11).

It is true that geothermal exploitation can interfere with natural surface manifestations. For instance, the activity of the geysers and boiling pools in the once famous Geysers Valley, close to the exploited area of Wairakei, has virtually ceased. The question of scenic amenities is one that can only be judged subjectively on a balance of considerations, by weighing the value of the energy won against that of touristic attractions and national heritage, a balance that must be largely emotional and cannot be strictly quantitative. The declaration of an area such as Yellowstone Park as a zone of outstanding beauty and interest, not to be exploited for the winning of energy, would be a value judgement having its protagonists and its opponents: there can be no absolute standards in such matters. There is a certain schizophrenic streak in human nature. We may hate the sight of electricity transmission lines extending over the countryside, but we would bitterly complain if our homes were deprived of electricity. We become demented by the noise of aircraft, forgetting the joys of travel that have been brought to us by the aeroplane; we object to the depressing sight of industrial areas and mining works, yet we enjoy their products. In so far as geothermal development is concerned, it is only fair to point out that we cannot do without energy, and that a geothermal power plant, which produces no smoke, has no unsightly chimney stacks, no ungainly coal- or ash-handling equipment, no coal storage yards or oil storage tanks, and no boiler house can be far less displeasing to the eye than a fuel-fired plant of the same capacity. For similar reasons, industrial establishments using geothermal heat are likely to be far less obtrusive than those which rely upon fuels. On balance, it may justly be claimed that geothermal exploitation is far less guilty of scenery spoliation than fuel combustion.

17.17 Ecology

This is a subject that has only recently begun to attract well-deserved attention. The discharge of chemicals into the air and into streams and rivers and thence into the ground water; small, but appreciable local changes of temperature and humidity; noise; a degree of deforestation; all these factors could, and possibly do, disturb the natural balance of nature prevailing in a thermal area before exploitation. The effects of geothermal exploitation upon local fauna and flora are already being studied in New Zealand, California and very possibly elsewhere. Protectors of wild life and of fisheries would do well to encourage more intensive research in this direction, although it is only fair to say that on present evidence there seems to be little cause for alarm. However, it is always possible that the development of certain fields hitherto untouched might have effects not yet observed elsewhere.

17.18 Conclusion

The comparatively recent upsurge of interest in geothermal development has roughly coincided with the genesis of a public sensitivity to the dangers threatening our environment. This is probably a happy coincidence, for the latter has acted as a timely check to a certain 'laissez faire' attitude that accompanied the earlier geothermal exploitation projects. This attitude, although relatively harmless when the exploitation of earth heat was in its infancy, could later become a serious hazard were it allowed to persist until this form of energy bears a major share – as it undoubtedly will – of the world's energy burden. A clear change of mood can now be discerned in geothermal circles from one of unreasoning optimism to one of sober understanding of the environmental aspects of geothermal exploitation. Gone is the pious belief that earth heat exploitation is entirely 'clean' and wholly guiltless of infecting the environment. Nevertheless, it is an undoubted fact that geothermal exploitation is far less culpable than fuel combustion of fouling the human nest, and that an antidote can be found to nearly every potential source of geothermal pollution. The more rapidly the use of earth heat takes over the energy burden from fuels, the better will it be for our human environment. Timely legislation in certain countries has enforced attention to this very important matter, even though it could perhaps have been introduced at a rather less drastic pace. The advances made in environmental studies in the 1970s have been impressive, and the good work may be expected to continue.

(Also of relevance to this Chapter are [169–172]).

18 Miscellaneous points

18.1 Systematic approach

Exploration for geothermal fields can be a costly affair, perhaps involving a few millions of US dollars, depending partly on luck and partly on the skills of the exploration team. To minimize the cost it is absolutely essential that development should be set about systematically. So often on past occasions countries have squandered their limited resources in looking for exploitable earth heat in an unsystematic way. They may have flitted from area to area, sinking a few holes here and a few holes there, in places chosen solely on the superficial evidence of surface manifestations which, as explained in Section 6.3, are not always a reliable guide to the exact location of a field and may even be entirely absent from an area overlying a valuable field. Such hit-or-miss methods are followed in the pious hope of making a lucky strike. International 'experts' may have been invited to pay short visits to thermal areas in the expectation that they will be able to assess the geothermal possibilities from a superficial inspection lasting only a day or two. Such unmethodical activities are not necessarily wholly wasted, for every hole sunk adds to the cumulative knowledge of an area, and every visiting expert may contribute some useful suggestions as to the next steps to be taken. It must be admitted that successes have been sometimes achieved by such haphazard methods. Nevertheless, without a systematic approach there will always be a risk that much money may be needlessly wasted. If exploration costs are to be kept to a minimum, and if good results are to be obtained quickly, assuming that a useful field *does* exist, it is vitally important that a logical system be followed from the outset. Even if no field exists at all, at least that negative fact can be established more quickly and at less cost by proceeding systematically.

A suggested methodology has been proposed [173] and is schematically illustrated in the process diagram shown in Fig. 68, which is more or less self-explanatory. Its main purpose is to concentrate first on the less expensive investigations and to postpone drilling operations, which absorb by far the

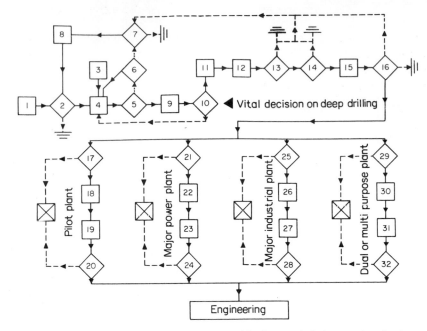

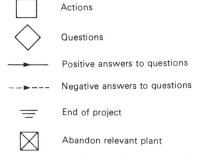

1. Inventory of alternative energy costs
2. Is geothermal energy likely to be competitive?
3. Collection and review of preliminary field data
4. Collection of more extensive field data
5. Are data enough for conjectural field model?
6. Are data sufficiently promising to proceed?
7. Is there an alternative area worth investigating?
8. Selection of alternative area
9. Conjecture of tentative working model
10. Are data sufficient to justify deep exploration drilling?
11. Minimum economic yield of bores
12. Deep exploratory drilling and borehole measurements
13. Are production bores likely to yield economic minimum?
14. Are corrosion problems soluble?
15. Estimation of energy potential of field
16. Are other engineering problems soluble?
17. Is a pilot plant required?
18. Preliminary design of pilot plant, with estimates
19. Re-appraisal of economics of pilot plant
20. Is pilot plant still economic?
21. Is field potential enough for major power plant (>20 MW)?
22. Preliminary design of major power plant, with estimates

23. Re-appraisal of economics of major power plant
24. Is major power plant still economic?
25. Is field potential enough for major industrial plant?
26. Preliminary design of major industrial plant, with estimates
27. Re-appraisal of economics of major industrial plant
28. Is major industrial plant still economic?
29. Is there a demand for waste heat from power plant?
30. Preliminary design of dual or multi-purpose plant, with estimates.
31. Economic study of dual or multi purpose plant
32. Is dual or multi purpose plant economic?

☐	Actions
◇	Questions
→	Positive answers to questions
--►---	Negative answers to questions
≑	End of project
⊠	Abandon relevant plant

Figure 68 Suggested process diagram for systematic approach to geothermal development. (From [173].)

greatest part of exploration costs, until a reasonably probable field model has been conceived which offers good chances of a high success ratio in the subsequent drillings. Actual exploration techniques are not described or specified either in Fig. 68 or in [173] because these are covered by the available literature: it is the *procedure* that is emphasized.

18.2 Legal questions

Before any country attempts to develop its geothermal resources it is most advisable that the legal position be fully clarified, and that suitable legislation be enacted as quickly as possible if the situation is not clear. Is the exploitation of earth heat to be a Government monopoly, or is it to be open to private enterprise? Is earth heat to be treated, legally, as a mineral, as ground water, or as neither? What are the rights of a land-owner under whose property lies an exploitable field, even though the field may be tapped (perhaps by taking advantage of subterranean flow patterns or perhaps by using directional bores, as in Fig. 18) from land owned by another party? Has the Government the right of compulsory acquisition, at a fair price, of land from which earth heat can supposedly be exploited? What of the rights to discharge unwanted fluids into water courses, or even to re-inject them into the substrata if the quality of the ground water could thereby be affected? What are fair rates for leasing rights? Are permissible pollution levels clearly defined, or the subject of total veto in certain cases? What redress has a land owner – e.g. of coffee plantations, of other crops or grazing lands – against pollution damage consequential to exploitation nearby (e.g. crop destruction from silica-laden spray)? What are the access rights across the land of third parties? Are drilling activities to be restricted to Government agencies, or may they be pursued by private enterprise under license subject to certain safety precautions? May a private individual exploit earth heat under his own land; and if so will a special license be needed? The questions are almost endless, but they must be comprehensively listed and answered with legal backing. Where geothermal legislation exists, as in California, existing acts can no doubt form a useful starting point for the guidance of other countries intent upon drafting their own legislation; but there could well be special problems peculiar to individual countries.

18.3 Training

Geothermal development calls for the services of many specialists versed in various disciplines – geology, geochemistry, geophysics, hydrogeology, engineering, economics, industry, district heating, husbandry, etc. Specialists in many of these fields are regularly being turned out by the universities and technical colleges of the world; but if geothermal development is to be fostered at an accelerating rate, these young people must be 'slanted' towards the geothermic application of their chosen discipline, and this requires at least some partial understanding of related disciplines as well as a deep understanding of their own. The purpose of the UNESCO geothermal review [4] was to encourage this process, but so much has happened since that was written that the breadth of necessary training has increased in recent years. Less than 30 years ago there were very few geothermal specialists in the world, but the proliferation of conferences, symposia and articles in the technical press bears witness to the accelerating awareness of the importance of geothermics. Carefully planned educational schemes will have to be prepared in many countries if the necessary legions of specialists are to be available for the expected 'geothermal renaissance' that lies shortly ahead. Already, valuable work is being done by the Istituto Internazionale per le Ricerche Geotermiche in Pisa, Italy, and by the Government of Japan who are initiating geothermal training courses with the support of UNESCO; but training facilities will have to be widely and rapidly extended in other countries too, if a manpower 'bottleneck' is not to hinder the steady growth of geothermal development.

18.4 Coordination of data

As with so many sciences, geothermics suffers rather from a plethora than from a shortage of information. The would-be student of the subject is apt to be bewildered by the sheer volume of literature available to him, and one of the purposes of this book is to ease his approach to the subject. Taking into consideration only the various conferences, symposia and the Review organized by the United Nations and UNESCO, the number of papers and Rapporteurs' summaries is somewhat forbidding:

Rome Conference on new sources of Energy (Geothermal papers only), 1961	80
Pisa Geothermal Symposium, 1970	209
San Francisco Symposium, 1975	315
UNESCO Geothermal Review, 1973	15
Total	619

However, this is only a fraction of what has been written and recorded on the subject of geothermal energy within the last 20 years or so. Various periodicals, such as *Geothermics* (published by the Istituto Internazionale per le Ricerche Geotermiche, Pisa, Italy) and the *Geothermal Energy Magazine* (published at West Covina, California) are devoted solely to the geothermal sciences, while many other technical periodicals often contain articles of geothermal interest. The proceedings of learned societies in many countries, and of national and international seminars abound in articles and discussions on earth heat, while even the popular press by no means neglects the subject. Several complete geothermal textbooks have also been published, mostly devoted to some particular facet of the subject.

The problem of codifying this huge volume of data, including test results from various sites, much of it of great value, some of it of ephemeral interest and some of little consequence, was recognized in 1974 when the International Geothermal Information Exchange Programme (IGIEP) was initiated [174]. The purpose of this organization is to provide for the prompt exchange and dissemination of new information and data. Two computerized centres have been established – one in Pisa, using the combined facilities of the International Institute of Geothermal Research (CNR) and the Italian National Research Council (CNUCE), and the other in the USA using the facilities of the Lawrence Berkeley Laboratory, California, and the US Geological Survey in Reston, Virginia. Since 1974, the coordinating group of IGIEP have initiated the following activities:

(i) GRID a computerized bibliography of geothermal literature at the Lawrence Berkeley Laboratory, California.
(ii) GEOTHERM a computerized data file on geothermal fields, wells, and other geothermal topics compiled by the US Geological Survey.
(iii) Implementation of GEOTHERM at the computer centre of the International Centre for Geothermal Research, Pisa, and at the Lawrence Berkely Laboratory.
(iv) Coordination of activities between the United States, New Zealand, Italy and other countries, and between the US Geological Survey and the Lawrence Berkeley Laboratory.

It will be seen that elaborate machinery already exists for the international exchange of geothermal information.

18.5 Field management

The art of field management may be described as that of extracting a maximum of heat of the required grade, in accordance with availability, in such a manner as to eke out the life of a field over a sufficiently long period to avoid large changes of pressure and temperature during the life of the exploitation plant – all as efficiently as possible and with a minimum of wastage. Over-exploitation of a field – that is, excessive fluid withdrawal – may pay an immediate but spurious dividend in the production of huge quantities of heat, which can be used for power generation or other purposes, but it will probably result in pressures falling so rapidly that a plant can no longer operate at its full initial rating after only a fraction of its potentially useful life has passed. This would mean that for the latter part of the plant's life it would have to operate at reduced capacity, which implies a waste of capital resources. By drawing on the field at a slower rate, a smaller plant might serve its full useful life of, say 25 years, with no decline – or at least with only minimal decline – of output during the latter part of its service. A second generation (and perhaps even a third or more) of plant could then follow, perhaps designed to operate at rather lower pressures than its predecessors. In this way a field could perhaps be run down over a century or more, during which it will have extracted far more useful energy than if the field had been over-exploited in the first place. Clearly the withdrawal of steam only, where a field is dry, is likely to run down a field less rapidly than the withdrawal of flashing hot water owing to the much higher enthalpy of the former and, where power generation is involved, its greater power potential.

   The choice of an optimum wellhead pressure (see Section 8.8) is of great importance in the good management of a field, as is the location of re-injection bores where the bore water of a wet field is returned to the aquifer. The depth of re-injection bores, in relation to that of the production bores, will also have considerable influence on field life, as it is desirable that the moderately hot re-injected water should sweep as much as possible of the virgin aquifer water ahead of it, while recovering as much as possible of the heat retained in the hot rocks formerly impregnated by the original hotter water. From time to time it could well be advisable to change the sites of re-injection bores (even if not forced to do so for reasons of chemical deposition) in order to avoid over-cooling part of a field while other parts of that field still remain much hotter. The 'model' of the field must constantly be kept under review, and the pattern of recharge water flows from beyond the confines of the field should be constantly studied, so that the approach of relatively cool waters may be 'headed off' by changing the points of re-injection as may be necessary.

   Constant logging of well characteristics, of downhole temperatures and pressures under static conditions; observations of geophysical and geochemical variations, of ground level changes and of the degree of activity of natural surface phenomena in the neighbourhood; and a continuous record of the quantities of fluids withdrawn and re-injected into the field; will all inform the enlightened earth scientist of what is happening in the aquifer belowground [169].

18.6 Outlook for geothermal energy prospects in countries lying outside the seismic belt

A reference to Fig. 9 will show that the chances of finding a hyperthermal field outside the seismic belt are virtually nil. Nevertheless it is always possible that *semi-thermal* fields may be discovered in almost any country. In Hungary semi-thermal waters are already being used to great advantage for space-heating and farming on an extensive scale, and the USSR claim to have vast semi-thermal fields beneath its territories. It seems probable that intensified surveys would disclose the existence of such fields in many countries. Even in volcanically tranquil Britain, where the most recent volcanic manifestation occurred 40–60 million years ago, hopes have been expressed of finding semi-thermal fields in the sedimentary basins of West Hampshire, the Bath/Bristol area and perhaps elsewhere [138].

The existence of *semi-thermal areas* in many countries far from the seismic belt is a known fact, and it seems probable that these areas betray, in many cases, the presence of hot rocks at not excessive depths. The granites of south-west and north-east England are cases in point, where fairly high temperature gradients have been observed. Many other countries have similar 'warm spots' beneath which hot rocks are believed to exist within commercial range of the conventional drill. The development of these semi-thermal areas awaits only the successful mastery of rock-fracturing techniques, in which promising progress is already being made at Los Alamos in New Mexico.

For the achievement of the ultimate goal, the recovery of very deep seated heat even in non-thermal areas, we must await not only the successful achievement of rock-fracturing but also new and cheaper methods of penetration to great depths – methods that will economically supersede the performance of conventional drilling by a large factor, (see Chapter 19). There may, however, be an earlier and small-scale partial development towards this goal brought about by the use of deep oil or gas drillings that have either failed to make a strike or that have become obsolete through the depletion of the fuel deposits which they once conducted to the surface. This, in fact, is already being achieved at Melun, near Paris (Section 11.2), where deep abortive oil drillings of 1500–2000 m depth are being used in a *non-thermal area* for circulating water through permeable zones to recover earth heat which is successfully being used for space-heating [102]. Water temperatures of 55–70° C are attained at Melun, and while these figures are rather lower than could be desired for space-heating, they are adequate at least for background heating which can be supplemented by relatively small quantities of fuel heat at times of exceptional cold. There must be many other abortive or disused oil drillings to great depth in other parts of the world, capable of exploitation. The proposal has already been made to use the North Sea oil and gas drillings for geothermal purposes after the oil and gas have run out [138].

19 ⃝ The future

19.1 The broadening horizon

Until 1973 it was unlikely that even the most optimistic of geothermal enthusiasts would ever have ventured to suggest that earth heat could one day supplant fossil fuels as the world's main source of energy. Even today, probably only a handful of people (including the author) believe in such a possibility. As recently as 1975 Worthington [175] expressed the opinion that geothermal energy was most unlikely ever to supply more than 10% of the energy needs of the United States – a country very richly endowed with natural geothermal resources.

The reason for the soft-pedalling of the potentialities of earth heat was that Nature appeared to have strictly rationed the supplies of this form of energy by making it available only in a limited number of hyper-thermal fields within the seismic belt and in more widespread, but less potent, semi-thermal fields elsewhere. The combined areas of fields of both classes probably form only a small fraction of the earth's total land surface and, although it was well known that gigantic quantities of heat lay beneath the outer crust at all points in the world, it seemed that this rich store of energy was locked away in a safe deposit, beyond the reach of man. The fuel crisis of 1973 turned a few men's thoughts to the possibilities of safe-breaking! In countries where the mere thought of geothermal development had hitherto seemed absurd, attention suddenly became focussed upon all unorthodox forms of energy and the apparent absurdity of really large scale geothermal development was called into question. In Britain the Department of Energy set up at Harwell an Energy Technology Support Unit (ETSU) to study unconventional energy sources, and one of the results of their deliberations was a report on the geothermal prospects for the United Kingdom [138] which, while rightly maintaining a cautious attitude, clearly showed that there could possibly be a modest future for geothermal energy in Britain. In 1974 the Commission of the European Communities convened two meetings in Milan, both attended by the author, for the purpose of roughing out a programme for the development of geothermal development in the member countries of the Community, in only one of which, Italy, had geothermal energy hitherto been exploited, and in most of which the geothermal outlook seemed to be bleak.

Meanwhile, the Americans had been devoting much thought and research effort to the problem of whether the availability of earth heat was truly limited, as had hitherto been tacitly assumed, to geothermal fields. The fact that thermal *areas*, which through lack of permeability could not be regarded as *fields*, were known to exist led them to examine the possibilities of creating artificial fields in these areas. If *natural* fields are a rarity, then clearly our task must be to create *artificial* fields; and where better could they be created than in the hot dry impermeable rocks which are known to exist in many places at depths within commercial reach to the drill?

As soon as the oil crisis of 1973 occurred, the author started to consider an even more ambitious goal, namely the recovery of very deep seated heat from *non-thermal* areas, so that immense reserves of heat could be tapped and made universally available at any point of our choice on the surface of the globe. His initial thoughts on this possibility were somewhat crude, as he then lacked the necessary knowledge of rock mechanics. He was also ignorant at that time of the valuable work being done by the University of California at Los Alamos, New Mexico, towards finding the two principal missing keys to the 'safe deposit' of deep seated earth heat – rock fracturing and new penetration methods.

In the space of three or four years our geothermal horizon suddenly became greatly widened from the mere search for natural thermal fields to the creation of artificial fields in hot rocks at moderate depth and even to the ultimate extraction of very deep seated heat at any point in the world.

19.2 Search for more natural fields

Although ambitious new projects are being conceived to exploit hot dry rocks at moderate depths and to extract very deep seated earth heat in non-thermal areas, and although the author is confident of our ultimate success in these tasks, he is fully conscious of the fact that success will not be achieved overnight. Meanwhile, the energy situation is steadily deteriorating (even though there may be a short deceptive easement in the 1980s when the flow of North Sea oil reaches its peak), and we must not relax in our search for the *natural* thermal fields – both hyper and semi-thermal – that Nature has rather grudgingly provided. First we must thoroughly examine all available worldwide records that have been collected by the earth scientists, and at the same time must *extend* our inventory of relevant knowledge by undertaking in likely zones as much exploratory field work as we can afford, in order to locate promising areas where fields of either class could perhaps be found. This is a task both for national governments and for the United

Nations. After locating promising areas, we should then undertake systematic and detailed exploration of all of them as rapidly as our funds, trained personnel and equipment permit. As previously emphasized, *time* is the essence of geothermal development. Doubtless many commercially exploitable fields will thus come to light quite rapidly, so that the curve of Fig. 1 will rise ever more steeply for at least two or three decades, so enabling earth heat to take over a much more significant share of the world's energy burden.

The process of exploiting natural fields, however vigorously pursued, could never provide the full answer to our energy problems, although it could well win for us a valuable time-gaining palliative to tide us over the comparatively few years during which we can discover ways of fulfilling our more ambitious aims of winning earth heat in far greater quantities by entirely new methods.

19.3 Rock-fracturing

It has already been emphasized that the fundamental difference between a thermal *area* and a thermal *field* is one of permeability. A thermal area is caused by the presence of large masses of hot dry rock at differing depths below the surface. To convert such an area into a *field* it is necessary to create permeability where none exists, and this can only be done by shattering, or at least fracturing, the hot dry rock so as to create voids and/or fissures through which water may be circulated to pick up the heat content of the rock and bring it to the surface for exploitation. Rocks are very poor thermal conductors, so if useful quantities of heat are to be collected at a useful rate it is necessary that the artificially created voids or fractures should have the following properties:

(i) The water/rock interface must have the greatest possible contact area.
(ii) The volume of voids and fissures must be large enough to ensure that the circulating water passes at low velocity over the hot rock surfaces so as to extract a maximum of heat during its passage.
(iii) The configuration of voids and fissures must be such as to offer a minimum resistance to fluid flow.

Ideally, the rock should be shattered into a rubble formation, which would meet all three of these *desiderata*. However, a mass of rubble, containing a high volumetric ratio of voids to solids, can only occur within a confined space of greater volume than that of the sum total of the rubble pieces – e.g. as with lumps of rock or coal contained in a hopper. How could such a condition be brought about in a solid mass of more or less homogeneous rock? It has been possible to do this ever since the advent of the nuclear bomb.

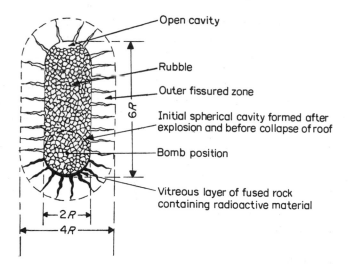

Figure 69   Rubble-filled 'chimney' formed by underground nuclear explosion (approximate proportions only are shown).

19.3.1 *Underground nuclear explosions.* If a nuclear bomb is exploded at depth in competent rock, the resulting underground formation will approximate to that illustrated in Fig. 69. The theory is that the intensely hot vaporized fission products expand within a split second of detonation to form a spherical cavity of such a size that the pressure within the cavity roughly balances the weight of the overburden plus an excess pressure depending upon the forces of shearing friction within the overburden. Temperatures within the cavity momentarily rise to millions of degrees, and pressures to several millions of atmospheres. Within seconds of the explosion, and persisting for minutes or even hours, the roof of the cavity starts to collapse; and the process continues until the rubble so formed rises to such a height that the roof is sufficiently supported thereby to prevent further rock-falls. The force of the explosion has created an 'empty' volume which changes shape from its initial spherical form to that of a rubble-filled 'chimney' as shown in Fig. 69. The sum total volume of voids in the chimney is equal to that of the initial spherical cavity, $4/3\pi R^3$. The relative proportions of the chimney, the shock-cracked surrounding zone and the initially formed spherical cavity approximately conform to Fig. 69 and also to the equation:

$$R = C.\frac{W^{1/3}}{(\rho h)^{1/4}}$$

where R = radius of cavity (and chimney) in metres,
          C = an empirically determined constant,

W = bomb capacity in kilotons. (1 kiloton = prompt release of 10^{12} calories),

ρ = mean density of overburden in g/cm³,

h = depth of bomb position in metres.

Not unexpectedly, this shows that the void volume at a given depth is directly proportional to the bomb capacity, for:

$$\text{volume} = \frac{4\pi R^3}{3} = \frac{4\pi}{3}\frac{C^3 W}{(h\rho)^{3/4}}.$$

Neither the formula nor the geometrical proportions of Fig. 69 are precise. The chimney height can vary according to the rock type and C can be known only to within $\pm 20\%$, though within much closer limits for a single rock type. The influence of rock type on C is shown in Table 14 [176]:

The uncertainties arise partly through inaccurate knowledge of ρ in a formation of different strata in different degrees of compaction, and partly because of different formation structures which can influence the frictional shearing resistance to the collapse of the entire cylindrical overburden into the initial spherical cavity. With the geometry shown in Fig. 69 the void volume would be 25%, assuming that the radiating peripheral cracks are formed by the space created by the explosion and not from the resulting roof collapse. These cracks are believed to form about 10% of the voids within the chimney, and since they would act as more or less stagnant pockets during circulation the *useful* proportion of voids would be about 0.9×25, or $22\frac{1}{2}\%$. In incompetent rock such as shales, rock salt, etc., the explosion would behave differently and less predictably.

Most of the radioactive fission products of long half-life would be trapped within a vitreous layer of molten rock which flows to the bottom of the chimney and there solidifies (Fig. 69). Obviously the bomb would have to be placed at sufficient depth to ensure adequate cover above the top of the chimney, lest the explosion should break the surface.

Nuclear explosions are a relatively cheap form of 'potted' energy, particularly if a hydrogen bomb is used. (Conventional explosives would be prohibitively expensive). Another great economic attraction of the nuclear explosion is that the price of the bomb rises only very slowly with the kiloton

Table 14 Influence of rock type on underground nuclear explosions.

Rock type	Average value of C
Alluvium	65.9
Tuff	78.1
Granite	60.6

capacity, so that the use of a big bomb would be far more economical in terms of heat gained or of power potential won, than for a small bomb. For example, in 1966 a 10 kiloton H-bomb cost US $350 000 whereas a 2000 kiloton bomb (200 times as powerful) cost less than US $600 000 [177]. Doubtless these costs have been hopelessly overtaken by inflation since then, but the proportions could well be valid still. The trouble with large bombs, however, is the resulting shock waves that could damage buildings and other surface structures. In densely populated areas their use would be ruled out, but in deserts or very thinly populated areas they might be practicable within electrical transmission range of the nearest population centre. It would be safe to detonate a 45 kiloton bomb at a distance of 19 km, or a 125 kiloton bomb at 48 km from the nearest unexpendable structure. The method could thus be practicable in some of the deserts of the western USA and doubtless in other parts of the world where cities are contiguous with deserts. It has also been pointed out [178] that off-shore nuclear explosions could safely be let off beneath the continental shelf – e.g. in the North Sea – at a safe distance from the coast, so it is not inconceivable that we may one day have off-shore power stations cable-connected to the land.

The possibility of shattering hot rocks in this way, and of extracting their heat for power generation purposes was analysed by Professor George Kennedy of the University of California, Los Angeles, as long ago as 1964 [179], no less than 9 years before the oil crisis of 1973. He pointed out that the heat energy of the bomb would account for about 20% of the total recoverable heat, the remaining 80% being truly geothermal, and he made out a good economic case for the suggested project. The idea has never been followed up, probably because, as with all matters concerning nuclear energy, opposition is encountered both on political grounds and from fear that despite the alleged sealing off of the more dangerous radioactive products in the vitreous layer at the base of the chimney, there could be a risk of radioactive contamination. Nevertheless, it is understood that the idea is far from 'dead', and that the US Atomic Energy Commission is still interested. The project may yet be revived.

It might be suggested that when the available heat from a rubble chimney starts to run down to an unacceptably low grade, further underground shots could be exploded adjacent to the original chimney, sequentially, so that the rubble-filled zones would merge together and the capacity of the artificial field extended more or less indefinitely. However, the vulnerability of the overlying exploitation plant to shock from subsequent explosions would more or less preclude this. It would be better at the outset to create either one very large chimney, or a contiguous series of smaller chimneys, depending upon the distance of the nearest vulnerable structure of importance, so as to last for at least two or three generations of exploitation plant.

19.3.2 *Hydro-fracturing*. Hydro-fracturing, or the underground splitting of rocks by means of hydraulic pressure, has been practised for many years in the petroleum industry as a means of increasing the permeability of a formation in order to enhance the yield of oil or gas from a well. Before considering some of the technical aspects of hydro-fracturing we should consider the ideal fracture pattern at which we are aiming, and to which the underground rubble-filled chimney created by a nuclear explosion approximates. Ideally we need a *labyrinth* possessing many parallel paths for the circulating fluid so as to meet as nearly as possible the three requirements mentioned in Section 19.3. If such a labyrinth could be obtained, it could be exploited in three possible ways, as illustrated in Fig. 70. Heat could be extracted by means of a 'once-through' open cycle with cold feed (Fig. 70a), by a closed cycle with reinjection and heat-exchanger (Fig. 70b) or, if a suitable vertically orientated labyrinth could be formed, by thermo-syphon (Fig. 70c). The third method would not only dispense with pumping: it could be used either with the open or closed cycle and could be formed either by two fairly closely spaced bores or even concentrically through a single bore. The closed cycle would be less vulnerable to chemical deposition than the open.

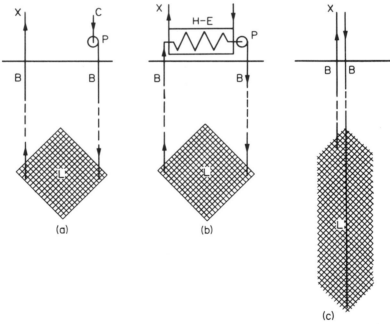

Figure 70 Extraction of heat from h t dry rocks by circulating water through artificially created underground labyrin' 1s. (a) Open cycle with cold feed (b) Closed cycle with reinjection and heat exchanger (c) Thermo-syphon.

With all forms of artificial labyrinth in hot rock there will always be an economic optimum depth at which the heat yield in terms of the winning costs is a maximum. We can then adjust the temperature of the heated water, up to a maximum somewhat short of the rock temperature, by varying the rate of circulation. We thus have a choice of winning a large quantity of low grade heat or a smaller quantity of higher grade heat. The choice would of course depend on the intended use for the heat, and the economics would involve the pumping costs, if any.

Unfortunately no one has yet succeeded in forming the ideal labyrinthine pattern of underground fissures by means of hydro-fracturing, but qualified success has nevertheless been obtained with hydro-fracturing and the subsequent circulation of heat-extracting water.

In a homogeneous mass of rock subject to no stresses other than those arising from its own weight, the compressive stress across a *horizontal* plane at any depth is equal to the lithostatic pressure, p_l arising from the weight of the overburden. This stress, in elastic competent rock, tends to make the material bulge out sideways; but this bulging is prevented by the presence of adjacent rock material trying to do exactly the same thing in the opposite direction. The result is a horizontal compressive stress across any *vertical* plane, having a value of $p_h = p_l \mu / (1 - \mu)$, where μ is Poisson's Ratio for the particular rock concerned. μ varies over a wide range from one type of rock to another.

When a bore is sunk vertically into competent rock, the horizontal, compressive stress at the rim of the hole is twice the horizontal stress p_h prevailing at the same depth in undisturbed rock, provided that tectonic pressures are absent. To effect a fracture *in a vertical plane* it is necessary to apply a pressure p_f at the bottom of the hole sufficient not only to relieve the rock of this doubled compressive stress but also to exceed this by the tensile breaking stress of the rock S_t. That is to say

$$\text{fracturing pressure } p_f = \frac{2\mu p_l}{1 - \mu} + S_t$$

(It will be noted that the diameter of the bore does not enter this equation. This is because we are not dealing with simple 'hoop' stress, but with stresses in thick cylinders – infinitely thick, in effect. The doubled compressive stress at the rim results from *strain* distribution, and a larger bore diameter would spread the greater total pressure in a different attenuating pattern from the bore rim outwards).

Taking 2.7 as a typical rock density, and neglecting rock compressibility (which in any case is insignificant), the lithostatic pressure p_l will be 0.2616 atm per metre of depth. Fig. 71 shows this lithostatic pressure in terms of depth as a straight line. Now let it be assumed that Poisson's Ratio (μ) for a particular rock type is 0.27, and that the tensile breaking stress of the rock, S_t is 50 atm. (735 p.s.i). Then

$$p_f = \left(2 \times \frac{0.27}{0.73} \times p_1\right) + 50 \text{ atm} = 0.74\, p_1 + 50 \text{ atm}.$$

This fracturing pressure line is also shown in Fig. 71 as *ABC*. It will cross the line of lithostatic pressure at point *B*, at a depth of 735 m and a pressure of 192 atm. This means that at depths exceeding 735 m the fracturing pressure will be *less* than the lithostatic pressure, while for shallower depths it will be *greater* than the lithostatic pressure. How is this to be interpreted?

First it is necessary to consider how a *horizontal* fracture could be formed by the action of hydraulic pressure. No rock is completely homogeneous and flawless. The presence even of a hair crack or slight indentation in a bore wall would admit hydraulic pressure capable of exerting an upward thrust; but this thrust would have to exceed the lithostatic pressure if the hair crack or indentation is to be extended, since it would be necessary to uplift the overburden. It is clear that so long as the value of p_f is less than p_1 no horizontal crack could ever be formed: instead, the lateral pressure on the bore walls will tend to force them apart, so that a *vertical*, or near-vertical crack will therefore form along the weakest plane.

Now to the left of point *B*, at shallower depths, the pressure p_f necessary to form a vertical fracture is greater than the lithostatic pressure p_1, so it might be thought that horizontal cracks would tend to form from hydro-fracture at shallow depths. However, in reasonably homogeneous rock it would be necessary to exert an upward thrust not only sufficient to overcome p_1 but to exceed it by the tensile strength of the rock. It can be seen from Fig. 71 that the excess of p_f over p_1 is never greater than S_t. Hence in homogeneous rock, vertical cracks would tend to form even at shallow depths. However, in non-homogeneous rocks such as stratified slate or shale formations, the value of S at right-angles to the cleavage planes is almost zero by comparison with the tensile strength to forces within the planes of the strata. With horizontally stratified rocks of this sort, the formation of horizontal cracks would therefore be possible at shallow depths.

The words 'horizontal' and 'vertical' should be treated liberally in the above context, as the direction of the strata in stratified rocks is usually inclined at some angle of tilt to the horizontal, and as the lack of perfect homogeneity in all rocks may often have the effect of imparting a 'grain' – i.e. the existence of more or less parallel planes fractionally weaker to tensile forces than others. Hence even 'vertical' cracks formed by hydro-fracture may have a slight tilt.

The general conclusion is that at the fairly great depths at which the fracturing of hot dry rocks is of interest, vertical or near-vertical cracking will nearly always result from the application of high hydraulic pressures, *unless* the value of μ is higher than 0.333, as will now be shown.

First let an example be considered of a rock whose value of μ is 0.31 and of S_t is 25 atm. In this case

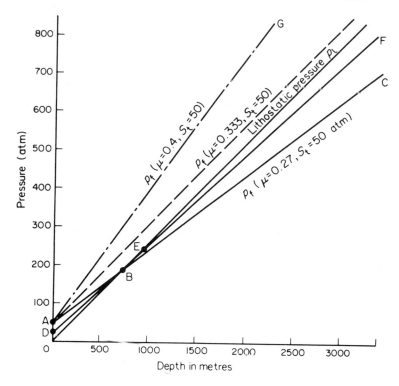

Figure 71   Hydro-fracturing theory.

$$P_f = \left(2 \times \frac{0.31}{0.69} \times p_1\right) + 25 = 0.899\, p_1 + 25.$$

This value of p_f is represented by the line *DEF* in Fig. 71, which crosses the lithostatic pressure line at a depth of 947 m and at a pressure of 248 atm. On reflection it will be clear that the greater the values of S and μ, the deeper will be the point at which the p_f and p_1 lines cross. Now what happens when $\mu = 0.333$ in value? The expression ... $2\mu p_f/(1 - \mu)$ then becomes equal to p_1; and no matter what the value of S_t may be the p_f line will lie parallel to the p_1 line and always *above* it, as shown (for example) by the broken line of Fig. 71. This implies that except in stratified rocks there will be a 50:50 chance that hydro-fracture could produce either approximately vertical or horizontal cracks. However, now suppose that μ is *greater* than 0.333 – e.g. 0.4. This could give a p_f line such as *AG* in Fig. 71, which diverges from the p_1 line at a fairly wide angle. In such circumstances, horizontal cracks could well form, as the cracking pressure is so much greater than the lithostatic pressure.

Values of S_t and μ can vary widely from place to place. S_t can be almost zero for stratified slates and shales (perpendicular to the cleavage planes), having quite low values for brittle rocks such as chalk, and attaining values as high as 50 to 75 atm or thereabouts for granites and other igneous rocks. Values of μ are even more capricious. In theory μ can never exceed 0.5, which is the value for water (since $0.5/(1-0.5)=1$, and equality between horizontal and vertical pressures is a property of liquids). Nevertheless, values as high as 0.8 have been observed in valleys close to the surface. This is because the lateral thrust due to the weight of neighbouring hills can overwhelm the overburden pressure due to the local rock in the floor of the valley. However, at the fairly great depths that are of interest in the hydro-fracturing of hot dry rocks, anomalies due to surface irregularities tend to disappear. Rocks having values of μ exceeding 0.333 are also rare (except as obtained from prepared samples in the laboratory, where test results are apt to differ appreciably from the properties of large buried rock masses). At Los Alamos in New Mexico, where the University of California (LASL, Los Alamos Scientific Laboratory) is conducting extensive rock-fracturing experiments in granite at depths of the order of 3 km, μ is found generally to lie between about 0.25 and 0.28, with a few exceptions outside those limits.

Up to now, consideration has here been confined to rocks in tectonic repose only, free from all stresses other than those induced by their own weight. In some parts of the world there will be induced tectonic stresses – tensile, compressive or shear – which may one day give rise to an earthquake, but which meanwhile can distort the rock-fracturing pattern from the simplified behaviour described above. Powerful horizontal compressive stresses in one direction, while increasing the value of p_f in a plane perpendicular thereto, can reduce the value of p_f in the plane parallel to the direction of the compressive forces. In general, tectonic forces can affect both the value of p_f (which could differ widely from the theoretical) and also the orientation of the resultant cracks from hydro-fracturing. It is even possible that tectonic repose is the exception rather than the rule. This could be due to the dragging forces of magmatic convection currents beneath the Moho. It is known that at Los Alamos, New Mexico, a very high horizontal stress is present. However, tectonic stresses would merely complicate, rather than invalidate, the theory of rock fracturing.

It is generally true to say that, with a few exceptions, hydro-fracturing will produce *vertical*, or near-vertical, cracks. These cracks are of lenticular shape and can be produced more or less symmetrically about the point at which the pressure is applied – the base of a cased bore – to diameters of some hundreds of metres and of widths of a very few millimetres at the centre, tapering off to zero at the periphery.

Lenticular cracks bear little resemblance to the ideal 'labyrinth' we are seeking, but it is believed that they could nevertheless enable us to extract rock heat at reasonable cost. At present the LASL experiments are mainly directed to the exploitation of lenticular cracks. The proposed method of exploiting these cracks is illustrated diagrammatically in Fig. 72. A vertical hole is sunk to the required depth into a mass of hot dry rock, and a lenticular

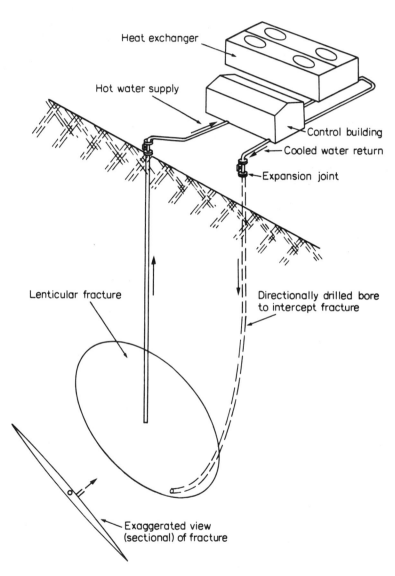

Figure 72 Conceptual diagram of man-made lenticular fracture in hot dry rock with circulating system. (After Fig. 1 of [180].)

fracture is formed by applying a high hydraulic pressure at the *un*cased base of the bore which is cased almost to the bottom. A second hole is then drilled *directionally* from a short distance away, aimed to strike the lower part of the lenticular crack. Water is then circulated in the manner shown – if possible by thermo-syphon action, otherwise by pumping. It is necessary to pressurize the circulating fluid so as to avoid the formation of steam, to which the crack pattern would offer excessive flow resistance owing to the high specific volume of steam and its consequent high velocities. Moreover, water has a better thermal capacity for heat absorption than steam.

For the experimental plant at Los Alamos the hot water rising from below-ground dissipates its heat to the atmosphere in heat exchangers, but for a working plant the heat would be usefully applied to some practical purpose. It is assumed that the cool water enters the bottom of the crack, is convected round it in a swirling motion, and ultimately returns to the surface up the original straight bore. It is possible that better results would be obtained if the lower part of the vertical pipe were blocked off and an opening cut in it nearer the top: this is a matter for empirical testing. Already (early 1977) qualified success has been obtained in that a circulating system has been established between two bores, though the flow impedance is still too high. A temperature of 200° C has been encountered at a depth of 9600 ft (about 290 m). This is equivalent to a gradient of about 64° C/km. Various proposals are to be tried out for improving the flow. As an alternative to continuous convective or pumped circulation, Professor R.W. Rex has suggested [181] a 'huff-puff' system of alternate injection of cold water and withdrawal of hot. It is far too early to judge the ultimate possibilities of the tentative crack system.

It is emphasized, however, that there are various secondary effects that are expected to favour the method of heat extraction:

(i) *Thermal stresses* resulting from the contact of cool water entering the fracture will cause the walls to shrink, thereby setting up a multitude of lateral cracks of uncertain extent. Although with a single lenticular fracture these lateral cracks would contain more or less stagnant water, they could possibly form the nucleus of a 'honeycomb' labyrinth. For if several lenticular fractures were created at the base of several closely spaced holes, the 'grain' of the rock would probably be such that the fractures would be roughly parallel to one another, and the lateral cracks could interconnect them. Thus a whole 'battery' of interconnected lenticular cracks could be built up side by side (Fig. 73). Furthermore, thermal stresses would tend to make lenticular fractures self-propagating – especially at the lower, hotter edge, where high temperature differences would give rise to large tensile splitting forces.

(ii) *Lower viscosity* of hot water would tend to increase the flow speed

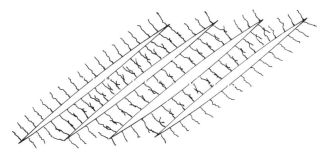

Figure 73 Possibility of establishing a labyrinth by means of parallel lenticular fractures and lateral cracks induced by thermal stresses.

and reduce the flow resistance at the lowest and hottest part of the fracture.

(iii) *Propping.* By injecting graded pellets of gravel or steel when applying the highest practicable dilating pressure, these pellets would fall to such a level as to act as props to sustain dilation after the removal or reduction of the hydraulic pressure. It is true that this would be difficult to contrive except in the lower part of the fracture, but the effect on reduced flow resistance could be appreciable.

(iv) *Wedge action.* The pressure required to extend a lenticular fracture would probably be substantially less than that needed to create one '*de novo*' owing to the wedge action at the narrow periphery of the fracture, much as a piece of cloth is more easily torn after making an incision with scissors, to concentrate the stress at one point. 'Moreover, although lateral (thermal stress) cracks are directed along less favoured planes than the 'parent' lenticular crack, it is quite possible that wedge action would cause them also to extend under high pressure so that a fairly extensive 3-dimensional fissure pattern would develop.'

(v) *Chemical leaching.* It is hoped that by circulating suitable solvents through the system, large volumes of solubles may be removed from the fracture, thus increasing the volume of contained fluid and effectively widening the flow passages to a small but useful extent.

At the Canborne School of Mines in Cornwall, England, a different approach is being made. It is argued that the steady application of a sufficiently high hydraulic pressure at the base of a hole must form a *single* fracture along the weakest plane of the rock (as determined by the inevitable irregularities and discontinuities in a structure that is never perfectly homogeneous). However, if a conventional explosion is detonated within a small confined space at the base of a bore, instantaneous pressures of 1 or 2 million p.s.i. will be generated which will cause multiple radial fractures to form in all directions, (Fig. 74a), because the inertia of the rock substance prevents

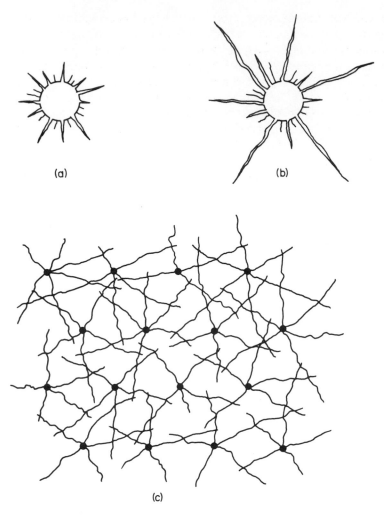

Figure 74 Formation of multiple radial fractures by means of conventional explosions at the base of bores, their subsequent selective extension by means of hydro-fracturing, and the ultimate build-up of a labyrinth. (a) Initial formation after explosion (b) Extension of cracks along weaker planes by hydro-fracturing (c) Network of bores with radial cracks to form 3-dimensional labyrinth.

it from moving sufficiently rapidly to provide an easy escape path through a single fracture for the high pressure gases from the explosion. Hence, other fractures will also form during the split second while the high pressures persist. Multiple radial fractures will in fact form and continue to grow until sufficient volume has been created to reduce the gas pressure to a value below that required for further fissure propagation. An advantage of this procedure is that small fragments of rock are forced violently into the radiating cracks at the moment of their formation, thus providing useful propping against subsequent collapse. After this initial 'star' formation has been created, the hydro-fracture technique may then be applied, and it is found that instead of a single crack being formed in one plane, as with straight hydro-fracturing, the applied pressure will extend *several* of the weakest cracks to considerable distances (Fig. 74b). This method suggests a possible means of building up an extensive labyrinth (Fig. 74c). Different pairs of bores near the periphery of the network could be used from time to time so as to vary the water flow paths and ensure that the greater part of the labyrinth is swept by the circulating fluid. Alternatively, radial flow between the centre and the periphery of the network could be arranged. A suitable system of pipework and valving could enable changes of flow pattern to be made at will. This method is in its infancy, and at present is being tried out experimentally at shallow depths – 30 to 40 m only – in order to study the fracture behaviour; but if the results are promising it is hoped to extend the tests to far greater depths.

Although Fig. 74c shows fissures crossing one another without change of direction, it is possible that the fracture pattern would be somewhat different. Where two fractures meet, extension could perhaps occur along one or more *lateral* cracks caused by thermal stress, in preference to straight continuation. On the other hand, since each crack would tend to be more or less circular in form, they would meet one another only along chords: cracks might tend to retain their generally circular shape even when the process of their extension is interrupted by intersection with another fissure. This will have to be determined empirically; but there would seem to be plenty of scope for the formation of elaborate labyrinthine passages both along and perpendicular to the main lenticular fractures.

A matter of great importance to the success of hydro-fracturing is the width of the fracture obtained, since this would clearly influence the flow resistance and the efficiency of heat extraction. It is virtually impossible to make precise calculations because rocks are never perfectly isotropic or perfectly elastic: they possess an element of plasticity. Harlow and Pracht [181] predicted in 1972 that in thick-bedded sedimentary rocks it should be possible to create disc-shaped cracks ideally of up to several kilometres in diameter and 'a few' centimetres wide at the centre, but experiments to date would seem to suggest that this was over-optimistic. The LASL experiments, in granite, had by late 1976 achieved a fracture of only 'a few millimetres' in width. Clearly, width is of greater importance with a single lenticular crack than for a complex labyrinth having many parallel flow paths. Much remains to be learned about attainable fracture widths and extents, but it should be noted that after the application of the fracturing pressure, when the crack is first formed, it may be filled with water and distended at a somewhat lower pressure; but on removal of the pressure the crack, except at great depths (see Section 19.6), will not completely reclose. An element of plastic deformation plus one of thermal contraction will tend to leave a thin crack permanently. As with an underground nuclear explosion, but of course on a far smaller scale, some 'free space' has been created within the rock. If the crack is flooded with water, hydrostatic pressure will help to retain, and slightly widen that space, but the width and volume can of course be increased by pressurizing. The greater the applied pressure, the lower will be the flow resistance, but the 'tougher' must be the surface heat exchangers. Too much pressurizing will simply *extend* the spread of the fracture without necessarily increasing its width. The energy stored in elastic dilation is theoretically recoverable (in part), though this is unlikely to be of practical value.

One source of concern in hydro-fracturing is water leakage through such minor permeability as may exist even in generally 'impermeable' rock. This could cause a serious loss of heat. However, if, after creating a fracture or series of fractures, leakages were very substantial, this could indicate that the impermeable layer has been penetrated to give access to an under-lying permeable zone that could form a potential aquifer which, if filled with water, could later serve as a 'natural' field. The problem would then become 'how long would it be necessary to wait before the water in the voids reaches equilibrium with the rock so that the newly created 'field' can be exploited?' This would be a function of the permeability of the newly opened up zone.

Geothermal hydro-fracturing is an art still in its infancy. Much information can be gleaned from [181–184], but more will depend upon the experimental work now in hand. Theorizing is limited by the fact that rocks do not behave like ideal isotropic substances. The presence of mountainous or hilly surface terrain, geological variations at depth, faults, discontinuities, lack of homogeneity, the presence of local 'arching' resulting from past subsidences and thus modifying local stress distribution; these and other factors necessarily place a limit upon the degree of theorizing possible. Nevertheless, some preliminary analysing (as advanced in this Section) helps to clarify thought and to give forewarning of the complications that may well arise in practice. Rock-fracturing is, however, a very promising line of investigation that could well lead to a tremendous extension of the quantity of exploitable earth heat available to us. It is also well to recognize that prospecting for hot dry rocks is comparatively simple – only high temperatures and a suitable lithology being required. Production drilling by conventional means would also be far simpler and cheaper than in 'wet' fields owing to the absence of corrosive fluids and the fact that circulation losses with their attendant troubles would scarcely arise [182]. Another point of over-riding importance is that hot dry rocks are believed to be our largest and most widely distributed energy source [182].

19.3.3 *Impact fracturing.* Yet another possible way of creating permeability in hot dry rocks is 'impact fracturing'. This method, which has not yet been tried out, was advanced by the author early in 1974 when, under the stimulus of the October 1973 oil crisis, he had been considering ways and means of making earth heat more accessible. The method was primarily intended for the recovery of very deep-seated heat in non-thermal areas; but it could equally well be applied for the exploitation of hot dry rocks at shallower depths. The proposed method is illustrated in Fig. 75. A bore, cased except near the lower extremity, is surmounted by a hydraulic accumu-

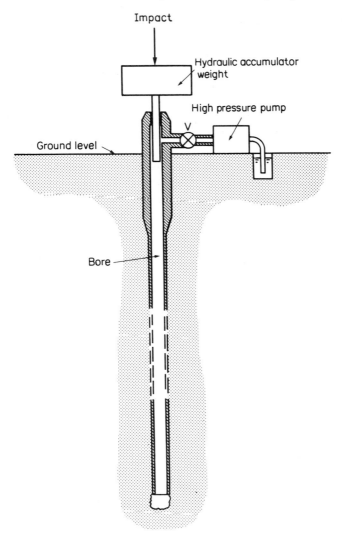

Figure 75 The author's proposal for rock-fracturing by means of dynamic impact.

lator, filled with water and pressurized by means of a pump until the accumulator weight is lifted. The lifting pressure should be perhaps about 90%
of the estimated hydro-fracturing pressure. A dynamic blow would then be
struck upon the top of the water column by means of some device resembling
a pile-driver, so that shock waves would travel to the base of the bore and
be reflected up and down at attenuating intensity. Very high instantaneous
pressures would be generated which would fracture the rock, and it is
expected that the steepness of the pressure rise would be such as to resemble
explosive action, so that the 'star' formation of Fig. 74a would be produced.
The process could be repeated as often as necessary. Before each blow is
struck, the pump would of course be isolated by closing the valve, V.
Immediately after each blow, water would flow into the newly created
cracks and the accumulator weight would fall onto its stops. By metering
the water that has to be pumped into the bore and cracks until the weight
is again lifted to its full extent, a measure of the volume of the cracks,
distended at the particular pressure of application, is obtained. Thereafter,
crack propagation could be effected either by straight hydro-fracturing or by
repeated dynamic impacts: in the latter case a pump of lower pressure
rating could be used, although possibly less extensive but more numerous
cracks would result. Only experience could show which would be the most
effective method of procedure.

Impact fracturing would demand very careful grouting between the bore
walls and the casing tubes, and also the use of a good ductile steel for the
casings, lest the very high instantaneous pressures should fracture the
casings. This is a ballistic problem, and it should not be beyond the wit of
man to ensure a sound shock-resisting casing, firmly bedded into the rock.
The difficulties should be no more formidable than those involved in the
design of a gun.

19.4 Improved penetration methods

The economic limitations of conventional rotary drilling have been stressed
in Section 7.15 and illustrated in Fig. 21, while novel methods of penetration
have been broadly mentioned in Section 7.17. Percussion drilling has certain
marked advantages over rotary drilling in dry rocks, principally because the
bit life can be extended to 12 or more times as long; but unfortunately the
method cannot be used in wet holes nor at great depths.

By far the most promising method for very deep penetration would seem
to be the 'subterrene' – a melt-penetration device [185, 186]. By applying
intense heat to a small local area, the rock may be fused so that the hot
penetrator can advance downwards, melting and displacing the fused rock
outwards in the process. Most igneous rocks melt at temperatures of about
1500 K – approximately the same as steel; so that melting penetrators must
be made of refractory metals such as molybdenum or tungsten which melt

at 2880 and 3650 K respectively, and which also have low creep rates at the rock-melting temperatures. The fused rock forms a dense, vitreous lining to the hole, intimately bonded with the surrounding rock (Fig. 76), having very great compressive strength capable of supporting the bore walls against collapse. Thus, no steel casings are required, with their attendant problems of jointing and grouting. The energy per metre of penetration in some igneous rocks, is more than double that required for conventional rotary drilling, but at great depths this is more than offset by the savings in casings and cementation and, more important, by the very great time savings in the avoidance of bit renewal, restringing and in correcting circulation losses. Moreover, as the penetrator enters hotter ground with increasing depth, less additional heat is needed per metre of penetration, so that the energy differential by comparison with conventional drilling improves. The result is that in place of the quasi-exponential cost/depth relationship shown in Fig. 21, the average cost per metre of penetration is expected actually to *decline* at great depths. Another feature of melt-drilling is that the penetration energy varies only slightly between one type of rock and another, so that the relative advantages of the method are far greater with very hard rocks such as granites and other igneous rocks likely to be found at great depths. The problem of increasing bit wear with rising temperature and increasing depth would be entirely eliminated; no mud or mud cooling system would be necessary; and directional drilling can be easily effected by controlling the temperature circumferentially around the penetrator. In that it does not rotate, the penetrator would be free from high torque and shock loads in deep wells. No elaborate surface area is required for all

Figure 76   Vitreous bore lining formed by melt-drilling, intimately bonded with the surrounding rock. (After Fig. D-11, Appendix D of [185].)

the stores, mud-cooling and other ancillary equipment needed for rotary drilling. Finally, vitreous encased continuous core can be extruded from the entire length of the hole. Hot fluids from any aquifer penetrated by melt-drilling can be captured by perforating the vitreous walls of the bore by means of explosive charges or other means of locally opening up the glass-like casing, but the method is more likely to be of real value in terrains where no natural aquifers exist, and where the object is to reach very deep-seated hot dry rocks.

Hitherto, electrically heated penetrators only have been used, and these have shown good promise. The ultimate intention is to use nuclear-heated heads which will be expendable, advancing at an increasing rate through the crust and finally entering the mantle and disappearing within the cores of the earth. In 1971 it was prophesied [185] that within 10 or 15 years the nuclear subterrene would be developed so as to be capable of producing holes of several metres diameter and several tens of kilometres in length or depth – either for purposes of tunnelling or of geothermal heat recovery.

In 1976 these great hopes were suddenly shattered by the unexpected abandonment of the American subterrene experimental programme at Los Alamos. As will be shown in Section 19.6, the real break-through in geo-thermal development that will enable it ultimately to compete with, and even surpass, fossil fuels as a principal (if not *the* principal) source of world energy, must largely depend upon the satisfactory development of a relatively cheap method of very deep penetration. Hence the abandonment of the subterrene programme can only be described as a disastrous decision. It seems probable that sheer necessity will sooner or later compel the readop-tion of the subterrene experiments; and it is to be hoped that if the financial burden is too great to be borne by the USA every effort should be made for the enterprise to be supported internationally by a concerted effort, since it would scarcely be an exaggeration to say that the future prosperity of the human race is at stake.

19.5 Direct tapping of volcanoes

The direct tapping of earth heat from active volcanoes has, in recent years, been the subject of much discussion, but not very much seems yet to have been achieved in the way of tangible results. Between 300 and 400 active volcanoes are known to exist in the world; and since many inactive ones may merely be dormant, the total number of volcanoes that *could* at some time become active may perhaps be measured in thousands. Volcanoes have the advantage over hydro-thermal fields in that their existence and location are self-evident, so that they do not have to be 'discovered', even though their hidden underground structure may have to be divined by exploration methods. A joint USA – Japanese seminar was held in Hawaii in February 1974, at which the possibilities of direct volcanic exploitation were discussed.

It has also been reported that the Russians plan to develop about 300 MW of power by boring nearly 10 000 ft – say 3 km – into the heart of the Avachinskaya Spoka volcano in Kamchatka. The proposed method of heat extraction is not known to the author. At the base of this volcano is a mass of hot magma at 1000–1200°C, at a depth of 3 to 5 km, and having a diameter of about 6 km. The volume of this magma has been estimated at about 100 km³ [18]. The heat content of this magmatic mass would be something of the order of 1% of the world's total measured fossil fuel reserves, which represents a gigantic concentration of energy within the comparatively small confines of a single volcano. Although this offers a great challenge for the winning of huge amounts of energy and at the same time of 'taming' a potentially dangerous volcano, the practical problems of recovering the heat must be formidable. The consistency of magma at such high temperatures is more or less plastic, so that 'fracturing' followed by water injection would scarcely be practicable. Any water injected into the hot mass would immediately form very high pressure steam which would probably escape by forcing a passage upwards towards the surface and thus eluding capture. Large refractory heat exchangers, buried in the magma, could offer a theoretical solution; but how would they be placed in position, and how would the problems of chemical scaling be satisfactorily dealt with? Doubtless, much thought will be devoted to these problems, and the Russian achievements in this field are eagerly awaited. Meanwhile there is a dearth of published information on this subject.

19.6 The ultimate goal: universal heat mining

The ultimate geothermal goal is, of course, to win earth heat wherever we wish – even in non-thermal areas. Fig. 77 shows the depths to which we would have to penetrate in order to reach various temperatures in zones of various mean temperature gradients. Of course in practice the temperature/depth relationship would never be a simple linear one as shown in Fig. 77 owing to changes of thermal conductivity with depth, but in terms of *average* thermal gradient the figure serves to show the order of magnitude of the required penetration depths to reach various required temperatures in non-thermal and semi-thermal areas. The shaded band represents a fair world average for non-thermal areas, from which it will be seen that to win low grade heat for such purposes as district heating or farming, and allowing for a reasonable temperature difference between the hot rock and the circulating water, it would be necessary to penetrate about $2\frac{1}{2}$ to 3 km; while for power generation we would have to go down to 6 or 7 km at least. (For more efficient power generation greater depths would be necessary – say 8 to 12 km – and an economic study would have to be made between the penetration costs and the thermal efficiency at which the extracted heat could be used).

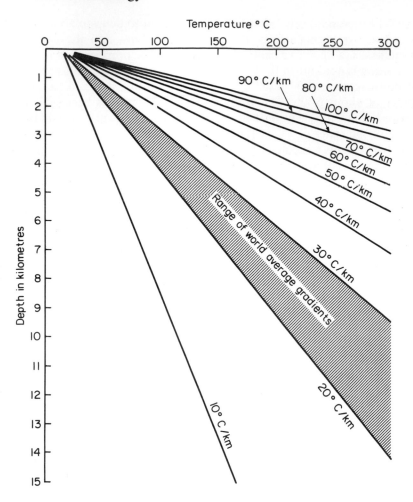

Figure 77 Temperature/depth relationship for non-thermal and semi-thermal areas for different average temperature gradients (assuming mean surface temperature to be 15° C).

Note: Shaded area is typical of non-thermal areas on land.

Such depths as $2\frac{1}{2}$ to 3 km, or even 8 to 12 km, can even now be penetrated by conventional drilling methods – but at a cost that would be unacceptable in terms of the grade of heat to be won at those depths in non-thermal areas. For the economic extraction of very deep-seated heat in non-thermal areas three prerequisites must first be achieved:

(i) The successful development of the nuclear subterrene or of some other cheap method of very deep penetration (See Section 19.4).

(ii) The successful mastery of rock-fracturing techniques (See Section 19.3).
(iii) The maintenance of excess pressures to dilate very deeply formed artificial cracks and fissures to permit the circulation of water through them.

The first two of these prerequisites have already been discussed at some length. The third is relatively simple. As is well known, fractured rocks and rubble can exist in a state of stability to moderate depths with a fairly high proportion of voids or permeability. As the depth increases, the rising lithostatic pressure tends to squeeze such voids out of existence, thus at first reducing the percentage of voids and ultimately – at perhaps 3 km or thereabouts and at greater depths – rendering the rock totally impermeable. It is this that accounts for the presence of bedrock beneath aquifers in hydro-thermal fields, as shown in Fig. 10 (Section 5.6). Hence, even after we have mastered the art of rock-fracturing to the required pattern, the man-made fractures would instantly reclose on the removal of the fracturing pressure. To recover earth heat at very great depths it will therefore be necessary not only to fracture the rock but also to keep the fractures permanently open by means of a dilating pressure so that water circulation may take place with a minimum of resistance to flow.

It may be asked 'If voids are squeezed out of existence at great depths, how can a bore be sunk to these depths without itself collapsing?' This is because even a plain uncased vertical cylindrical hole could survive to far greater depths than a flat or amorphous fissure, owing to the 'arch' effect of the hole walls which are in compression $(2p_h)$. If the hole were supported from within by hydrostatic pressure and also strengthened with a tough vitreous lining, it could doubtless resist collapse to immense depths. Even if water could not be admitted during the actual penetration process, high pressure inert gases or even steam could be admitted to give internal support.

It has already been shown that with rare exceptions hydro-fracturing will form vertical, or near-vertical cracks (Section 19.3.2). Typical fracturing pressures have been illustrated in Fig. 71. However, once a vertical fracture has been formed, the pressure exerted by one wall of a crack against the other (on reclosure) will be less than half the fracturing pressure: in fact it will, assuming no plastic deformation or thermal shrinkage, have a maximum value of $\mu p_1/(1-\mu)$ or p_h.

In Fig. 78, OA represents the lithostatic pressure p_1, for an assumed average rock density of 2.7; BC the fracturing pressure (for $\mu = 0.27$ and $S_t = 50$ atm); and OD the maximum pressure exerted horizontally by one wall of an empty vertical crack upon the other. The slightly curved line OE represents the hydrostatic pressure, p_w at depth, allowing for the falling water density at higher temperatures and assuming a temperature gradient of $25°$ C/km (as shown on the upper horizontal scale of Fig. 78). The point

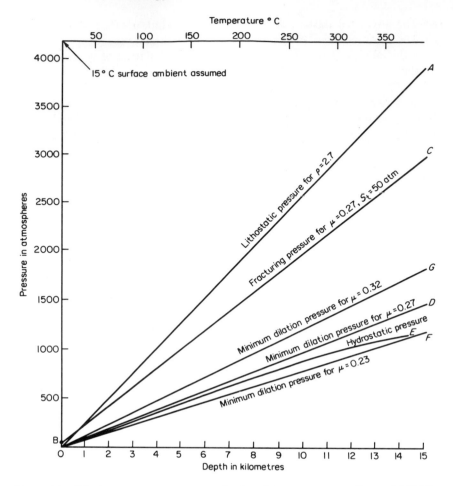

Figure 78 Dilation pressures at great depths. (Temperature gradient of 25° C/km assumed.)

E is of interest in that it indicates the depth at which, under the assumed conditions, water would reach the critical temperature of 374° C. At greater depths an upper column of water would be supported upon a lower cushion of superheated steam.

Now it can be seen from Fig. 78 that for the assumed temperature gradient, and for rock of the characteristics represented by the line *OD*, hydrostatic pressure alone would provide the greater part of the necessary pressure to relieve the walls of a vertical crack from pressing against one another – i.e. to *start* the dilation process. In fact the excess pressure that would have to be applied to start the dilation process would be as in Table 15.

Table 15   Minimum excess pressure required to start dilation process ($\mu = 0.27$, specific gravity $= 2.7$).

Depth (km)	Temperature (°C)	Proportion of minimum dilation pressure provided by hydrostatic pressure (as a percentage)	Minimum excess pressure required	
			atm	% of hydrostatic pressure
3.0	90	98.5	4.3	1.5
6.2	170	95.8	25.0	4.3
13.4	350	85.7	185.4	16.7

The reason for the rising proportion of excess pressure with increasing depth is of course due to the slight curvature of the hydrostatic pressure line *OE*, which in turn is due to the falling density of hot water. For the example taken, ($\mu = 0.27$) only quite modest excess pressures are needed to start off the process of dilation. For lower values of Poisson's Ratio – e.g. $\mu = 0.23$ – as represented by the line *OF* it would be possible for hydrostatic pressure to provide rather *more* than the minimum dilation pressure. For certain intermediate values of μ – e.g. 0.25 – hydrostatic pressure would be more than sufficient to start dilation except at very great depths.

For rocks having high values of μ – e.g. 0.32, as represented by the p_h line *OG* – hydrostatic pressure, though still providing the greater part of the minimum dilation pressure, would fall substantially short of what is needed.

In order to dilate a vertical fracture to the maximum possible extent, an excess pressure, over and above the hydrostatic, would have to be exerted to achieve a total pressure just short of that which would cause the fracture to extend, rather than to widen. The extension pressure, owing to wedge action referred to in Section 19.3.2.(iv), would probably be appreciably less than the original fracturing pressure. That is to say, for $\mu = 0.27$ (p_f line *BC*, p_h line *OD* in Fig. 78) the fracture extension pressures would probably lie well below the line *BC*. In fact it should be possible to extend a crack laterally and upwards (but only to a limited extent downwards owing to the rising lithostatic pressure) by pumping alone even if, as might well be so with impact fracturing, the pressure capacity of the pump were substantially lower than p_f for a cylindrical hole. The limit of extension would probably be fixed by the ratio of the maximum crack width to the distance from the bore to the crack extremity: that is to say, the extension pressure would depend on the geometrical *proportions* of a lenticular crack. Nevertheless, quite substantial excess pressures over the hydrostatic pressure might be required in order to obtain maximum dilation without fracture extension.

At Los Alamos a dilation pressure of about 88 atmospheres has been found possible at a depth of about 3 km – i.e. perhaps 25 to 30% of p_h.

The dilation of *horizontal* fractures would of course call for much higher pressures, slightly exceeding the lithostatic pressures. Thus for the extraction of very deep-seated heat it would seem that we shall probably have to rely upon vertical or near-vertical cracks only, rather than upon complex labyrinths of vertical and horizontal cracks.

To summarize, maximum dilation pressure will depend upon the value of μ and will thus vary widely from rock type to rock type. Nevertheless, a large proportion of this pressure would usually be provided by hydrostatic pressure alone. This, of course, is a very rough generalization, but for most rocks it would be no overstatement. No great difficulty should therefore be experienced in effecting dilation. The more formidable obstacles to be overcome in order to achieve universal heat-mining are therefore the successful mastery of rock-fracturing and of cheap deep penetration. It should not require an excessive degree of optimism to expect that these obstacles will be surmounted within two or three decades or even less.

A problem could arise when first initiating fractures at the base of a melt-formed bore. Clearly a bore must be cased all the way from the surface to within a short distance of the bottom, where the fracturing pressure is to be applied: otherwise fracturing would occur in the upper part of the bore. If a melt-formed bore has a vitreous lining the whole way down to the bottom it would be necessary to remove or penetrate that lining over a short distance from the bottom so as to expose the rock itself to fracture. At very great depths it might be difficult to support cutting or fusing equipment for this purpose owing to the high tensile stresses induced in the means of suspension. Hence it would probably be necessary to use explosives for gaining access to the virgin rock. The radial firing of bullets (see Section 7.11) is a technique that could possibly be used to penetrate the base of the vitreous lining so as to admit pressurised water into the very substance of the rock. Alternatively it might be necessary to use a straight explosive beneath a plug so as to shatter the lower part of the lining. When it comes to applying the fracturing pressure, either by pumping alone or by impact, it must be ensured that the vitreous lining remains intact over all its length except at the lowest extremity. It is true that this lining is claimed to be 'as tough as steel', but whether this is true of tension as well as compression seems doubtful. However, it is probable that the tremendous compaction of the vitreous lining and contiguous rock would give a more monolithic formation than would be possible with steel casings and cement grout, and would induce much higher compressive stresses than in the case of a simple cylindrical hole. Hence the initial circumferential compression would probably greatly exceed $2p_h$ (as with a simple hole) and would extend to a greater distance outward from the bore. In other words, the stress distribution would be very different from that of a simple drilled hole. The resistance to bursting under the influence of fracturing pressure would be very great, and a state of tensile hoop stress would probably be confined to the upper part of the bore only. It is possible that this upper part might need a protective sleeve of some sort. this is a point that will require careful investigation. It should also be remembered that once the true cylindrical shape of the base of the bore has been disturbed by explosive action it is probable that the fracturing pressure would be reduced by wedge action below the theoretical value of $2p_h + t$, so that the tensile stresses induced in the upper part of the bore lining would also be reduced accordingly.

It should be noted that the heading of this section refers to 'heat-mining'. This phrase is used advisedly, for the effect of a successful outcome would be to extract a vast, but *finite* quantity of *stored* heat which, except after the passage of great spans of future time, when the extracted heat could be replenished by conduction from the deeper recesses of the earth or by radioactivity from some distance away from the points of exploitation, should not be regarded as 'renewable'. Nevertheless, as already deduced in Section 4.1, it would seem that although it is not possible to fix a precise figure to the amount of extractable energy from deep earth-heat mining, we could at least be assured of satisfying all our likely energy needs for many centuries. That is all we can reasonably hope for.

19.7 Flexibility of use of artificial fields

A great advantage of artificially created fields, whether formed in hyper-thermal or semi-thermal areas underlain by hot rocks at slight or moderate depth or whether extracted from very great depth in non-thermal areas, is that the argument in favour of their use for base load application is less strong than with natural fields. By varying the rate of fluid circulation, artificial fields could be used equally well for peaking or for medium plant factor loadings without any wastage of heat. For power purposes, this would mean that an artificial field could perhaps supply economically any electrical system of whatever load factor. For non-power purposes the rate of heat extraction could be varied to suit the demand exactly.

20 Epilogue: the frame of the geothermal picture

20.1 Rationale

The broad theme of this book is that the development of geothermal energy is desirable and that it should be encouraged to expand as rapidly as possible. That is the picture, and the book has to some extent analysed the composition and structure of that picture. Still, what of the frame into which the picture is to be set? Surely that frame must be the overall world energy problem, a brief study of which will provide a suitable setting for the geothermal picture. This final chapter, in fact, seeks to show *why* there is a pressing need to develop the use of earth heat with a sense of great urgency. Geothermal development is not being advocated just for the sake of growth, as such, or because it is an interesting and novel idea: it is being urged as a way of trying to avert a human disaster of terrifying proportions.

20.2 Energy hunger

About 175 years ago Malthus foretold that the human race would founder on the twin rocks of population growth and food scarcity. Basically, he was one of the first men to foresee the obvious consequences of living on a finite planet. He prophesied that widespread war, famine, pestilence, misery and vice would inevitably be the lot of man before the end of the nineteenth century. The fact that human prosperity had broadly improved, rather than diminished, by the early twentieth century caused Malthus to be largely discredited as an alarmist prophet. Nevertheless, there was little wrong with his general thesis: it was only his time scale that proved to be unduly pessimistic. He should not be blamed for failing to foresee, in the early days of the Regency, the astonishing effects of mechanical transport and colonialization which so immensely increased the areas of the world from which food and raw materials could be drawn in the nineteenth century. In short, the arguments of Malthus were sound, but the early universal disasters which he foretold were mercifully postponed by the unexpected discovery of vast new resources.

Ever since the dawn of the Industrial Revolution, man's appetite for *energy* has grown far more rapidly than his appetite for food, and there is no lack of neo-Malthusians who foretell the early coming of a world energy famine. The gloomy prognostications of these prophets of doom are understandable when one pauses to note the following facts:

(i)   From 1900 to 1950 the world's annual energy demands rose by a factor of about 3.6, equivalent to an average annual growth rate of rather more than $2\frac{1}{2}\%$. From 1950 to 1960, in only one-fifth of that time, they rose at an average annual growth rate of about 4%; and from 1960 to 1970 by about $5\frac{1}{2}\%$ p.a. (compound). These growth rates are hyper-exponential, and give cause for alarm.

(ii) As late as 1974 (the latest year for which published world statistics are available) no less than 97.6% of the world's energy needs were supplied by fossil fuels. The remaining 2.4% came mainly from hydropower, a small amount of nuclear fission and a miniscule quantity from geothermal energy. We are still almost wholly dependent upon fossil fuels for our energy.

(iii) The World Energy Conference Survey of Energy Resources, 1974, quoted a figure of *measured* reserves of fossil fuels (excluding nuclear resources) equivalent to 139 years supply at the world consumption level at 1973 (the last year before the pattern of world energy growth became disturbed by the Middle East oil crisis and the sympathetic price rises of other fossil fuels that followed).

(iv) Now 139 years may sound reassuring to those who pursue the philosophies of 'sufficient unto the day is the evil thereof', but at best that is only about five generations from 1973, and could carry us only into the early part of the 22nd century: in practice our reprieve would be far less, as will now be shown.

(v)   If *growth* were to continue at anything approaching the rates to which we have become accustomed, the 139 years would shrink in a spectacular fashion, as Table 16 will show.

Table 16   Life of (1974) measured reserves of energy for calculated growth rates.

Growth rate p.a. (%)	Life of measured reserves (as from 1973; in years)
nil	139
1	87
2	67
3	55
4	48
5	42
6	38

These figures will serve to show that unless we can virtually halt growth altogether the outlook is very far from satisfactory.

(vi) This is not all. As stocks become depleted, prices will soar and wars could perhaps break out in a free-for-all scramble for what remains. There could well be chaos before the end of this century if we are to continue to rely upon fossil fuels as our main sources of energy.

(vii) Although 'measured reserves' are probably a very cautious figure which grossly under-estimates the ultimately available fossil fuel resources, the benefits to be gained from higher figures are largely offset by the evils of continued growth unless the growth be very small, zero or even negative, as can be seen from Table 17.

(viii) Yet another argument used, with good reason, by the pessimists, is that not all fossil fuels will be burnt: much will be needed as feed stock for industry.

(ix) In certain countries the coal miners seem to be pricing themselves out of the market. Certainly the winning of coal has failed to keep pace with the rising energy demand, and in some countries the rate of coal production has actually fallen.

(x) The continued growth of fossil fuel combustion may so pollute the earth as to spoil the environment, menace public health, and perhaps even make the world uninhabitable.

(xi) Whereas for several decades we were discovering new deposits of fossil fuels more rapidly than we were consuming them, the reverse is now true: we are consuming them more rapidly than we are discovering new sources.

These eleven points broadly reflect the arguments of the pessimists – or as some would call them, the scaremongers. Even if they do not emulate the precision of Bishop Ussher's extrapolations into the past, the pessimists

Table 17   Life of possible reserves of energy for various growth rates.

Annual growth rate (compound) p.a. (%)	Life in years, assuming that ultimate fossil fuel resources are:		
	1974 measured reserves	*10 times, the 1974 measured reserves*	*100 times, the 1974 measured reserves*
nil	139	1390	13900
1	87	271	496
2	67	170	284
3	55	127	204
4	48	103	161
5	42	87	134
6	38	76	115

make a rough stab at the early part of the next century as the likely time when doom will catch up with us.

20.3 Need for timely action

Now what have the optimists to say? How do they propose to act? Their first duty is to recognize that the arguments of the pessimists cannot be lightly dismissed, for they contain plenty of sound sense. Disaster will not be averted by Micawberism: we must *act*, and act vigorously and urgently; otherwise the pessimists may well prove to be right. Complacency is a very common human failing. Too often have we drifted into devastating wars through failure to grasp a nettle in time. Too often do we read in the press letters and reports of speeches – sometimes from eminent men of science who should know better – giving assurance that there is plenty of energy available and that it always will be so, and accusing those who say otherwise of being scaremongers.

Two main lessons are to be learned from the points enumerated in Section 20.2:

(a) We just cannot afford to sustain growth at anything approaching the levels to which we have become accustomed since 1950. There is no way of avoiding, sooner or later (and it had better be sooner) a reduction in growth rates. This reduction must be judiciously applied. Too drastic reduction could impose intolerable social stresses: too slow reduction could fail to avert the disasters foretold by the pessimists. Depending upon the outcome of (b) which follows, we may perhaps have to reduce the growth rate to zero within the foreseeable future, or even reverse it to negative values. Certainly anything approaching *exponential* growth, even at a very low rate, cannot for long be tolerated; for such growth is a divergent series that expands at an ever-increasing rate towards infinity, whereas we live on a planet with large, but *finite* resources.
(b) We cannot for much longer rely almost entirely upon fossil fuels for our main sources of energy. Herein lies our best hope of mitigating the horrors that would otherwise follow from (a) if tackled too rapidly; for at long last we are becoming aware of the importance of 'non-fuel' energy sources.

It cannot be over-emphasized that the principal enemy is not so much energy *demand* as energy *growth*, as can be seen from Tables 16 and 17. The author makes no apology for repeating the well known legend of the Chinese traveller who, some centuries ago, introduced to the Moghul Emperor in Delhi the game of chess. The Emperor was so delighted with the game that he invited the Chinese to name his own reward. The Chinese said that he would gratefully accept one grain of rice for the first square of the chessboard, two for the second, four for the third, and so on – doubling each time. The Emperor thought him mad to demand such a trifling reward, until he discovered that the whole world did not contain enough rice to satisfy the request. The effects of exponential growth – and even more, of *hyper*-exponential growth such as has occurred in the first 70 years of this century – are not unlike those grains of rice.

Politicians and economists are constantly preaching the need for growth as a necessary adjunct to prosperity. This may well be true for developing nations who have not yet attained a high standard of living, but for affluent countries it could prove disastrous in the long term, however attractive its short term benefits may appear; for it amounts to the mortgaging of a comfortable present against a painful future.

It is a curious fact that until 1973 no difficulty, other than temporary and local ones, was experienced in matching energy supplies to the rapidly growing demand. A belief in quasi-exponential energy growth had become almost an Article of Faith. Prognostications of electricity demands (which incidentally had been growing more rapidly than total energy demands), based on simple exponential growth modified by a small 'regressive factor', proved to be astonishingly accurate. By 1973, thanks to the general availability of abundant and cheap energy, the human way of life had so accustomed itself to hyper-exponential growth that dislocation, or even disaster, seemed to threaten us if our voracious energy appetite could not be fully satisfied. Then came the events of late 1973, which shattered our complacency and showed that dislocation could indeed occur almost overnight at the crack of the Arab whip. The days of cheap energy had apparently gone forever. Astronomical rises in the price of oil were soon followed by sympathetic rises in the prices of other energy sources. National balances of payment were thrown into disorder and many countries were beset by serious economic troubles. It is still too early to gain a proper perspective of the results of the 1973 fuel crisis (it takes about 3 years for world energy statistics to be compiled and a few more years before short term 'wobbles' in the demand pattern can become smoothed out). Nevertheless, it is clear that this demand pattern was undoubtedly disturbed by the events of 1973, although it is equally clear that energy demand is still growing (but at a reduced rate): a decline in oil consumption has been more than offset by increased uses of other energy forms.

If our principal enemy is growth, it is equally clear that our potential allies are new forms of energy. To stave off disaster we must weaken the enemy and rally our potential allies. Those are our two urgent tasks that will brook no delay.

20.4 Curtailment of growth
It is easy to talk of growth curtailment, but far from easy to bring it about. There is of course great scope for effecting economies: enormous quantities

of energy are at present being needlessly wasted. Profligacy is a luxury no longer to be afforded. Apart from the avoidance of waste, which need hurt no one, it must be recognized that curtailment of energy growth could be a very painful process; for man is an industrial animal, and industrial growth has been achieved only at the cost of energy growth. A too sudden reduction in energy growth rate could well have serious social consequences – even to the extent of political revolutions (which of course would not solve the dilemma but are apt to serve as an emotional vent to people subjected to unaccustomed restraints). Democracy, as we know it, could well be an early casualty of too rapid growth reduction. The brakes should be applied as gently as we can afford, so that people can become accustomed to the necessary changes to their habits that must be an inevitable corollary to a declining energy growth rate. In the last quarter century, what was formerly regarded as a luxury has come to be regarded as an essential, or even as a right. That attitude must gradually be reversed, as has been successfully done in times of war, and a more proper distinction must be restored as between necessities and luxuries. It is reassuring that President Carter seems to be very conscious of the urgent need to curb our appetite for energy. It is not inconceivable that if this process is not too rapidly applied it could somewhat paradoxically improve the quality of life; for there is little to show that the peoples of those countries that enjoy a high material standard of life are any happier than their forefathers whose demands were far simpler.

Another theoretical partial solution to the energy problem lies in reducing, halting, or even reversing population growth. However, to effect this painlessly would take several generations of intense propaganda. Deeply ingrained religious and social attitudes would have to be surmounted, and the process would inevitably be slow. We lack the time to rely on this solution alone, although every effort in this direction could be helpful.

20.5 Development of new energy sources and alternatives to fossil fuels

Our greatest source of hope of averting disaster is that just as Malthus fell down in his prophesies of early impending chaos by failing to take into account new developments which, to do him justice, he could not have been expected to foresee, so may there well be new developments in the near future *which can already be discerned* (so we lack the excuse of being unable to foresee them) that will drive away the spectre of an energy famine. It is clear that however successful we may be in curtailing growth, there can be no *permanent* solution in that direction. The only ultimate key to the energy problem lies in the development of new energy sources on a scale hitherto unimagined. It is therefore useful and instructive to take stock of the options apparently available to us.

20.5.1 *Hydro-power*. Of all conceivable alternative energy sources to fossil fuels, the first that naturally comes to mind is hydro-power, which is clean and renewable and often brings secondary benefits like irrigation and flood control. The World Energy Conference Survey of 1974 gave an estimate of the world's total exploitable hydro-resources at 3.7099×10^7 TJ p.a., equivalent to about 16% of the world's total energy consumption in 1973. In that year only about 1/8 of these hydro-resources had been developed – i.e. about 2% of the world's (then) total energy needs. However, it is clear that even if fully developed, hydro-power could never bring about our salvation. It could only supply about 1/6 of our *present* needs, and this proportion will fall if growth continues.

20.5.2 *Other renewable energy sources*. Wave, wind, solar and tidal energy are now receiving much deserved attention and may yet contribute quite large shares of our total energy needs. However, it is doubtful whether, even collectively, they will ever be able to offer more than a very useful palliative. Moreover, with the exception of tidal power, they are likely to involve the use of obstructive, ungainly and unsightly structures that would be aesthetically offensive.

20.5.3 *Nuclear fission*. Here of course we have a very valuable potential source of energy which, with the use of breeder reactors and of thorium, could more than treble our 'measured reserves'. This could give us valuable breathing space. However, from Table 17 it will be seen that even a trebling of resources would not gain us a great deal of time unless we first succeed in reducing the growth rate to a very low annual percentage. Moreover, there is a very powerful anti-nuclear lobby, and also a certain reluctance in high places to make decisions. These together are raising serious doubts as to the extent to which this energy source will ultimately be used. That there are possible environmental, and even terrorist, hazards cannot be denied, but these are perhaps no greater than those associated with off-shore oil rigs: in any case, little that is worthwhile has ever been accomplished without some element of calculated risk.

20.5.4 *Nuclear fusion*. Controlled nuclear fusion would indeed open the floodgates of available energy, so that we could afford to relax somewhat in our otherwise vitally urgent task of reducing the energy growth rate. So far the mastery of this vast energy source has eluded us, and although it is devoutly to be hoped that our efforts will ultimately be crowned with success, we cannot afford to place all our eggs in one basket.

20.5.5 *Geothermal energy*. This really is our last remaining option. Great strides have been made in recent years, both in exploration and in exploitation; while valuable research work is in progress that will almost certainly extend enormously the accessibility to earth heat. Herein would seem to lie our best hopes.

20.6 Mutual inter-dependence of growth curtailment and the development of new energy sources

The more rapidly we can develop new energy sources, the less urgent becomes the need to reduce the energy growth rate, and *vice versa*. The abundant availability of new energy sources within the near future would greatly help to absorb the shock of reducing the growth rate; while conversely, the more rapidly we can reduce the growth rate, the more time will be available within which we must develop new energy sources. Both remedies should be tackled simultaneously, and success in either will reduce the urgency of the other. What we cannot possibly afford is to fail in both endeavours; for that could lead only to inescapable disaster – the end of Industrial Man and perhaps a reversion to barbarism, if not to the total extinction of Mankind.

20.7 World energy: the probable shape of things to come

Fig. 79 illustrates the sort of pattern we may expect with world energy development. Except for Curve A, which is precise, it is purely qualitative and makes no claim to accuracy as to the future time scale: it is the *shapes* of Curves B and C that are of significance.

If we take as the unit of the vertical scale the world energy consumption (8538 million tons coal equivalent) in 1973 – the last year before the Middle East oil crisis really started to take effect – Curve A represents the *actual* annual world energy consumption from the beginning of the century until 1973, and the *extrapolated* world energy consumption thereafter at an assumed exponential growth rate of 5% p.a. This figure of 5% is arbitrary, and has been chosen simply by way of illustration. It has already been pointed out that the mean annual growth rate in the 1960s was $5\frac{1}{2}$%. It is true that from 1970 to 1973 there was a slight recession to rather less than 5% p.a. growth, but this could have been merely due to inevitable short term fluctuations lacking long term significance. The chosen figure of 5% p.a. is slightly less than the average growth rate from 1960 to 1973, and has been taken in order to show the utter impossibility of maintaining for long growth rates anywhere approaching those to which we had become accustomed by 1973.

Curve B shows the actual growth in the supply of fossil fuels from 1900 to 1973. It lies so close to Curve A that it is almost indistinguishable from it, as these fuels supplied nearly all the world's energy needs during the 73 years under consideration. The effect of the huge increases in fossil fuel prices since 1973 are not likely to cause an actual *fall* in fuel production for some time owing to the pressures of still rising demands, but they will

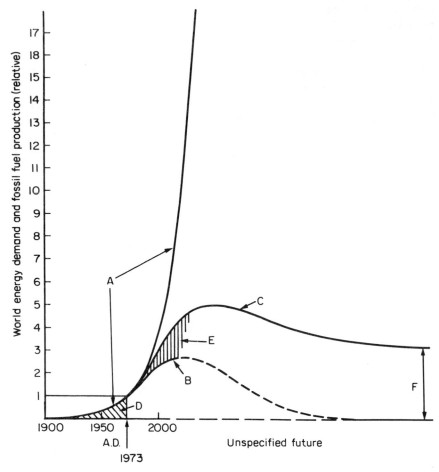

A, Actual world energy supply and demand until 1973,
   and extrapolated future demand assuming 5% annual
   exponential growth.
B, Probable *shape* (not to scale) of future fossil fuel
   production
C, Probable *shape* (not to scale) of future world energy
   demand if disaster is to be avoided
D, Total world energy consumption from 1900 to 1974
E, Energy deficit between world energy demand and
   fossil fuel availability, to be met from new energy
   sources and economies
F, Ultimate stability when world energy demand can
   be met from renewable sources alone

Figure 79   Probable pattern of future world energy supply and demand (see [187].)
   Note: The vertical scale is in multiples of the world energy demand in 1973–the last
year to be scarcely affected by the Middle East oil crisis.

almost certainly lead to a falling off of the growth rate of production. Moreover, it is the avowed policy of some of the oil-producing countries to keep more of their reserves in the ground as an insurance against future contingencies. Against these retarding factors, new oil and gas fields are now coming into production, and high prices are likely to stimulate the exploitation of other more exotic fuels such as oil shales and tar sands Before very long fossil fuels will inevitably begin to show signs of approaching exhaustion after one or two short-lived 'bonanzas' have been played out. Also, as new energy forms are developed to an increasing extent, market forces will start to retard the growth in demand for fossil fuels. On the whole, it seems likely that the pattern of fossil fuel production will follow some such shape as the extrapolated part of Curve B after 1973. Production could continue to rise at a declining pace to reach a peak – perhaps in the early part of the next century – and then gradually decline to approach zero almost asymptotically at some remote and unspecified future date when all the accessible stocks of fossil fuels have become exhausted. Against the background of history, the bulk of the world's fossil fuels will have been consumed within perhaps a couple of centuries – a mere instant in eternity – with very attenuated production over longer periods on either side of the 'hump'. The area D approximately represents the total fuel consumption during the first 73 years of this century – which will differ little from the total fuel consumed from the dawn of history until 1973. It represents about 15% of the 'measured reserves' referred to in Chapter 4 (excluding nuclear resources), but probably a far smaller percentage of the fossil fuel stocks now 'inferred' and possibly ultimately recoverable. In fact according to the World Power Conference 1968 estimates of world energy resources a total believed stock of fossil fuels, proved *and* inferred, of 8.8×10^{12} tons coal equivalent was quoted. This figure is almost 8 times the 'measured' reserves shown in Table IX-2 of the 1974 World Energy Conference Survey of Energy Reserves. So the shaded area D of Fig. 79 could perhaps represent only about 2% of the ultimately recoverable fossil fuel resources of the world.

Now it is quite obvious that we could not possibly hope to fill the gap between Curves A and B for more than a very short time, since the two curves will soon diverge from one another so rapidly that the gap between them would become unmanageable, with Curve A soaring away merrily towards infinity and curve B bending further away from it. Even if we could *discover* new energy sources at a rate undreamed of, there must be a limit to the rate at which we could construct the necessary equipment for harnessing them and converting them into the form required. For instance, in 50 years time the energy demand at a sustained exponential growth of 5% p.a. would have increased about $11\frac{1}{2}$ times whereas the population, which for quite a long time has been growing at a fairly steady rate of 2% p.a. would have increased only about 2.7 times. Hence the production of exploitation equipment *per capita* would have to increase by about $4\frac{1}{4}$ times. After 100 years this factor would have risen to more than 18 times. Clearly such a trend could not possibly be maintained. Moreover, it is often overlooked that all artifacts, including energy exploitation equipment, themselves consume energy in the process of manufacture – sometimes on an immense scale. There could be a danger that the energy absorbed in making exploitation equipment would be so great that the net gain of energy usable for other purposes might be negligible.

Thus it is abundantly clear that Curve A, or anything approaching it is doomed as totally impracticable. The demand curve must be forcibly 'bent' into some shape such as Curve C in Fig. 79. Curve C could be permitted to rise above Curve B only to the extent that we can supplement fossil fuels with new forms of energy. The shaded area E represents these new energy sources, and *it is within this area that the importance of geothermal energy will lie*. This clearly shows the inter-dependence between growth curtailment and the development of new energy sources (see Section 20.6). As drawn in Fig. 79, it will be seen that Curve C need depart from Curve A only fairly slowly at first, provided that we can develop area E fairly rapidly. If we cannot do this, Curve C can lie only slightly above Curve B, and this could be very painful. Obviously, if E can be expanded rapidly it will give us much more time in which to adapt ourselves to a new, and drastically re-shaped, demand Curve C; whereas if Curve C can be bent over quickly, smaller values of E would be acceptable.

It is possible that the maximum height to which Curve C can be allowed to grow will be limited not so much by our ingenuity in winning new energy sources as by an approach towards the ultimate of environmental acceptance. Certainly geothermal energy would permit a far higher peak value to Curve C than would fuel combustion.

Ultimately, in the very remote future, we shall have to learn to live on renewable energy sources only – F in Fig. 79. When we have learned to do that, and to limit our population accordingly, there will no longer be a world energy problem.

Our future lies largely in our own hands [187]. Mankind has shown himself to be particularly adept in failing to learn from the lessons of history and in blundering from crisis to crisis through his inability to see ahead. The purpose of this chapter is to show that to some extent we *can* see ahead, and that by taking timely action we may avert the disasters that may otherwise be inevitable. The obstacles that lie in our path to a brilliant energy future are probably no greater than those which were successfully overcome within a few years in attaining mastery over nuclear fission and space travel. All that is needed is a sense of urgency, and our problems will vanish.

Despite the evidence that this sense of urgency is now being felt by a growing number of people, we must be on our guard against complacency which may arise in the 1980s when a temporary glut of energy could engender a false sense of security, as North Sea oil flows in spate. The future of energy supplies is the problem of Mankind, and not only of a small group of countries.

This chapter is intended to be no more than a warning: it makes no claim to being a precise prophecy. As already remarked in Chapter 4, long term prognostications are a vain pursuit. At all costs we should not attempt to extrapolate the future into the realms of absurdity! Who can possibly foresee how Mankind will have developed in tens of thousands of years? He may by then have learned to harness energy from sources as yet not dreamed of. He may have learned so to modify his way of life as to be able to do with far less energy – *per capita* and *in toto* – than at present. There may have occurred tectonic or solar cataclysms that will have destroyed or drastically reduced human and animal life. He may even (not altogether improbably) have destroyed himself by continuing to follow his foolish policies, perhaps at the same time annihilating all other forms of life on this planet, or perhaps leaving other species to survive, thus putting back the evolutionary clock by aeons.

One thing is certain: the next few decades will be critical for the human race. We may perhaps be faced with a Morton's Fork of discomfort on the one hand and disaster on the other; but by timely action – and there are encouraging signs of an awareness of the need for such action – we may weather the storm successfully. In any case, babies already born will see incredible changes in the human way of life before they are elderly, or even middle-aged.

References*

The first four of the following publications have been printed as composite documents, and contain a wealth of specialized papers and articles, many of which are included in this reference list. To avoid tedious repetition, these publications are referred to in other references by the following abbreviations:

LASL: The Los Alamos Scientific Laboratory of the University of California.

Rome: 1 Proceedings of the United Nations Conference on New Sources of Energy, Rome, August 1961, Vols **2** and **3**. United Nations: Geneva. E/CONF 35/3 and E/CONF 35/4. Document sales Nos 63.1.36 and 63.1.37.

Pisa: 2 Proceedings of the United Nations Symposium on the Development and Utilization of Geothermal Resources, Pisa, Sept/Oct 1970. Published in *Geothermics* special issue, Vols **1** and **2**.

San Francisco: 3 Proceedings of the Second United Nations Symposium on the Development and Use of Geothermal Resources, San Francisco, USA May 1975, Vols **1**, **2** and **3**. Lawrence Berkeley Laboratory, University of California.

UNESCO: 4 Geothermal Energy, a review of research and development. Earth Sciences **12** (1973). UNESCO: Paris.

5 Einarsson, S.S. (1961). Proposed 15 MW geothermal power station at Hveragerdi, Iceland. *Paper G/9, Rome*, **3**, 354.

6 Reynolds, G. (1970). Cooling with geothermal heat. *Pisa*, **2**, Part 2, 1658.

7 Einarsson, S.S. (1975). Geothermal space heating and cooling. *San Francisco*, **3**, 2117.

8 Líndal, B. (1973). Industrial and other applications of geothermal energy. *UNESCO*, 135.

9 Kunze, J. F., Richardson, A.S., Hollenbaugh, K.M., Nichols, C.R. and Mink, L.L. (1975). Non-electric utilization project, Boise, Idaho. *San Francisco*, **3**, 2141.

10 Kunze, J.F., Miller, L.G. and Whitbeck, J.F. (1975). Moderate temperature utilization project in the Raft River valley. *San Francisco*, **3**, 2021.

11 Reistad, G.M. (1975). Potential for non-electrical applications of geothermal energy and their place in the national economy. *San Francisco*, **3**, 2155.

12 Bullard, Sir Edward, FRS. (1969). The origin of the oceans. *Scientific American*, **221**, 68.

*When the manuscript for this book was completed the author, in his capacity as Rapporteur, had only had access to sections v, viii and ix of *San Francisco*.

13 Bullard, Sir Edward, FRS. (1969). The origin of the oceans. *Scientific American*, **221**, 69.

14 Tamrazian, G.P. (1970). Continental drift and thermal fields. *Pisa*, **2**, Part 2, 1212.

15 Bullard, Sir Edward, FRS. (1973). Basic Theories, *UNESCO*, 19.

16 Decius, L.C. (1961). Geological environment of hyperthermal areas in continental United States and suggested methods of prospecting them for geothermal power. *Paper G/48, Rome*, **2**, 166.

17 Calami, A. and Ceron, P. (1970). Air convection within Montaña del Fuego, Lanzarote Island, Canary Archipelago. *Pisa*, **2**, Part 1, 611.

18 Tikhonov, A.N. and Dvorov, I.M. (1970). Development of research and utilization of geothermal resources in the USSR. *Pisa*, **2**, Part 2, 1072.

19 Facca, G. and Tonani, F. (1967). The self-sealing geothermal field. *Bull. Volcanologique*, **30**, 271.

20 Facca, G. (1973). The structure and behaviour of geothermal fields. *UNESCO*, 61.

21 Chierici, A. (1961). Planning of a geothermoelectric power plant: technical and economic principles. *Paper G/62, Rome*, **3**, 299.

22 House, P.A., Johnson, P.M. and Towse, D.F. (1975). Potential power generation and gas production from Gulf Coast geopressured reservoirs. *San Francisco*, **3**, 2001.

23 Banwell, C.J. (1961). Geothermal drillholes: physical investigations. *Paper G/53, Rome*, **2**, 60.

24 McNitt, J. R. (1973). The rôle of geology and hydrology in geothermal exploration. *UNESCO*, 33.

25 McNitt, J.R. (1970). The geologic environment of geothermal fields as a guide to exploration. *Pisa*, **1**, 24.

26 Craig, H. Boato, G. and White, D. (1956). The isotopic geochemistry of thermal waters. *National Research Council on Nuclear Science Service Report No. 19*, 29–44.

27 Hulston, J.R. (1961). Isotope geology in the hydrothermal areas of New Zealand. *Paper G/31, Rome*, **2**, 259.

28 Banwell, C.J. (1973). Geophysical methods in geothermal exploration. *UNESCO*, 41.

29 Banwell, C.J. (1970). Geophysical techniques in geothermal exploration. *Pisa*, **1**, 32.

30 Dorbin, M.B. (1960). *Introduction to geophysical prospecting* McGraw Hill: New York.

31 Sestini, G. (1970). Heat flow measurement in non-homogeneous terrains. Its application to geothermal areas. *Pisa*, **2**, Part 1, 424.

32 Lee, W.H.K. and Uyeda, S. (1965). Terrestial heat flow. *Geophysical Monograph*. **8**, American Geophysical Union: Washington D.C.

33 Burgassi, R., Battini, F., and Mouton, J. (1961). Prospection géothermique pour la recherche des forces endogènes. *Paper G/61, Rome*, **2**, 134.

34 Meidav, T. (1970). Application of electrical resistivity and gravimetry in deep geothermal exploration. *Pisa*, **2**, Part 1, 303.

35 Gorhan, H.L. (1971). Geophysics *versus* drilling in engineering geology and hydrogeology. *Queensland Government Mining Journal*.

36 Lumb, J.T. and Macdonald W.J.P. (1970). Near-surface resistivity surveys of geothermal areas using the electromagnetic method. *Pisa*, **2**, Part 1, 311.

37 Clacy, G.R.T. (1973). Geothermal ground noise, amplitude and frequency spectra in the New Zealand volcanic region. *J. Geophys. Res.*, 5377.

38 Brune, J.N. and Allen, C.R. (1967). A micro-earthquake survey of the San Andreas fault system in southern California. *Bull. Seismol. Soc. Am.* **57**, 277.

39 Ward, P.L., Pálmason, G., Drake, C. and Oliver, J. (1968). The microseismicity of Iceland and its relation to the regional tectonics. *49th Annual Meeting of the American Geophysical Union.*

40 Pálmason, G., Friedman, J.D., Williams, R.S.Jr., Jonnsson, J. and Saemundsson, K. (1970). Aerial infra-red surveys of Reykjanes and Torfajokull thermal areas, with a section on cost of exploration surveys. *Pisa*, **2**, Part 1, 399.

41 Schirra, W Jr. (1976). National Geothermal Conference, Palm Springs. *Geothermal Energy*, **4**, No. 6, 32.

42 Sigvaldason, G.E. (1973). Geochemical methods in geothermal exploration. *UNESCO*, 49.

43 Craig, S.B. (1961). Geothermal drilling practices at Wairakei, New Zealand. *Paper G/14, Rome*, **3**, 121.

44 Matsuo, K. (1973). Drilling for geothermal steam and hot water. UNESCO, 73.

45 Cigni, U. and Giovannoni, A. (1970). Planning methods in geothermal drilling. *Pisa*, **2**, Part 1, 725.

46 Cigni, U. (1970). Machinery and equipment for harnessing of endogenous fluid. *Pisa*, **2**, Part 1, 704.

47 Giovannoni, A. (1970). Drilling Technology. *Pisa*, **1**, 81.

48 Woods, D.I. (1961). Drilling mud in geothermal drilling. *Paper G/21, Rome*, **3**, 270.

49 Fabbri, F. and Vidali, M. (1970). Drilling mud in geothermal wells. *Pisa*, **2**, Part 1, 735.

50 Contini, R. and Cigni, U. (1961). Air drilling in geothermal bores. *Paper G/70, Rome*, **3**, 89.

51 Dench, N.D. (1970). Casing string design for geothermal wells. *Pisa*, **2**, Part 2, 1485.

52 Fabbri, F. and Giovannoni, A. (1970). Cements and cementation in geothermal well drilling. *Pisa*, **2**, Part 1, 742.

53 Bolton, R.S. (1961). Blowout prevention and other aspects of safety in geothermal steam drilling. *Paper G/43, Rome*, **3**, 78.

54 Innes, I.A. (1961). Management, in relation to measurements, and bore maintenance of an operating geothermal steam field. *Paper G/15, Rome*, **3**, 208.

55 Stilwell, W.B. (1970). Drilling practices and equipment in use at Wairakei. *Pisa*, **2**, Part 1, 714.

56 Brunetti, V and Mezzetti, E. (1970). On some troubles most frequently occurring in geothermal drilling. *Pisa*, **2**, Part 1, 751.

57 Smith, J.H. (1961). Casing failures in geothermal bores at Wairakei. *Paper G/44, Rome*, **3**, 254.

58 Budd, C.F. (1973). Producing geothermal steam at the Geysers field. In *Geothermal Energy Resources, Production and Stimulation*, (edited by P. Kruger and C. Otte). Stanford University Press.

59 Hunt, A.M. (1961). The measurement of borehole discharges, downhole temperatures and pressures, and surface heat flows at Wairakei. *Paper G/19, Rome*, **3**, 196.

60 James, R. (1970). Factors controlling borehole performance. *Pisa*, **2**, Part 2, 1502.

61 James, R. (1967). Optimum wellhead pressure for geothermal power. *N. Zealand Engineering*, **22**, No. 6, 221–8.

62 Bangma, P. (1961). The development and performance of a steam-water separator for use on geothermal bores. *Paper G/13, Rome*, **3**, 60.

63 Armstead, H.C.H. and Shaw, J.R. (1970). The control and safety of geothermal installations. *Pisa*, **2**, Part 1, 848.

64 Dench, N.D. (1961). Silencers for geothermal bore discharge. *Paper G/18, Rome*, **3**, 134.

65 Pollastri, G. (1970). Design and construction of steam pipelines. *Pisa*, **2**, Part 1, 780.

66 James, R. (1968). Pipeline transmission of steam-water mixtures for geothermal power. *New Zealand Engineering*, **23**, 55–61.

67 Armstead, H.C.H. (1968). The extraction of power from hot water. Seventh World Power Conference, Moscow, August, 1968, Paper 174, Section C4.

68 Haldane, T.G.N. and Armstead, H.C.H. (1962). The geothermal power development at Wairakei, New Zealand. *Proc. Inst. mechan. Eng.*, **176**, No. 23, 603.

69 Takahashi, Y., Hayashida, T., Soezima, S., Aramaki, S. and Soda, M. (1970). An experiment on pipeline transportation of steam-water mixtures at Otake geothermal field. *Pisa*, **2**, Part 1, 882.

70 Aikawa, K. and Soda, M. (1975). Advanced design in Hatchobaru geothermal power station. *San Francisco*, **3**, 1881.

71 Dal Secco, A. (1970). Turbo-compressors for geothermal plants. *Pisa*, **2**, Part 1, 819.

72 Moskvicheva, V.N. and Popov, A.E. (1970). Geothermal power plant on the Paratunka River. *Pisa*, **2**, Part 2, 1567.

73 Rollet, A. (1957). Centrale géothermique de Kiabukwa: lecons tirées de quatre années d'exploitation. *Bull. Acad. r. Sci. Col., Bruxelles*, **3**, 1246.

74 Hansen, A. (1961). Thermal cycles for geothermal sites and turbine installation at the Geysers power plant, California. *Paper G/41, Rome*, **3**, 365.

75 Harrowell, R.V. (1975). Private communication.

76 Wigley, D.M. (1970). Recovery of flash steam from hot bore water. *Pisa*, **2**, Part 2, 1588.

77 Wood, B. (1973). Geothermal power, *UNESCO*, 109.

78 Armstead, H.C.H. (1970). Utilization of steam and high enthalpy water (for electric power generation and other purposes. *Pisa*, **1**, 106.

79 Dal Secco, A. (1975). Geothermal plants: gas removal from jet condensers. *San Francisco*, **3**, 1943.

80 James, R. (1970). Power station strategy. *Pisa*, **2**, Part 2, 1676.

81 Armstead, H.C.H. (1970). Geothermal power for non-base load purposes. *Pisa*, **2**, Part 1, 936.

82 Ciapica, I. (1970). Present development of turbines for geothermal applications. *Pisa*, **2**, Part 1, 834.

83 Ricci, G. and Viviani, G. (1970). Maintenance operations in geothermal power plants. *Pisa*, **2**, Part 1, 839.

84 Matthew, P. (1975). Geothermal operating experience, Geysers power plant, *San Francisco*, **3**, 2049.

85 Palmer, T.D., Howard, J.H. and Lande, D.P. (1975). Geothermal development of the Salton Trough, California and Mexico. *Lawrence Livermore Laboratory Document*, UCRL-51775.

86 Austin, A.L. (1975). Prospects for advances in energy conversion technologies for geothermal energy development. *San Francisco*, **3**, 1925.

87 Armstead, H.C.H. (1975). Basic design of a cheap wellhead non-condensing turbine. *San Francisco*, **3**, 1905.

88 Pessina, S., Rumi, O., Silvestri, M. and Sotgia, G. (1970). Gravimetric loop for the generation of electrical power from low temperature water. *Pisa*, **2**, Part 1, 901.

89 Daman, E.L. (1966). *Electrogasdynamic power generation.* Gourdine Systems, Inc.: New Jersey.

90 Naymanov, O.S. (1970). A pilot geothermoelectric power station in Pauzhetka, Kamchatka. *Pisa*, **2**, Part 2, 1560.

91 Alger, T.W. (1975). Performance of two-phase nozzles for total-flow geothermal impulse turbines. *San Francisco*, **3**, 1889.

92 Armstead, H.C.H. (1975). Some unusual ways of developing power from a geothermal field. San Francisco, **3**, 1897.

93 Charropin, P., Despois, J., Fauconnier, J-.C. and Nougarède, F. (1976). Paper presented at the first International Total Energy Congress, Copenhagen, *Total Energy*, p. 460. Miller Freeman Publications: San Francisco

94 Lindal, B. (1976). Private communication.

95 Boldizsár, T. (1976). Private communication.

96 *Japan Geothermal Energy Association.* (1974). Geothermal energy utilization in Japan, 1974. Official brochure.

97 Mashiko, Y. and Hirano, Y. (1970). New supply systems of thermal waters to a wide area in Japan. *Pisa*, **2**, Part 2, 1592.

98 Shannon, R.J. (1975). Geothermal heating of Government buildings in Rotorua. *San Francisco*, **3**, 2165.

99 Dvorov, I.M. and Ledentsova, N.A. (1975). Utilization of geothermal water for domestic heating and hot water supply. *San Francisco*, **3**, 2109.

100 Lund, J.W., Culver, G.G. and Svanevik, L.S. (1975). Utilization of intermediate temperature geothermal water in Klamath Falls, Oregon. *San Francisco*, **3**, 2147.

101 Behl, S.C., Jegadeesan, K. and Reddy, D.S. (1975). Some aspects of the utilization of geothermal fluids in the North-West Himalayas. *San Francisco*, **3**, 2083.

102 Coulbois, P., and Hérault, J-P. (1975). Conditions for the competitive use of geothermal energy in home heating. *San Francisco*, **3**, 2104.

103 Arnórsson, S., Ragnars, K., Benediktsson, S, Gislason, G. and Thórhallsson, S., with Björnsson, S., Grönvold, K. and Lindal, B. Exploitation of saline high temperature water for space heating. *San Francisco*, **3**, 2077.

104 Boldizsár, T. (1970). Geothermal energy production from porous sediments in Hungary. *Pisa*, **2**, Part 1, 99.

105 Einarsson, S.S. (1973). Geothermal district heating. *UNESCO*, 123.

106 Kremnjov, O.A., Zhuravlenko, V.J. and Shurtshkov, A.V. (1970). Technical–economic estimation of geothermal resources. *Pisa*, **2**, Part 2, 1688.

107 Lokchine, B.A. and Dvorov, I.M. (1970). Applications expérimentales et industrielles de l'energie géothermique en URSS. *Pisa*, **2**, Part 2, 1079.

108 Lindal, B. (1961). The extraction of salt from sea water by multiple-effect evaporators using natural steam. *Paper G/27, Rome*, **3**, 479.

109 Mizutani, Y. (1961). Salt production by geothermal energy in Japan. *Paper G/7, Rome*, **3**, 483.

110 Komagata, S., Iga, H., Nakamura, H. and Minohara, Y. (1970). The status of geothermal utilization in Japan. *Pisa*, **2**, Part 1, 185.
111 Lindal, B. (1970). The production of chemicals from brine and seawater, using geothermal energy. *Pisa*, **2**, Part 1, 910.
112 Thiagarajan, B. (1967). Report of a UNIDO mission on the manufacture of chemicals from seawater in Iceland, UNIDO: Vienna.
113 Kennedy, A.M. (1961). The recovery of lithium and other minerals from geothermal water at Wairakei. *Paper G/56, Rome*, **3**, 502.
114 Werner, H.H. (1970). Contribution to the mineral extraction from supersaturated geothermal brines, Salton Sea area, California. *Pisa*, **2**, Part 2, 1651.
115 Mazzoni, A. (1948). The steam vents of Tuscany and the Larderello plant. pp. 59–75, Bologna, Amonina Arts Grafiche.
116 Lenzi, D. (1961). Utilization de l'énergie géothermique pour la production de l'acide borique et des sous-produits contenus dans les 'soffioni' de Larderello. *Paper G/39, Rome*, **3**, 512.
117 Valfells, A. (1970). Heavy water production with geothermal steam. *Pisa*, **2**, Part 1, 896.
118 Shreve, R.N. (1956). *Chemical process industries* (2nd Edition) McGraw Hill: New York.
119 Ludviksson, V. (1970). Nýting jardhitans (The application of natural heat). *National Research Council Report 70–3*, Reykjavik.
120 Burrows, W. (1970). Geothermal energy resources for heating and associated applications in Rotorua and surrounding areas. *Pisa*, **2**, Part 2, 1662.
121 Einarsson, S.S. (1970). Utilization of low enthalpy water for space heating, industrial, agricultural and other uses. *Pisa*, **1**, 112.
122 Lindal, B. (1961). Greenhouses by geothermal heating in Iceland. *Paper G/32, Rome*, **3**, 476.
123 Gutman, P.W. (1975). Geothermal hydroponics. *San Francisco*, **3**, 2217.
124 Dragone, G. and Rumi, O. (1970). Pilot greenhouse for the utilization of low temperature waters. *Pisa*, **2**, Part 1, 918.
125 Hallsson, S. (1970). Drying seaweeds and grass by geothermal energy. *Timarit Verkfraed. Islands*, **4**.
126 Kerr, R.N., Bangma, R., Cooke, W.L., Furness, F.G. and Vamos, G. (1961). Recent developments in New Zealand in the utilization of geothermal energy for heating purposes. *Paper G/52, Rome*, **3**, 456.
127 Armstead, H.C.H. (1967). Fresh water from geothermal fluids. *International Water for Peace conference, Paper P/673, Washington, D.C.*
128 Armstead, H.C.H. and Rhodes, C. (1970). Desalination by geothermal means. *3rd International Symposium on Fresh Water from the Sea, Dubrovnik. Proc.* **3**, 451–9. Athens.
129 De Anda, L.F., Reyes, S.C. and Tolivia, M.E. (1970). Production of fresh water from the endogenous steam of Cerro Prieto geothermal field. *Pisa*, **2**, Part 2, 1632.
130 Chiostri, E. (1975). Geothermal resources for heat treatment. *San Francisco*, **3**, 2094.
131 Cooke, W.L. (1970). Some methods of dealing with low enthalpy water in the Rotorua area of New Zealand. *Pisa*, **2**, Part 2, 1670.
132 Sigurdsson, H. (1961). Reykjavik Municipal district heating service and utilization

of geothermal energy for domestic heating. *Paper G/45, Rome*, **3**, 486.

133 Bodvarsson, G. (1961). Utilization of geothermal energy for heating purposes and combined schemes involving power generation, heating and/or by-products. *Paper GR/5(G), Rome*, **3**, 429.

134 Wong, C.M. (1970). Geothermal energy and desalination: partners in progress. *Pisa*, **2**, Part 1, 892.

135 Leardini, T. (1970). Economie de l'energie géothermique. *Pisa*, **2**, Part 1, 958.

136 Armstead, H.C.H. (1973). Geothermal economics. *UNESCO*, 161.

137 Bodvarsson, G. and Zoëga, J. (1961). Production and distribution of natural heat for domestic and industrial heating in Iceland. *Paper G/37, Rome*, **3**, 449.

138 Garnish, J.D. (1976). Geothermal energy: the case for research in the United Kingdom. Report (Energy paper No. 9) prepared for the Dept. of Energy by the Energy Technology Support Unit, Harwell. HMSO: London.

139 Dan, F.J., Hersam, D.E., Kho, S.K. and Krumland, L.R. (1975). Development of a typical generating unit at the Geyscrs geothermal project – a case study. *San Francisco*, **3**, 1949.

140 Worthington, J.D. (1976). Private communication.

141 Facca, G. and Ten Dam, A. (1964). *Geothermal power economics*. Worldwide Geothermal Exploration Co.

142 Leardini, T. (1976). Private communication.

143 Ragnars, K., Saemundsson, K., Benediktsson, S. and Einarsson, S.S. (1970). Development of the Namafjall area, Northern Iceland. *Pisa*, **2**, Part 1, 925.

144 Iga, H. (1976). Personal communication.

145 Nakamura, S. (1970). Economics of geothermal electric power generation at Matsukawa. *Pisa*, **2**, Part 2, 1715.

146 Bowen, R.G. and Groh, E.A. (1971). Geothermal – Earth's primordial energy. *Technology Review*, 42–48.

147 Thórhallsson, S, Ragnars, K., Arnórsson, S. and Kristmannsdóttir, H. (1975). Rapid scaling of silica in two district heating systems. *San Francisco*, **2**, 1445.

148 Usui, T. and Aikawa, K. (1970). Engineering and design features of the Otake geothermal power plant. *Pisa*, **2**, Part 2, 1533.

149 Bruce, A.W. (1970). Engineering aspects of a geothermal power plant. *Pisa*, **2**, Part 2, 1516.

150 Allegrini, G. and Benvenuti, G. (1970). Corrosion characteristics and geothermal power plant protection; (collateral processes of abrasion, erosion and scaling.) *Pisa*, **2**, Part 1, 865.

151 Hermannsson, S. (1970). Corrosion of metals and the forming of a protective coating on the inside of pipes carrying thermal waters used by the Reykjavik Municipal District Heating Service. *Pisa*, **2**, Part 2 1602.

152 Marshall, T. and Braithwaite, W.R. (1973). Corrosion control in geothermal systems. *UNESCO*, p. 151.

153 Ozawa, T. and Fujii, Y. (1970). A phenomenon of scaling in production wells and the geothermal power plant in the Matsukawa area. *Pisa*, **2**, Part 2, 1613.

154 Tolivia, M.E. (1970). Corrosion measurements in a geothermal environment. *Pisa*, **2**, Part 2, 1596.

155 Einarsson, S.S., Vides, R.A. and Cuéllar, G. (1975). Disposal of geothermal waste water by re-injection. *San Francisco*, **2**, 1349.

156 Chasteen, A.J. (1975). Geothermal steam condensate re-injection. *San Francisco*, **2**, 1335.

157 Gringarten, A.C. and Sauty, J.P. (1975). The effect of re-injection on the temperature of a geothermal reservoir used for urban heating. *San Francisco*, **2**, 1370.

158 Kubota, K. and Aosaki, K. (1975). Re-injection of geothermal hot water at the
Otake geothermal field. *San Francisco*, **2**, 1379.

159 Reed, M.J. and Campbell, G, (1975). Environmental impact of development in the
Geysers geothermal field, U.S.A. *San Francisco*, **2**, 1399.

160 Axtmann, R.C. (1975). Chemical aspects of the environmental impact of geothermal power. San Francisco, **2**, 1323.

161 Allen, G.W. and McCluer H.K. (1975). Abatement of hydrogen sulphide emissions
from the Geysers geothermal power plant. *San Francisco*, **2**, 1313.

162 Rothbaum, H.P. and Anderton, B.H. (1975). Removal of silica and arsenic from
geothermal discharge waters by precipitation of useful calcium silicates. *San
Francisco*, **2**, 1417.

163 Anderson, S.O. (1975). Environmental impacts of geothermal resource development on commercial agriculture: a case study of land use conflict. *San Francisco*,
2, 1317.

164 Guiza, J.L. (1975). Power generation at Cerro Prieto geothermal field. *San Francisco*, **3**, 1976.

165 Mercado, G.S. (1975). Cerro Prieto geothermoelectric project: pollution and
basic protection. *San Francisco*, **2**, 1394.

166 Swanberg, C.A. (1975). Physical aspects of pollution related to geothermal energy
development. *San Francisco*, **2**, 1435.

167 Axtmann, R.C. (1975). Environmental impact of a geothermal power plant.
Science, **187**. No. 4179.

168 Stilwell, W.B., Hall, W.K. and Tawhai, J. (1975). Ground movement in New
Zealand geothermal fields. *San Francisco*, **2**, 1427.

169 Bolton, R.S. (1973). Management of a geothermal field. *UNESCO*, 175.

170 Armstead, H.C.H. (1975). Environmental factors and waste disposal. *San Francisco*, **1**, p. lxxxvii.

171 Cuéllar, G. (1975). Behaviour of silica in geothermal waste waters. *San Francisco*,
2, 1343.

172 Hatton, J.W. (1970). Ground subsidence of a geothermal field during exploitation.
Pisa, **2**, Part 2, 1294.

173 Armstead, H.C.H., Gorhan, H.L. and Müller, H. (1974). Systematic approach to
geothermal development. *Geothermics*, **3**, No. 2.

174 Clark, A.L., Calkins, J.A., Tongiorgi, E. and Stefanelli, E. (1975). A report on the
International Geothermal Information Exchange Programme, 1974–75. *San
Francisco*, **1**, 67.

175 Worthington, J.D. (1975). Geothermal Development in Chapter 9 of *Status Report–
Energy Resources and Technology*, a report of the *ad hoc* Committee on Energy
Resources and Technology, Atomic Industrial Form, Inc.

176 Boardman, C.R., Rabb, D.D. and McArthur, R.D. (1964). Contained nuclear
detonations in four media – geological factors in cavity and chimney formation.
3rd Ploughshare Symposium, University of California, Davis, Cal. April, 1964.

177 Witherspoon, P.A. (1966). Economics of nuclear explosives in developing underground gas storage. University of California, Lawrence Radiation Laboratory,
Livermore, California, Documant UCRL-14877.

178 Parker, K. (1974). Personal communication.

179 Kennedy, G.C. (1964). A proposal for a nuclear power programme. 3rd Plough-

share Symposium, University of California, Davis, Cal. April, 1964.

180 Pettitt, R.A. (1976). Environmental monitoring for the hot dry rock geothermal energy development project. LASL Document LA-6504-SR Status Report. UC-11 and UC-66e

181 Harlow, F.H. and Pracht, W.E. (1972). A theoretical study of geothermal energy extraction. *J. Geophys. Res.* **77**, 7041.

182 Smith, M. (1973). Geothermal Energy. LASL Document LA-5289-MS, Informal Report UC-34.

183 Harrison, E., Kieschnick, W.F.Jr. and McGuire, W.J. (1953). The mechanics of fracture induction and extension. *Petroleum Trans.*, AIME.

184 Blair, A.G., Tester, J.W. and Mortensen, J.J. (1976). LASL Hot dry rock geothermal project, *Progress Report* LA-6525-PR. UC-66a.

185 Robinson, E.S., Rowley, J.C., Potter, R.M., Armstrong, D.E., McInteer, B.B., Mills, R.L. and Smith, M.C. (Editor). (1971). A preliminary study of the nuclear subterrene. LASL Document LA-4547. UC-38.

186 Altseimer, J.H. (1975). Geothermal well technology and potential applications of subterrene devices – a status review. *San Francisco*, **2**, 1453.

187 Armstead, H.C.H. (1975). World Energy: the shape of things to come. *Energy International*, **12**, No. 1, 13.

188 Barbier, E. and Fanelli, M. (1975). Relationships as shown in ERTS satellite images between main fractures and geothermal manifestations in Italy. *San Francisco*, **2**, 883.

189 Comisión Federal de Electricidad, México. (1971). Official brochure on the Cerro Prieto geothermal development.

190 Barton, D.B. (1970). Current status of geothermal power plants at the Geysers, Sonoma County, California. *Pisa*, **2**, Part 2, 1552.

191 Finney, J.P., Miller, F.J., and Mills, D.B. (1972). Geothermal power project of Pacific Gas & Electric Co. at the Geysers, California, presented at the summer meeting of the IEEE Power Engineering Society, July, 1972.

192 Villa, F.P. (1975). Geothermal plants in Italy: their evolution and problems. *San Francisco*, **3**, 2061.

193 Ente Nazionale per l'energia elettrica (ENEL). Larderello and Monte Amiata: electric power by endogenous steam. Official brochure.

194 Sato, H. (1970). On Matsukawa geothermal power plant. *Pisa*, **2**, Part 2, 1546.

195 New Zealand Government. (1970). Wairakei: power from the Earth. The story of the Wairakei geothermal project. *Official brochure.*

196 Minohara, Y. and Sekioka, M. (1975). Geothermal utilization in the Atagawa tropical garden and Alligator farm: an example of successful geothermal utilization. *San Francisco*, **3**, 2237.

197 Meidev, T., Sanyal, S., and Facca, G. (1977). An update of world geothermal development. *Geotherm. Energy Mag.*, **5**, May, 30.

198 Garnish, J.D. (1977). Geothermal energy usage in the Paris Basin. Report ETSU N2/77. Energy Technology Support Unit, Harwell.

199 Sheinbaum, T. (1978). Geothermal well stimulation with a secondary fluid. *Geothermal Energy Mag.* **6**, No. 1, 33.

200 Contini, R. (1961). Methods of exploitation of geothermal energy and the equipment required. Paper G/71. *Rome*, **3**, 111.

Index